Cereals in Breadmaking

FOOD SCIENCE AND TECHNOLOGY

A Series of Monographs, Textbooks, and Reference Books

1. Flavor Research: Principles and Techniques, *R. Teranishi, I. Hornstein P. Issenberg, and E. L. Wick*
2. Principles of Enzymology for the Food Sciences, *John R. Whitaker*
3. Low-Temperature Preservation of Foods and Living Matter, *Owen R Fennema, William D. Powrie, and Elmer H. Marth*
4. Principles of Food Science
   Part I: Food Chemistry, *edited by Owen R. Fennema*
   Part II: Physical Methods of Food Preservation, *Marcus Karel, Owen R Fennema, and Daryl B. Lund*
5. Food Emulsions, *edited by Stig E. Friberg*
6. Nutritional and Safety Aspects of Food Processing, *edited by Steven R Tannenbaum*
7. Flavor Research: Recent Advances, *edited by R. Teranishi, Robert A Flath, and Hiroshi Sugisawa*
8. Computer-Aided Techniques in Food Technology, *edited by Israel Sagu*
9. Handbook of Tropical Foods, *edited by Harvey T. Chan*
10. Antimicrobials in Foods, *edited by Alfred Larry Branen and P. Michael Davidson*
11. Food Constituents and Food Residues: Their Chromatographic Determination, *edited by James F. Lawrence*
12. Aspartame: Physiology and Biochemistry, *edited by Lewis D. Stegin and L. J. Filer, Jr.*
13. Handbook of Vitamins: Nutritional, Biochemical, and Clinical Aspects *edited by Lawrence J. Machlin*
14. Starch Conversion Technology, *edited by G. M. A. van Beynum and J. A. Roels*
15. Food Chemistry: Second Edition, Revised and Expanded, *edited b Owen R. Fennema*

16. Sensory Evaluation of Food: Statistical Methods and Procedures, *Michael O'Mahony*
17. Alternative Sweetners, *edited by Lyn O'Brien Nabors and Robert C. Gelardi*
18. Citrus Fruits and Their Products: Analysis and Technology, *S. V. Ting and Russell L. Rouseff*
19. Engineering Properties of Foods, *edited by M. A. Rao and S. S. H. Rizvi*
20. Umami: A Basic Taste, *edited by Yojiro Kawamura and Morley R. Kare*
21. Food Biotechnology, *edited by Dietrich Knorr*
22. Food Texture: Instrumental and Sensory Measurement, *edited by Howard R. Moskowitz*
23. Seafoods and Fish Oils in Human Health and Disease, *John E. Kinsella*
24. Postharvest Physiology of Vegetables, *edited by J. Weichmann*
25. Handbook of Dietary Fiber: An Applied Approach, *Mark L. Dreher*
26. Food Toxicology, Parts A and B, *Jose M. Concon*
27. Modern Carbohydrate Chemistry, *Roger W. Binkley*
28. Trace Minerals in Foods, *edited by Kenneth T. Smith*
29. Protein Quality and the Effects of Processing, *edited by R. Dixon Phillips and John W. Finley*
30. Adulteration of Fruit Juice Beverages, *edited by Steven Nagy, John A. Attaway, and Martha E. Rhodes*
31. Foodborne Bacterial Pathogens, *edited by Michael P. Doyle*
32. Legumes: Chemistry, Technology, and Human Nutrition, *edited by Ruth H. Matthews*
33. Industrialization of Indigenous Fermented Foods, *edited by Keith H. Steinkraus*
34. International Food Regulation Handbook: Policy · Science · Law, *edited by Roger D. Middlekauff and Philippe Shubik*
35. Food Additives, *edited by A. Larry Branen, P. Michael Davidson, and Seppo Salminen*
36. Safety of Irradiated Foods, *J. F. Diehl*
37. Omega-3 Fatty Acids in Health and Disease, *edited by Robert S. Lees and Marcus Karel*
38. Food Emulsions: Second Edition, Revised and Expanded, *edited by Kåre Larsson and Stig E. Friberg*
39. Seafood: Effects of Technology on Nutrition, *George M. Pigott and Barbee W. Tucker*
40. Handbook of Vitamins: Second Edition, Revised and Expanded, *edited by Lawrence J. Machlin*
41. Handbook of Cereal Science and Technology, *Klaus J. Lorenz and Karel Kulp*
42. Food Processing Operations and Scale-Up, *Kenneth J. Valentas, Leon Levine, and J. Peter Clark*
43. Fish Quality Control by Computer Vision, *edited by L. F. Pau and R. Olafsson*
44. Volatile Compounds in Foods and Beverages, *edited by Henk Maarse*

45. Instrumental Methods for Quality Assurance in Foods, *edited by Daniel Y. C. Fung and Richard F. Matthews*
46. Listeria, Listeriosis, and Food Safety, *Elliot T. Ryser and Elmer H. Marth*
47. Acesulfame-K, *edited by D. G. Mayer and F. H. Kemper*
48. Alternative Sweeteners: Second Edition, Revised and Expanded, *edited by Lyn O'Brien Nabors and Robert C. Gelardi*
49. Food Extrusion Science and Technology, *edited by Jozef L. Kokini, Chi-Tang Ho, and Mukund V. Karwe*
50. Surimi Technology, *edited by Tyre C. Lanier and Chong M. Lee*
51. Handbook of Food Engineering, *edited by Dennis R. Heldman and Daryl B. Lund*
52. Food Analysis by HPLC, *edited by Leo M. L. Nollet*
53. Fatty Acids in Foods and Their Health Implications, *edited by Ching Kuang Chow*
54. *Clostridium botulinum*: Ecology and Control in Foods, *edited by Andreas H. W. Hauschild and Karen L. Dodds*
55. Cereals in Breadmaking: A Molecular Colloidal Approach, *Ann-Charlotte Eliasson and Kåre Larsson*
56. Low-Calorie Foods Handbook, *edited by Aaron M. Altschul*

Additional Volumes in Preparation

Antimicrobials in Foods: Second Edition, Revised and Expanded, *edited by P. Michael Davidson and Alfred Larry Branen*

Cereals in Breadmaking

A Molecular Colloidal Approach

Ann-Charlotte Eliasson
Kåre Larsson

Food Technology
Chemical Center
University of Lund
Lund, Sweden

CRC Press
Taylor & Francis Group
Boca Raton London New York

CRC Press is an imprint of the
Taylor & Francis Group, an **informa** business

Library of Congress Cataloging-in-Publication Data

Eliasson, Ann-Charlotte.
Cereals in breadmaking : a molecular colloidal approach / Ann-Charlotte Eliasson, Kåre Larsson.
p. cm. — (Food science and technology ; 55)
Includes bibliographical references and index.
ISBN 0-8247-8816-8
1. Bread—Analysis. 2. Grain—Analysis. 3. Colloids. I. Larsson, Kåre, . II. Title. III. Series: Food science and technology (Marcel Dekker, Inc.) : 55.
TX558.B7E45 1993
664'.752—dc20 92-40313
CIP

Preface

Bread has a unique status among foods based on its long history in all parts of the world. It is not surprising that in many languages the word for bread means foods in general, as we know from the Bible. The science of breadmaking is a challenge in food research. There is scarcely any other type of food in which the progression of changes from the initial raw material to the final product is more complicated. It is also remarkable that this art was refined to perfection thousands of years ago, without any knowledge of the chemistry, physics, or microbiology involved.

The effort to understand the chemistry of the baking process has a relatively short history. When the nature of proteins was revealed through work by The Svedberg and others, it became possible to study wheat proteins, aqueous gel properties fractions of wheat flour components. Remarkable achievements were reported by Haugaard and Johnson in 1930. Their work deserves particular attention among earlier studies, as being the first attempt at a colloidal point of view.

Research over the last three decades has explained many of the unique properties of wheat storage proteins, which in turn determine the baking properties of wheat flour. Most recent studies have focused on separation and characterization of these proteins and, in particular, relating protein quality to breadmaking performance. This is certainly one important piece of information needed to provide a general basis for plant breeding with baking performance and bread quality in mind. A long-neglected field of knowledge, however, is the surface and colloid chemistry involved in the wheat flour–water interaction—from dough mixing until fixation of the bread structure in connection with the starch gelatinization in the oven. Colloid science is in fact the only discipline that provides the tools to describe this process completely; if we try to understand these phenomena only through chemical/analytical knowledge of all the components, we can't see the forest for the trees.

Our laboratory has applied a colloidal approach to bread technology. New results are presented in this volume on the gluten structure in bulk and at interfaces, on phase separation in the dough, and on the transition from a foam to a pore system. On the basis of these results and recent progress in the understanding of shape in biomolecular assemblies derived from forces and curvature at interfaces, it is possible to present a complete structural model of the breadmaking process. As a background to this model it was natural to review the present molecular and colloidal knowledge of the chain from wheat grain to bread. As an introduction, the basic concepts of colloid science are presented with a minimum of theory.

The scientific literature on breadmaking is concentrated in journals and textbooks on cereals. Wheat is by far the most important crop for breadmaking and is given the most coverage in this book. We have included research on oats, rye, maize, and other grains in Chapters 4 and 7.

Bread research in our department started in about 1975 with a Ph.D. thesis by T. Carlson, supervised by one of us (K. Larsson), which was followed by many others. We are grateful for these contributions and to the Swedish Council for Forestry and Agricultural Research and to the Cerealia Foundation R&D, Stockholm, for supporting the cereal research in our department.

This volume is intended to be a useful textbook for advanced studies in cereal technology. We also hope the book will be read by scientists active in cereal research world-wide.

Ann-Charlotte Eliasson
Kåre Larsson

Contents

Cereals in Breadmaking

1

Basic Concepts of Surface and Colloid Chemistry

I. INTRODUCTION

An ordinary light microscope can be used to study particles down to micrometer size. The world of colloids exists between this size and the size of individual molecules, down to about 10 Å. Consider one phase dispersed as particles in another phase, thus forming a continuous medium. With successively smaller and smaller particles within this size range, eventually a limit will be reached at which the two phases become one. An example of such a single phase is a micellar solution, which consists of amphiphilic molecules associated into an aggregate. It is a colloidal solution, and we call it an association colloid. The individual molecules, however, form a molecular solution. In this small size range of colloids we can also find various macromolecules, such as proteins, in solution.

If wheat flour is stirred in an excess of water, the proteins will cover the entire range of sizes of colloidal dispersions. Some of these proteins are insoluble in water but can still be dispersed. The dispersed particles, however, tend to collect into aggregates, and larger aggregates will form a sediment together with starch when the stirring ceases. The stable water solution will therefore contain only small particles—monomers of the water-soluble fraction of proteins. An important feature characterizing colloidal dispersions is the size of the interface between the particles and the continuous phase, the total surface area of the particles. The fat globule in milk, for example, exposes an interface between fat and water of almost 10 m^2 per gram of milk. It is obvious that the nature of this interface plays a central role in our understanding of numerous properties of milk products, such as the separation of cream, the whipping of cream, and the formation of butter by phase inversion.

The importance of surface structure for colloidal phenomena is the reason for the close relation between surface science and colloidal science. This chapter presents the fundamental aspects, on the colloidal level, of the changes of the wheat endosperm from milling via dough mixing, fermentation, and heating to the final bread. Various crucial factors involved in baking properties—for example, the formation of gas cells and the gas-holding capacity of the dough—become obvious once we understand the underlying laws of surface and colloid science.

II. INTERFACES—SURFACE ENERGY AND SURFACE TENSION

Molecules in a liquid or in the bulk of a solid have neighbors in all three dimensions to interact with. In the surface layer, however, the molecules can interact only downwards toward the bulk or laterally within the surface plane. This is the reason for the existence of a *surface free energy*, which is the energy needed to increase the surface (or to bring molecules from the bulk to the surface layer). The surface free energy of a pure water surface toward air is about 73 mJ/m^2, whereas it is much lower for nonpolar liquids such as triglyceride oil (about 30 mJ/m^2). The surface energy is determined experimentally via the *surface tension* (γ), which is the force in millinewtons per unit length in meters that acts against increasing the area.

A drop hanging at the edge of a tube remains there because of surface tension (the mass of the drop is balanced by the surface tension), and this phenomenon provides one method to determine surface tension. Another method, described in the next section, is used to study the effect of surface-active molecules on the surface free energy. A third method is to use a capillary, provided that the walls are completely wetted by the liquid. The capillary force that causes the liquid to rise through the capillary tube is due to surface tension balancing the mass of liquid in the capillary above the level of the liquid outside the capillary.

III. SURFACE-ACTIVE MOLECULES—SELF-ASSEMBLY IN WATER AND MONOMOLECULAR SURFACE FILMS

In living tissues two groups of molecules tend to accumulate at interfaces: lipids and proteins. They are therefore classified as *surface-active molecules*, or *amphiphiles*. The driving force is their ability to reduce the surface energy between water and a gas phase or the interfacial energy between oil and water (or between solids and oil or water). Surface-active simple molecules such as lipids have one hydrophilic region; the rest of the molecule is hydrophobic. At an oil/water interface, the molecules of a monomolecular amphiphilic film will be oriented with their hydrophilic (water-attracting) region toward the water and their hydrophobic (water-repelling) region toward the oil.

Amphiphilic molecules form organized structures in bulk water by self-assembly. They do so because of hydrophobic effect, which is a tendency of the hydrocarbon region of the molecule to avoid water contact as it breaks the hydrogen bonds between water molecules. The structures obtained by such self-assembly of lipid molecules in water are further described below.

Suppose we have a pure water surface in a trough as shown in Fig. 1 with barriers defining the area. If a surface-active substance, for example, a long-chain fatty acid is spread from a solvent between the barriers, it is possible to record the pressure (π) versus molecular area of the fatty acid (A). The π–A isotherm is known from extensive studies of "insoluble" lipid monolayers to reflect the structure of the monomolecular film, as illustrated in Fig. 2. There are, in fact, the same structures as in the crystalline state of the lipid (solid state of the monolayer), in the lamellar liquid crystalline state (liquid condensed monolayer), or in the gaseous state. These corresponding lipid structures are described in later sections.

The surface film pressure π acts against the surface tension. Thus the surface film pressure is identical to the reduction in surface tension:

$$\pi = \Delta\gamma \tag{1}$$

Any protein in an aqueous solution will tend to go to the surface toward air, and the conformation and orientation it adopts will reduce the interfacial energy. A typical value of the reduction in interfacial tension when an excess of protein is present is about 20 mN/m. The absorption of proteins is usually irreversible, contrary to the behavior of lipids. This is due mainly to the size of the molecule. The protein molecule is often unfolded, and many water-insoluble segments of the peptide chain will be above the water surface. To undergo desorption, these segments would have to detach from the surface at the same time, which is very unlikely.

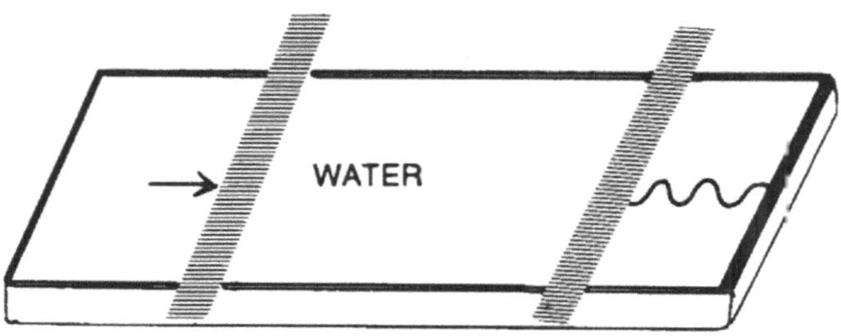

Fig. 1 Surface balance. One barrier can be moved to vary the area, whereas the other can record the presence of the surface film by the spreading pressure of the monolayer.

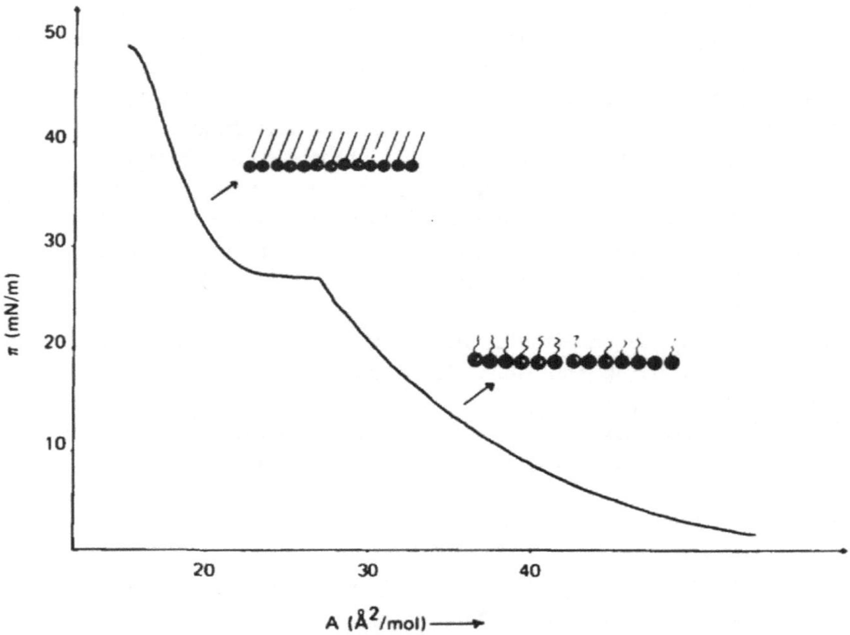

Fig. 2 Monolayer phases of a simple amphiphilic substance, one with lipid chains and the other (at high pressure) with crystalline chains. Polar heads of the molecules are indicated by circles, and the hydrocarbon chains by their axes.

Surface tension can be studied under various conditions using the surface balance, and even complex systems such as wheat lipids and wheat proteins can be successfully analyzed by spreading the corresponding monolayers, as will be described.

IV. INTERFACES OF SOLIDS—CRITICAL SURFACE TENSION OF WETTING

The surface energy of solids cannot be determined easily. A lot can be learned, however, from wetting properties. If a drop of liquid is put on the surface of a solid, it can either spread spontaneously, exhibiting complete wetting, or form a certain contact angle (θ) as shown in Fig. 3. The equilibrium conditions are given by the equation relating the interfacial tensions between the liquid and solid, between solid and gas, and between gas and liquid.

By studying the wetting properties using different polar/nonpolar liquids, it is possible to get indirect information on the surface properties of solids. A

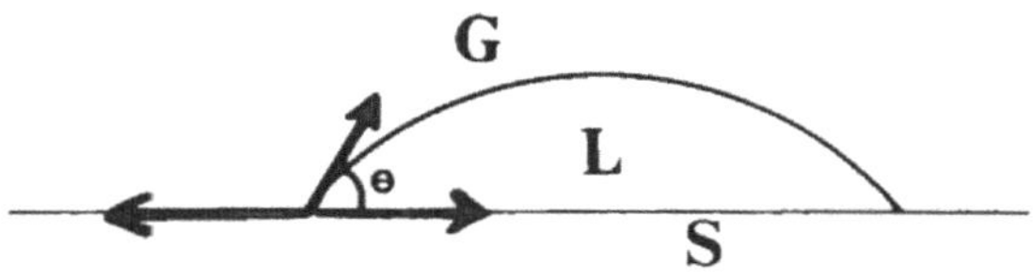

$$\gamma_{GS} = \gamma_{GL} \cos\theta + \gamma_{LS}$$

Fig. 3 Interfacial tensions (γ) determining the contact angle (θ) of a droplet on a solid.

particularly useful concept is the critical surface tension of wetting, which is derived by plotting the cosine of the contact angle versus surface tension as shown in Fig. 4. The extrapolation to wetting gives the critical surface tension of wetting (γc).

Solid surfaces can be classified as low-energy surfaces (e.g., polyethylene, which has a surface energy $\pi \approx 30$ mJ/m^2) or high-energy surfaces. Pure metals have values of a few thousand millijoules per square meter in surface free energy, whereas the values for metal oxides are considerably lower. A liquid with a certain surface tension will always spread and wet a surface with a higher surface free energy.

V. THE ELECTRIC DOUBLE LAYER

When particles contain charged groups or have an adsorbed monolayer of an ionic surface-active substance that forms an interface toward water, there will also be an adsorbed layer of counterions close to the surface. The nature of this electric double layer is of utmost importance to an understanding of the stability of colloidal dispersions in water. The distribution of charges as reflected by the surface potential is illustrated in Fig. 5.

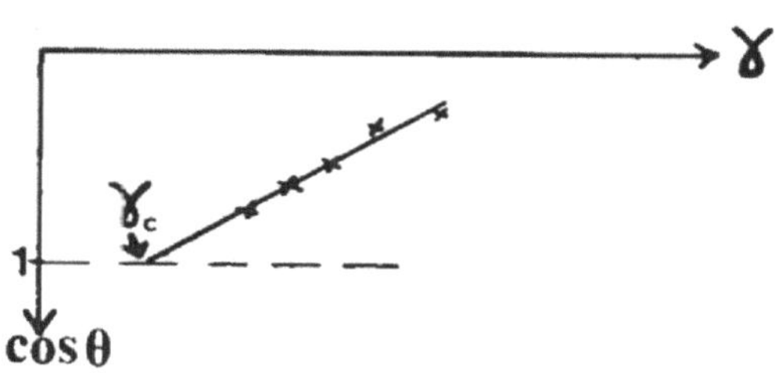

Fig. 4 Critical surface tension (γc) of wetting. Different liquids are used, and their contact angles (θ) recorded. The plot of cos θ versus surface tension gives the critical surface tension of wetting by extrapolation to contact angle 0°.

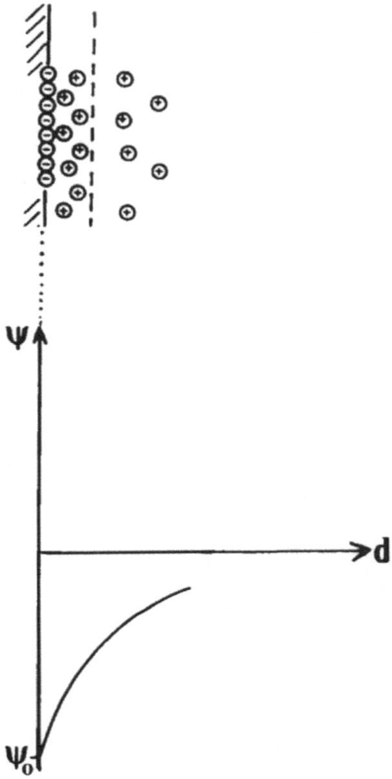

Fig. 5 Electric double layers as illustrated, of ionic distribution outside a surface, exposing negative groups.

The surface potential (ψ) has an exponential falloff

$$\psi = \psi\, e^{-\kappa d} \tag{2}$$

where d is distance from the surface and κ is directly proportional to the charge of the counterion and the square root of the concentration of the electrolyte in water. Thus added salt will compress the double layer. The repulsive force from this electric double layer is the basis for colloidal stability, as shown below.

A. The DLVO Theory

Deryaguin and Landau in the Soviet Union and Verwey and Overbeck in The Netherlands presented a simple theory more than half a century ago on the forces stabilizing colloidal dispersions. This theory, known as the DLVO theory, is still most useful. It describes the interaction energy between equal particles as a sum of their van der Waals attractive forces and their electrostatic repulsion. The van der Waals attractive forces are due to the difference in polarizability between the electron clouds of the particles and the liquid phase. When the distance d between the particles is considerably less than the particle diameter, this force is equal to H/d^6, where H is the Hamaker constant. The electrostatic repulsion is due to the electric double layer outside the particles as described in the previous section. The electrostatic interaction can be defined by $\epsilon^{-d/l}/d$, where l is the Debye–Hückel screening length. The sum of these forces describes the colloidal stability; a typical curve is shown in Fig. 6.

When two particles approach each other from a long distance, they will first experience very weak forces and then successively a minimum in interaction. Under special conditions this minimum (the "secondary minimum") is so deep that the particles form an agglomerate, separated by this distance. To come closer to each other, the particles must overcome a barrier, which, of course, is directly related to the degree of electrostatic repulsion. The electrostatic repulsion can be successively screened if we add more and more salt. This is the reason we can flocculate colloidal dispersions with salt (the particles attach; the "primary minimum" is reached).

B. Steric Stabilization of Colloids

There is sometimes an additional factor that can contribute to colloidal stability, particularly in food systems. This is a repulsive force due to polymer molecules

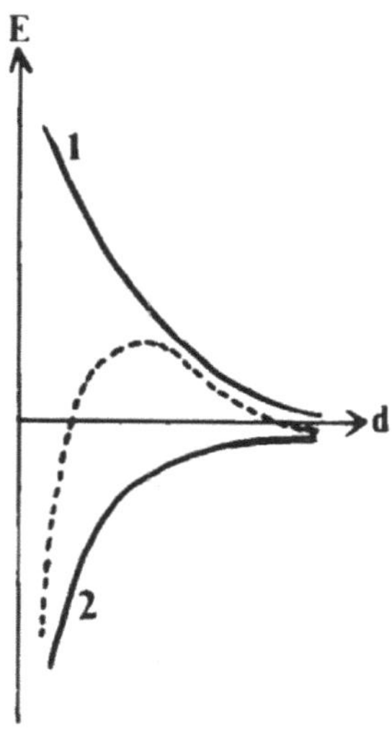

Fig. 6 Interaction energy curve between two particles according to the DLVO theory (dashed curve). The underlying electrostatic repulsion and van der Waals attraction curves are also shown.

adsorbed on the surfaces of the particles. If a flexible polymer is amphiphilic, it will tend to adsorb on hydrophobic particles in a hydrophilic medium, such as an oil-in-water emulsion. Such a polymer chain can extend far out from the particle surface, and particle collision must then involve conformational changes or removal of the polymer chain at the contact region. It is therefore natural to term this factor *steric stabilization*.

Adsorbed polymers may sometimes have the opposite effect and flocculate a dispersion. This can take place if the polymer chain can bridge two or more particles. Block polymers (polymers in which chain segments of two or more types alternate) often have the capability of stabilizing or flocculating a dispersion, depending upon the conditions. For example, blocks with negatively charged groups will produce an electrostatic repulsion in addition to the steric stabilization. If calcium ions are added to a dispersion stabilized in this way, the ionic cross-linking can lead to flocculation.

VI. MONOMOLECULAR FILMS AT THE AIR/WATER AND OIL/WATER INTERFACES

A. The Monolayer Structure

At the beginning of this century, Langmuir was able to reveal the nature of amphiphilic monolayers on water. The most intensive period of monolayer research took place during the decade before World War II. Interpretations of π–A isotherms were published before the characteristic crystalline and liquid crystalline structural features were known. Much of the later monolayer work was based on these early concepts, without relating results to three-dimensional lipid phases. Reports of molecular arrangement in monolayers are therefore often misleading, and the nomenclature used is inadequate. A structural description of monolayer phases was given in Fig. 2, and some general additional features will be considered here.

When the molecules in an amphiphilic monolayer are far apart, they lie more or less flat on the water surface, distributed like a gas in two dimensions. In analogy with the general gas law in three dimensions, this type of monolayer will exhibit a pressure equal to kT, where k is Boltzmann's constant and T is the temperature (in kelvin). Experimental values of monomolecular film pressures at very low values (below 0.1 mN/m) have been used to calculate the molecular weight of the film, and the values obtained are in good agreement with known values, demonstrating that the gas law gives a good description.

When a gaseous monolayer is compressed, a transition into the *liquid-expanded* phase usually takes place. This phase has been the subject of many controversies. Most of this discussion, however, took place before the structure of the lamellar liquid phase was known. From lipid–water studies it is known that the cross-sectional area per chain is always near 30 Å^2 in the lamellar liquid crystalline phase. Comparisons between monolayer and bulk structure in monoglycerides, lecithins, and galactolipids show this agreement in molecular area and also in temperature transitions. The rheological properties of this monolayer phase are also consistent with a liquid type of disorder. The formation of the liquid-expanded phase means that a gaseous phase condenses to a coherent film in the liquid state. Further compression can give phases with extended hydrocarbon chains tilted in relation to the surface. Such phases are called *liquid-condensed (L2)* phases, which is a misleading term, as they are in fact crystalline.

Many such phases are possible, as the angle of tilt of the chains can vary. The smallest possible molecular area is obtained when the extended zigzag chains are vertical. Phases with crystalline chains show linear π–A curves with well-defined plateau values in the π–A isotherms at phase transitions. Owing to hydrocarbon chain-packing alternatives, only certain discrete tilt values are possible. The geometric relations existing between different monolayer phases and three-dimensional polymorphic forms indicate the occurrence of the same structures in monolayers as in crystals. Considering polar lipids, the interaction with water can take place the same way in a monolayer phase as in the lamellar liquid crystalline lipid–water phase.

Polar lipids give two types of π–A isotherms. One type, exhibited by egg yolk lecithin, for example, is obtained above the chain melting temperature in the actual lipid–water system. Under these conditions there is only one monolayer phase, which usually exists up to a film pressure of about 40 mN/m. The structure is thus considered to be the same as in the lamellar liquid crystalline phase. The other type of π–A isotherm is obtained below the chain melting point of the lipid, and the monolayer structure is considered to be the same as in the gel phase. However, below the chain transition temperature, the monolayer phase with liquid chains can also be formed at low surface pressure. About a certain film pressure, this monolayer phase transforms into one with crystalline chains. The lamellar liquid crystalline phase is shown in Fig. 8, and the gel phase differs only by having crystalline chains.

The *equilibrium spreading pressure* (ESP) of monolayers from polar lipids has the same value as the plateau pressure, defining the transition between the form with liquid chains and that with crystalline chains. Above the chain melting temperature, the ESP value is the same as the collapse pressure of the monolayer phase with liquid chains. Monolayers of lipid on a water surface can be transferred to a solid, and lipid multilayers can be built up on the solid this way. Thus a hydrophobic material that is dipped through a lipid monolayer, down into water, will transfer the monolayer to give it a hydrophilic surface at the interface with the water. Another monolayer is transferred as the material is moved up from the water to give it a hydrophobic surface in air. Repeated dipping will thus give multilayers. Lipid monolayers on a water surface can also be compressed to multilayers under certain conditions.

B. Oil/Water Interfaces

Because of experimental difficulties, there are very few studies of polar lipids at the oil/water interface. In general terms it can be stated that at this interface also, polar lipids tend to form monolayers with structural properties very similar to those at the air/water interface. An important difference, however, is that a *separate phase* often builds up at the interface. In a system of a vegetable oil, water, and phosphatidylcholine, a lamellar liquid crystalline phase will build up at the interface, exposing a hydrophobic side toward the oil and a hydrophilic side toward the water. This also seems to be a common mechanism in emulsification.

C. Mechanical Properties of Monolayers at Interfaces

The pressure–area isotherm, as shown in Fig. 2, also demonstrates certain mechanical properties of the film at the air/water interface. Thus the slope is a direct measure of the monolayer compressibility. By expanding the film it is possible to detect hysteresis effects, and by repeated compression–expansion cycles chemical changes within the monolayers can eventually be observed. Such reactions within wheat protein monolayers will be demonstrated in Chapter 2.

Another technique used to investigate monolayer viscoelastic properties is based on wave damping. There are basically three types of waves that form on a water surface. Those of the shortest wavelength are called capillary waves and are due to thermal molecular mobility. Above them in wavelength are the Faraday waves, with wavelengths in the centimeter range. Wavelengths longer than this, such as the waves on the surface of a lake, are called gravitational waves. Faraday waves with frequencies of about 5 Hz are particularly useful in the investigation of monolayer properties, and an application with wheat storage proteins is demonstrated in Chapter 2. A summary of the technique is given here. The pioneering work in this field was done by Miles and by Lucassen-Reynders and Lucassen (see [1]). The damping of capillary waves, with shorter wavelength range than Faraday waves, has also been studied.

Standing Faraday waves can be generated experimentally by vertical oscillations of a circular cylinder containing the liquid. In the studies on wheat proteins reported in Chapter 2, a beaker of radius $R = 3.3$ cm was used [2]. When the oscillations had an amplitude a of 0.046 cm and a frequency of 5 Hz, standing waves exhibiting the (0,1) mode were obtained (see Fig. 7).

The wave damping to be expected has been derived theoretically, both in the case of a clean water surface and when the surface is covered by a monolayer of surface-active molecules. The monolayer damping effect involves surface elasticity and surface viscosity, thus expressing monolayer mechanical properties.

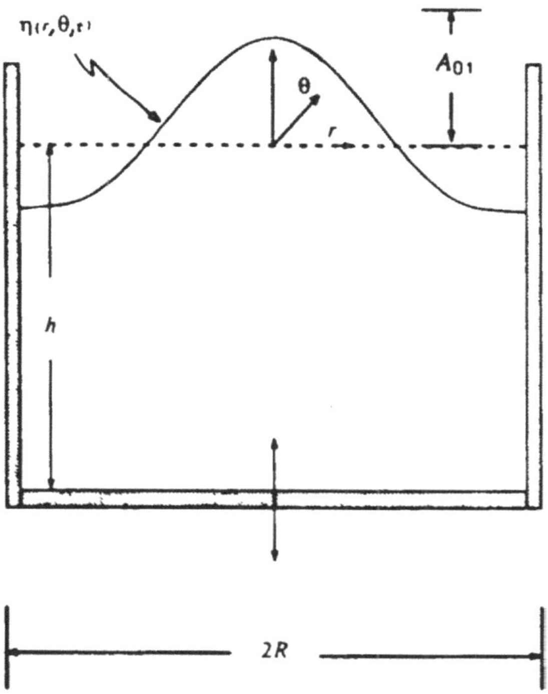

Fig. 7 Experimental setup to produce standing Faraday waves on water [2]. The amplitude A_{01} of the (0,1) mode is shown. The surface displacement is given by $\eta(r,\theta,t)$.

VII. SELF-ASSEMBLY OF LIPID MOLECULES IN WATER

The association of polar lipids in an aqueous environment results in remarkable phases with long-range order although they are disordered like liquids over atomic distances. These phases are therefore *liquid crystalline* in character. About three decades ago the liquid nature of the hydrocarbon chains was shown to be the fundamental feature in the structure of liquid crystalline phases. This made possible the determination of the structures of the most common liquid crystalline phases.

A. Lamellar and Hexagonal Phases

When polar lipids in the solid states are heated, they either melt directly or are transformed into liquid crystalline phases. A liquid crystalline phase obtained by heating is classified as *thermotropic*. More common in polar lipids, however, is that liquid crystalline phases are formed in the presence of water. Above a hydrocarbon chain melting temperature, water penetrates the polar region, and a lamellar lipid–water structure is formed in which water layers alternate with lipid bilayers. The structure of this phase, the L_α-phase, is shown in Fig. 8.

Two other structures, which together with the lamellar phase are the most important liquid crystalline phases in lipid–water systems, are shown in Fig. 9. The two hexagonal structures shown are termed H_I and H_{II} phases. The H_I phase consists of infinite cylinders of lipid molecules, which expose their polar groups on the surface, and consists of a core of hydrocarbon chains in a disordered liquidlike conformation. These rods are hexagonally arranged in a water medium. The H_I phase is very rare among lipids, whereas H_{II} is common. The H_{II} is the inverse of H_I. Thus water forms rods, and hydrocarbon chains form a continuous medium with the polar groups at the water/hydrocarbon chain interface.

B. Cubic Phases

Cubic lipid–water phases have been observed in the temperature and composition ranges between H_{II} and L_α, between L_α and H_I, and between H_I and the micellar solution. First the structure of the cubic phase of soaps was determined, which was reported to consist of two networks of rods formed by the polar groups embedded in a hydrocarbon chain matrix. This rod-system structure was proposed to be the general structure for the cubic lipid–water phases.

Studies of the cubic monoglyceride–water phase indicated that the structure was closely related to that of the lamellar phase. It was concluded from nuclear magnetic resonance (NMR) diffusion measurements that the local environment of the lipid molecules was similar to that of the lamellar phase. Another

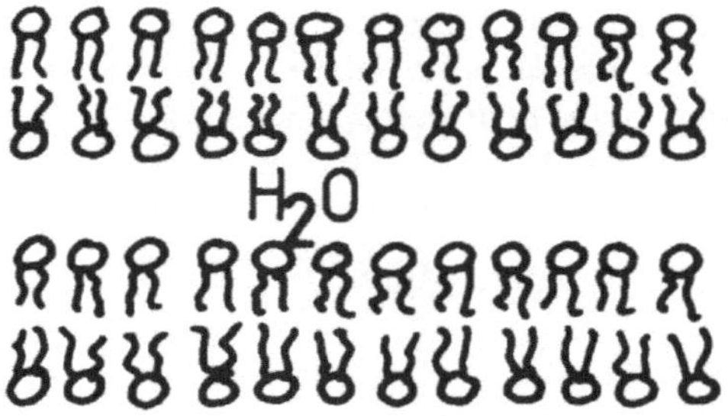

Fig. 8 Structure of the lamellar liquid crystalline phase. The polar heads (circles) and the disordered chains of the lipid molecules are shown.

Fig. 9 Structures of the two hexagonal liquid crystalline phases (indicated as in Fig. 8). (Top) A three-dimensional section of the reversed hexagonal structure.

conclusion from the NMR measurements was that the structure was continuous with respect to both lipid and water. The structure was proposed to consist of lipid bilayer units connected into a network. Later it was realized [3] that such a structure is consistent with an infinite periodic minimal surface (IPMS). The methyl end-group gap of the bilayer is identical to the IPMS, and the infinite lipid bilayer separates two identical water channel systems. In fact, monoolein forms two such cubic IPMS phases, one with a body-centered lattice at low water content and the other with a primitive lattice at high water content. One structure is shown in Fig. 10. There are three fundamental IPMS structures with cubic symmetry, and these three structures differ by the shape factor discussed below.

Fig. 10 Structure of the lipid bilayer in a cubic lipid–water phase. The lipid bilayer forms an infinite periodic minimal surface, and this plastic model is identical to such a surface. There are two continuous water channel systems, one on each side of the surface.

C. Phase Diagrams

Polar lipid mixtures of biological origin behave like one component with regard to the phase rule, although the lipid can be a mixture of hundreds of homologues. Phase equilibria in binary lipid–water systems are illustrated by composition–temperature phase diagrams, and one example is shown in Fig. 11. The lamellar liquid crystalline phase dominates. There are two types of lipid–water phase diagrams. The first, shown in Fig. 11, is obtained from polar lipids, which are "insoluble" in water. An ordinary phospholipid has a solubility of about 10^{-10} mol/L, and its behavior is representative for phase diagrams of polar lipids.

A second type of binary system is obtained when the lipid is soluble as micelles in water. Examples of such lipids are fatty acid salts and lysolecithin. Such a system does not occur among wheat lipids or surfactants used in bread-making. When the lipid concentration in the micellar solution is increased, spherical micelles are usually transformed into rod-shaped micelles. At still higher lipid concentrations, the lipid cylinders are hexagonally arranged, and a liquid crystalline phase (H_I, as shown in Fig. 9) is formed. The lamellar liquid

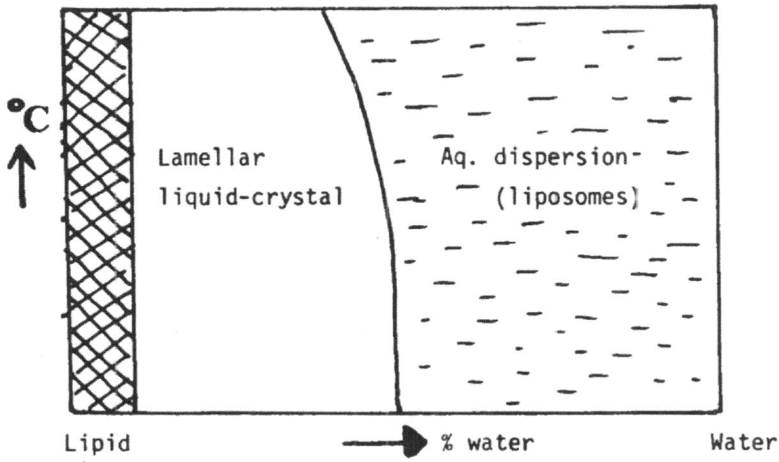

Fig. 11 Binary phase diagram of a typical polar lipid, such as a phospholipid, and water. The dominating one-phase region is formed by the lamellar liquid crystalline phase. Two-phase regions are shaded.

crystalline phase is usually formed in the region between the hexagonal phase and this anhydrous lipid.

Ternary systems consisting of polar lipids, water, and triglyceride oil will be discussed in connection with cereal lipids. The term L1 is used for an ordinary micellar solution and L2 for the reversed type of liquid phase, that is, water aggregates in a continuous hydrocarbon chain medium. According to the definition of microemulsions, the L2 phase in an aqueous system of a polar lipid and triglyceride oil is a *microemulsion*, which thus is an alternative term for the phase. Structural studies of such systems indicate that a close relationship exists between the L2 phase structure and that of the corresponding liquid crystalline phase obtained by cooling (in wheat lipids, the lamellar liquid crystalline phase).

When water is added to the lamellar liquid crystalline phase above its limit of swelling, *liposomes* can be formed mechanically—for example, by stirring. A liposome consists of spherically concentric lipid bilayers alternating with water layers. The structure is similar to the lamellar liquid crystalline phase, the difference being that the bilayers are curved into a closed aggregate. Liposomes can be up to several micrometers in diameter, with the innermost layer having a minimal diameter of a few hundred angstroms. A liposomal dispersion is kinetically stable, like an emulsion.

The wedge shape of the amphiphile molecule (expressed as a shape factor calculated from molecular dimensions) determines the phase that is formed with water. Once the wedge shape is known, it is possible to predict the different amphiphile–water structures on the basis of the corresponding curvature of the bilayer [3].

An increase in temperature will increase the disorder along the hydrocarbon chain, which means that the shape will become more wedgelike. The transitions with temperature will therefore follow the sequence lamellar → cubic → reversed hexagonal. The effect of water content is more complex. In the case of nonionic lipids, the disorder will increase with water content; thus the structural effects will be similar to those obtained by an increase in temperature. Ionic lipids, on the other hand, will exhibit the phase transition sequence reversed hexagonal → cubic → lamellar when the water content is increased. This behavior is related to an increased lateral repulsion between the polar heads with hydration and thus a reduced wedge shape (smaller angle).

VIII. WATER AND THE HYDRATION FORCE

To understand the unique properties of water, such as its ability to dissociate salts or the exceptional density of the liquid state compared to the solid (ice floating on water), it is necessary to start with the structure of the water molecule (see Fig. 12).

Structure in water or in ice is determined by the strong tendency of water molecules to associate by forming hydrogen bonds. The four directions from each molecule, as shown in Fig. 12, will give a very open structure, and this ideal arrangement exists in ice. Close packing of spheres, which occurs in metals, for example, shows 12 neighbors. When water is formed from ice, the increased disorder allows the water molecules to dynamically fill the empty

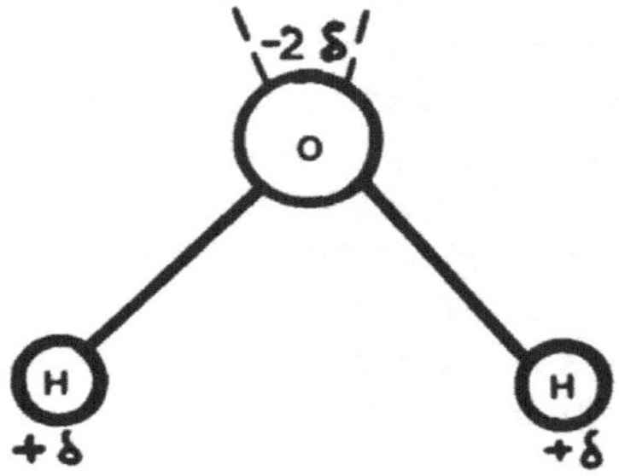

Fig. 12 A water molecule with its tetrahedral arrangement of the two hydrogen atoms and the directions of the two electron clouds available for forming intermolecular hydrogen bonds. The charge distribution is also indicated.

space in the ice lattice. Thus the density increases. It should be pointed out, however, that the detailed structure of water still is unknown.

As indicated by the charges on the water molecule shown in Fig. 12, the molecule is a strong dipole. This is the reason water can dissolve various ionic substances by dissociation. When sodium chloride, for example, is dissolved, the water shell surrounding the chloride ions will have the positively charged hydrogens directed toward the ions, and the sodium ions will be surrounded by water molecules whose oxygen side is turned toward them.

From studies of effects of osmotic stress on the lamellar liquid crystalline phase, the term "hydration force" was coined to explain repulsive forces between phospholipid bilayers. At distances of up to 30 Å between bilayers, the hydration force is the dominating force, independent of type of polar head groups. This repulsion is responsible for the spontaneous swelling of lipids in water. The force has an exponential falloff, with a decay distance of 2–3 Å, comparable to the size of the water molecule. The force is usually described as

$$F = p_0 e^{-d/l} \tag{3}$$

Its value can be obtained experimentally by plotting the hydration pressure (p) versus water-layer thickness (d); P_0 is the pressure at zero thickness. The hydration pressure can be obtained by osmotic stress, exposing a lipid–water phase to a polymer solution with solutes so large that they cannot penetrate the water layer between lipid bilayers. The value of d can be followed by X-ray diffraction. This force has also been demonstrated in other systems that hydrate and swell.

Finally, it should be pointed out that the hydration force has also been characterized by numerous measurements using surface-force apparatus.

IX. COLLOIDAL ASPECTS OF PROTEINS

To understand the behavior of wheat proteins during baking we must consider their colloidal properties, and a natural starting point is a description of protein conformation in water solution.

Through systematic work by protein engineering and crystal structure determinations, the folding mechanisms of protein molecules are being gradually revealed. A major factor is the hydrophobic versus hydrophilic nature of the amino acid side chains. As a simple first approximation, proteins in water resemble micelles, with more polar amino acid sequences covering the surface and more nonpolar ones forming a core protected from water contact. As the polar character is so important to the folding of the peptide chain, the different amino acids are ranked in this respect in Table 1. From the top down, the 20 amino acids are listed according to increasing polarity. The exact order is, of course, tentative, as it depends on various factors in the environment, such as pH and the presence of counterions.

Table 1 Amino Acids Ranked According to Polarity

Isoleucine	Nonpolar
Leucine	↓
Valine	
Phenylalanine	
Methionine	
Alanine	
Proline	
Threonine	
Cysteine	
Tryptophan	
Histidine	Medium polarity
Tyrosine	↓
Glycine serine	
Aspartate	
Asparagine	
Glutamate	
Glutamine	
Lysine	
Arginine	Polar

The organization of the peptide chain into the final structure can be described on the following six levels:

Primary structure: amino acid sequence along the peptide chain
Secondary structure: peptide backbone conformation (α-helix, β-sheet, β-turn, random coil)
Supersecondary structure: aggregates of secondary structure
Domains: separate globular regions within the molecule
Tertiary structure: three-dimensional arrangement of all atoms in the molecule
Quaternary structure: structural units consisting of associated molecules

The driving force toward the final assembly of molecules is determined by the primary structure, but there are still no methods to predict the folding of a particular peptide chain. The structure is determined by hydrophobic interaction and various types of interactions between polar groups: ionic interaction, hydrogen bonding, salt bridging, and dipole–dipole interaction. There is only one type of covalent bridge between peptide chains: the disulfide bridge, which is of utmost importance in breadmaking.

Physical properties are of fundamental importance with regard to the effects of proteins on food texture and other external characteristics. It is usual to define such overall properties as *functional properties*. Common functional properties are gel formation/stabilization, foam stabilization, and emulsification.

As soon as the temperature is increased, the individual amino acid residues along the peptide chain start to increase in thermal mobility, which, of course, involves not only vibrations of the individual atoms but also coordinated movement of the individual segments. The space they require will thus change, and ultimately a drastic reconformation must take place to account for these requirements.

Denaturation is an all-or-none reaction due to these changes, and it is similar to a phase transition. Tanford (see recommended textbook list at the end of the chapter) defined this major change as starting from the native organization and taking place without the rupture of any primary covalent bonds. (Disulfide bridges, however, are often opened during the unfolding process.) Denaturation caused by heat as described here can also be caused by other factors such as surface exposure, oil, ethanol, pH changes, or exposure to a solvent. A remarkable feature in the insoluble wheat storage proteins is their lack of denaturation, in the strict meaning of the term.

Going back to the micellar model of the protein structure, the hydrophilic amino acid residues, which face outwards, must be able to cover the surface. If there are not enough hydrophilic groups to cover the surface, association is a solution, and caseins are assembled in that manner. If, on the other hand, there are too many hydrophilic groups to correspond to a spherical shape, the protein will probably not exhibit a globular shape but will instead form fibers (with a larger surface-to-volume ratio).

X. COLLOIDAL STRUCTURES—GELS, DISPERSIONS, POLYPHASIC AQUEOUS SYSTEMS, AND FOAMS

The basic types of colloidal structures are various types of *dispersions*, such as *suspensions* (a solid phase dispersed in a liquid continuous phase), *emulsions* (a liquid phase dispersed in another liquid phase), and *foams* (gas cells in a liquid). All these types of structures occur in breadmaking. Furthermore, the water phase is "immobilized" into a *gel*, which is another important type of structure that occurs in colloidal systems. Details of the characteristics of gels will be evident from the subsequent chapters; we discuss only some general structural features here.

All of us have a "feeling" for what a gel is, although a strict definition is lacking. *Gels* are three-dimensional networks that immobilize large amounts of water into a viscoelastic state. There are two types of aqueous gels:

1. *Particle gels* consist of colloidal aggregates (the shape of the elements can range from spheres to rods and disks, and the size is of the same order of magnitude as the wavelength of visible light).
2. *Macromolecular gels* consist of polymer chains covalently bound or "physically" associated.

Both types can be identified by their rheological behavior. Type 1 is usually not transparent, whereas type 2 is. The two major gel-forming constituents in bread and other foods are proteins and polysaccharides.

The most common mechanism of protein gel formation is thermal molecular change in connection with denaturation. It is usually a two-step process.

$$xP_n \rightarrow xP_d \rightarrow (P_d)_x \qquad (4)$$

The first step is the unfolding of a native protein molecule, P_n, to give the denatured protein, P_d. Then there is an association step, resulting in the gel.

Other mechanisms of gel formation may involve the formation of intermolecular disulfide (-S-S-) bridges. If we compare the disulfide bridge with a hydrogen bond, the first is about 335–375 kJ/mol in energy, and the second about 8–42 kJ/mol.

A gel should also exhibit *cohesiveness;* if two pieces with the same composition are put in contact, they should fuse, eliminating any interface at the contact region. If, on the other hand, the interface does not disappear at sites of contact, the material is not a true gel. We then use the term *coagel*. The egg white of a boiled egg is an example of a coagel. Like a gel, a coagel contains a large amount of immobilized water and shows viscoelastic behavior, but it is not a real gel, as it lacks cohesiveness. The gluten gel is successively transformed into a coagel by heating, as will be considered in detail in Chapter 2.

Polysaccharides often exhibit a gel structure. In breadmaking we are concerned with two types: starch and pentosans. Their gel mechanisms are discussed in Chapter 2.

When a wheat flour dough is mixed, the naturally present nonpolar lipids in the form of an oil are emulsified. The dough is thus an oil-in-water *emulsion*. In certain fat-based baked products such as cakes, we can even have the reverse type of emulsion in the batter—an emulsion of water with wheat flour and other ingredients in a continuous oil (fat) phase. Both types of emulsions are stabilized by an interfacial film of amphiphilic molecules. Thus the emulsion droplets are covered by a monolayer, exposing polar regions toward water and hydrocarbons toward the oil phase. Sometimes the interface even consists of multilayers; for example, it may include a layer from a liposome.

A special case of an emulsion occurring during the baking process consists of water droplets dispersed in a water phase. This situation, known as ***two-phase separation***, or coascervation, is due to the fact that one polymer can form an aqueous solution whereas another polymer forms such an unstable solution that it separates. In fact, a third, fourth, etc. phase may separate out, forming what we call an ***aqueous polyphasic system***. If, for example, amylose and amylopectin are mixed in water, phase separation can be observed, with the appearance of an upper amylose-rich phase and a bottom amylopectin-rich phase.

A *foam* consists of close-packed gas cells encapsulated by lamellae of a liquid. To stabilize the foam structure, some surface-active substance is needed. A simple soap foam, with which we all are familiar, is shown in Fig. 13. The fatty acid salts form outside monolayers, and the core of the lamellae consists of water (with the rest of the soap molecules eventually entering into solution). The surfactant surface performs many functions, including reduction of surface tension, mechanical stabilization, and the provision of repulsive forces for stabilization. All foams look basically the same, even the protein-based foam in dough, as is further discussed in Chapter 2.

XI. FLOW PROPERTIES

The flow behavior (rheology) of the different phases that occur during breadmaking is significant in many respects, as is evident from Chapters 5 and 7 in particular. The fundamental aspects of rheology are summarized in Chapter 2 in connection with wheat proteins. We describe here a related phenomenon: *lubrication*.

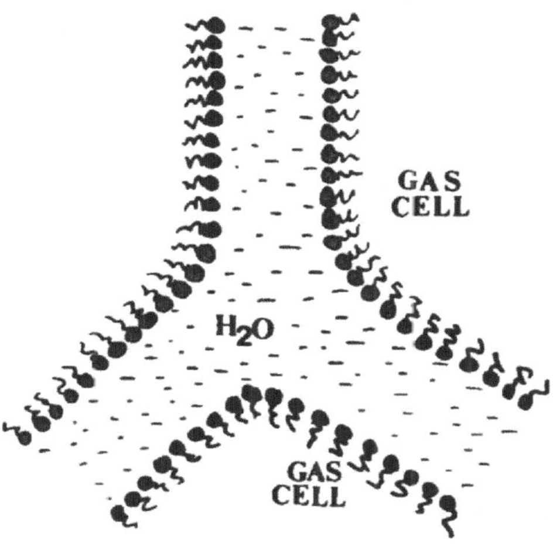

Fig. 13 The structure of a foam. The surfactant molecules are indicated as in the lipid structures shown earlier.

Friction was described by Leonardo da Vinci as a force tending to oppose movement. The friction (F) between two solids moving relative to one another is equal to

$$F = \mu W \qquad (5)$$

where μ is the coefficient of friction and W is the load or the force at the contact region normal to the direction of movement. Friction can also be regarded as the force needed to induce movement, which is equivalent to adhesion. Adhesion is a product of the contact area and the shear strength of the junction. Often, however, a deformation occurs in the contact zone, and plastic and/or elastic deformation gives a contact area that is almost linearly related to the total load.

When there is a liquid film between the solid surfaces and no deformation occurs, ***hydrodynamic lubrication*** is obtained. The friction is then related only to the viscosity of the fluid. An obvious requirement is that the surface roughness be less than the film thickness. The film thickness is related to the viscosity (η) and the movement velocity (v) according to the equation

$$\text{Thickness} = (\text{constant})(\eta v)^{2/3} \qquad (6)$$

From this equation it is possible to estimate when contact will occur. This situation is defined as ***boundary lubrication***. The important feature of this situation is that although there is direct contact between the two solids, the friction can be much lower than what should be expected from the coefficient of friction between the solids. The reason is that a lubricant monolayer on each solid takes up part of the load. The force needed to shear the solids depends on the area of solid–solid contact in relation to the area where there is direct contact between the lubricant monolayers. Usually the latter area dominates, and the shear force is then mainly determined by the lubricant film.

In breadmaking, the shear forces between the starch granules involved in dough mixing, fermentation, and heating expansion are closely related to the lubricating effects provided by the protein gel phase as well as by lipid–water phases. The lamellar lipid–water phase $L\alpha$ has obvious advantages in lubrication. Effects of phospholipids on dough rheology, which will be discussed later, are certainly related to their lubrication effect.

XII. POROUS STRUCTURES

A. Relevance to the Structure of Dough and Bread

When dough is fermented, the colloidal structure is transformed from a dispersion of gas cells into a foam, and then a pore system is formed. Pores form an open network, in contrast to a foam, which consists of closed cells. The final bread has a typical pore structure. Porous structures can be characterized by the pore network geometry and the intrafacial curvature. Living tissues provide many examples of porous structures; one is spongy bone. Mineralization is one case in which the interface between organic and inorganic matrices is sometimes an ideal interface, a minimal surface [3]. To have a basis for further discussion, it is necessary to start with a description of curvature.

B. Curvature of a Surface

The curvature at any point on a surface is defined by the principal radii of curvature, R_1 and R_2. If we imagine a circle with a tangential contact at that point, and this circle is allowed to rotate with its radius coinciding with the vector normal to the surface (see Fig. 14), the radius will have to vary in order to keep the tangential contact. The maximum and minimum values of these radii are the two principal radii of curvature R_1 and R_2. If we consider the surface formed by the soap film in Fig. 14 the radius has either a + or – sign depending on whether the rotating circle is perpendicular or parallel to the rings. There are two significant factors, the average curvature H,

$$H = 1/R_1 + 1/R_2 \tag{7}$$

and the Gaussian curvature G,

$$G = 1/R_1 R_2 \tag{8}$$

C. Minimal Surfaces

Long ago it was shown that the condition of minimization of a surface spanning a given boundary is that the average curvature be zero everywhere.

$$1/R_1 + 1/R_2 = 0 \tag{9}$$

This means that the surface is as convex as it is concave at all points. Many experiments on minimal surfaces were made using soap films (which always exhibit the minimal area) before the various surfaces could be described analytically.

A simple example is the catenoid shown in Fig. 14, which is obtained when a soap film is allowed to span the space between two circular threads.

The Gaussian curvature of a gas bubble

$$G = 1/R_1 R_2 > 0 \tag{10}$$

as the two radii have the same sign (occur on the same side of the surface). The Gaussian curvature of a minimal surface is always negative, except at so-called flat points (where one or both radii are infinite), when $G=0$. The value of G is directly related to adsoprtion forces. Thus, active sites in enzymes and

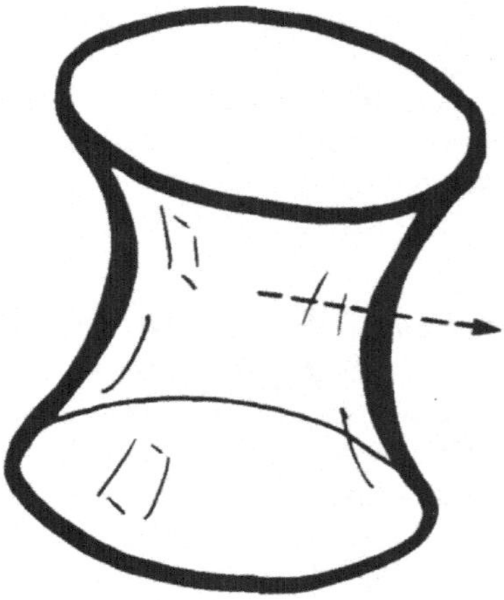

Fig. 14 The catenoid formed by a soap film. The two rings are shown, and the soap film between them is indicated. The dotted arrow represents the normal vector.

catalytic centers on inorganic catalysts are usually locations on the surface where there is a maximum in the absolute value of *G* (see [3].

D. Sphere Model of a Pore System

The packing of spheres, whether equal or different in size, has frequently been used to model porous materials. Thus a kind of porosity parameter is given by the average coordination number of spheres. Even when equal spheres are considered, the situation is very complicated from a theoretical point of view. The highest coordination number possible is 12 (corresponding to cubic or hexagonal close packing of spheres, as in metal lattices), and the lowest is 3 (in order to form a contact in three dimensions). Furthermore, the geometry of contacts between the spheres can vary, depending on distances between the centers and the coordination geometry. These models consider the space between spheres as the pore system, as many porous materials are based on particles, which, for example, are sintered into a ceramic. In a dough, however, it seems more natural to consider a pore system as consisting of fused spheres, the approach taken below.

E. Network Model of Pores

A more recent approach than the sphere model is a lattice network model of the porous material. This has been related to flow properties according to the so-called percolation theory. This theory explains the experimentally observed sudden increase in penetration or diffusion through the pore system above a critical value of network branching.

F. A Constant-Curvature Surface—A Model of Pore Structures Formed from a Foam

The pore system of a dough is quite different from that of solids such as ceramic materials, which have been most investigated. Earlier theories on pore structure are therefore not relevant to a dough or bread. A major difference is the mirror symmetry of the center plane between gas cells or gas pores, a structure quite different from the gas/solid interface in a conventional solid porous material. A new approach, presented below, is based on recent knowledge of infinite periodic surfaces.

Gas cells introduced during dough mixing are spherical, and not until the fermentation has proceeded above the geometrical limit of close packing of actual spheres do we obtain polyhedra and an overall structure corresponding to that of a foam. An ordinary soap foam like a dough contains polyhedra of various shapes and sizes. The most common is the tetrakaidecahedron, first described by Kelvin as the polyhedron with the smallest ratio of surface to volume [4]. It consists of 14 faces (as is evident from its name), six squares and eight hexagons, as shown in Fig. 15. These polyhedra can be regularly packed in space corresponding to the ideal structure of a foam, as shown in Fig. 16.

When the dough is baked to a bread, heating in the oven results in a fast expansion, called *oven spring*, and the foam lattice is then successively transformed into a pore structure. Thus, at a critical point as described by percolation theory, the pore system becomes open for gas transport. This sudden increase in

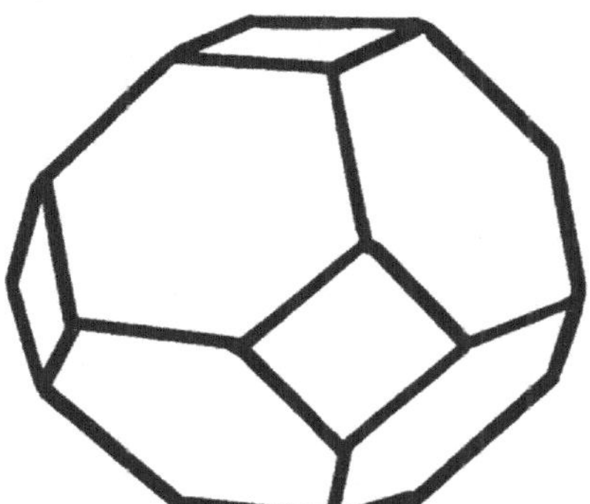

Fig. 15 A tetrakaidecahedron.

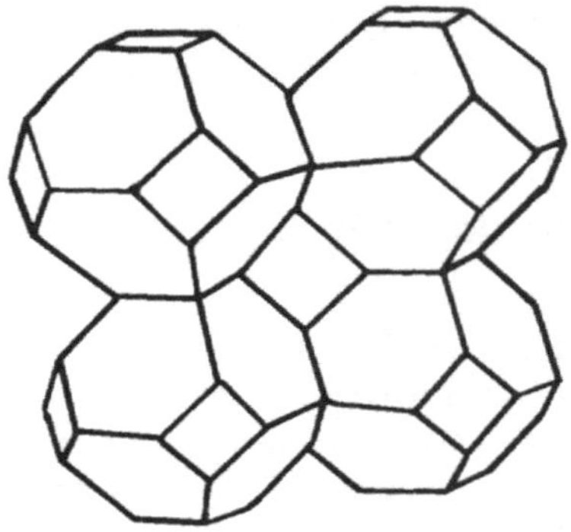

Fig. 16 Ideal foam structure formed by packing of tetrakaidecahedra.

carbon dioxide leakage seems to occur at temperatures around that of the gelatinization of starch, and the corresponding change in mechanical properties of the pore walls (from a gel type to a glass type) might "trigger" this transition from a closed-pore to an open-pore system. The expansion of the bread stops when the pore system becomes open. Also, the open-pore structure is the reason the final bread does not contract when it is cooled but is able to keep its shape from the oven.

In considering the details of the pore geometry, it is obvious that the pressure in each gas cell and in each pore is constant. This tells us directly that the pore surface must exhibit an average curvature everywhere that is constant and positive. Thus,

$$H = 1/R_1 + 1/R_2 = \text{constant} > 0 \tag{11}$$

This is a direct consequence of the Laplace–Young equation (further discussed in Chapter 5):

$$\Delta p = \gamma(1/R_1 + 1/R_2) \tag{12}$$

where Δp is the pressure difference over the interfacial and γ is the interface tension.

Infinite periodic minimal surfaces with constant average curvature are related to the minimal surfaces discussed at the beginning of this section. It is easy to imagine a surface that is parallel to the minimal surface, and if the curvature does not differ too much from zero, the constant average surface will be close to the parallel surface. Let us start with the idealized case of a pore system that is identical to the minimal surface. Remember the cubic lipid–water phases discussed in Section VII. B; they are also minimal surfaces. We can now relate the pore structure to the foam structure from which it is formed. Two adjacent foam cells are shown in Fig. 17. When all the polyhedra forming the foam fuse, the minimal surface shown is formed. This type of idealized pore system will, in fact, consist of two independent networks with no connections. If there is a pressure gradient over the interface between these two pore networks, the surface will be a constant-curvature periodic surface as shown at the bottom of the figure.

Two other simple periodic surfaces are clear candidates for models for pore systems. One is the so-called diamond surface [3], where the pore network follows the diamond lattice. The most interesting surface, however, is the *gyroid*, which is shown in Chapter 6. We have tried to relate the dough pore system to the basic periodic minimal surfaces, and we consider this surface to be most relevant. It can be described as the most effective packing of helical pore networks.

G. Physical Properties of Porous Materials

Pores play an important role in various adsorption phenomena, which can even lead to the formation of a thin liquid film from a vapor (capillary condensation). The first theoretical treatment by Thomson and later by Kelvin gave the relation between the equilibrium pressure over a liquid concave surface (P) and the saturated vapor pressure (P_o):

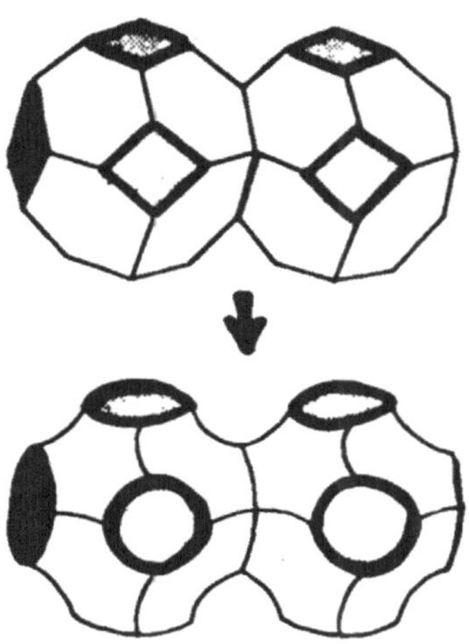

Fig. 17 Two foam cells, which by fusion can form a pore with constant average curvature.

$$\ln(P/P_0) = \gamma v/RT(1/R_1 + 1/R_2) \tag{13}$$

where γ is the surface tension and v is the molar volume of the adsorbed liquid.

Pores are usually classified as follows:

Micropores—pore size about 20 Å
Mesopores—pore size about 20–500 Å
Macropores—pore size above 500 Å

As micropores are the same size as the molecules of various actual vapors, there is a strong tendency for a liquid condensate to form in these pores.

Adsorption behavior of porous materials can be described by the isotherms shown in Fig. 18. The middle curve is most representative of the pore structure

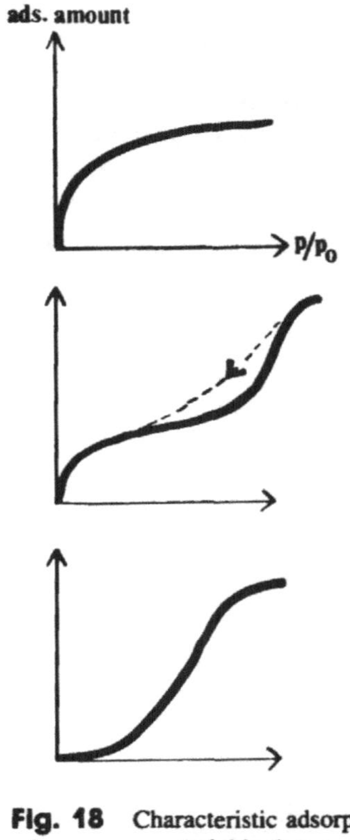

Fig. 18 Characteristic adsorption isotherms of different types of porous materials. A microporous material is shown at the top, a mesoporous "normal" material in the middle, and a weakly adsorbing mesoporous material at the bottom. Adsorption is given as a function of the vapor pressure of water.

of bread. Many aspects of the adsorption properties of the pore system are relevant to bread, as will be shown later.

TEXTBOOKS RECOMMENDED FOR FURTHER STUDY

Israelachvili, J., *Intermolecular and Surface Forces*, Academic, London, 1985.

Kruyt, H. R., Ed., *Colloid Science*, Vols. I and II, Elsevier, Amsterdam, 1952.

MacRitchie, F., *Chemistry at Interfaces*, Academic, London, 1991.

Matijević, E., *Surface and Colloid Science*, Vol. 5, Wiley, New York, 1972. See also the other volumes in the series.

Tanford, C., *The Hydrophobic Effect: Formation of Micelles and Biophysical Membranes*, 2nd ed., Wiley, New York, 1980.

REFERENCES

1. Gaines, L. G. J. (1966). *Insoluble Monolayers*, Wiley, New York.
2. Henderson, D., Larsson, K., and Rao, Y. K. (1991). A study of wheat storage protein monolayers by Faraday wave damping, *Langmuir* 7:2731.
3. Andersson, S., Hyde, S. T., Larsson, K., and Licin, S. (1988). Minimal surfaces and structures: from inorganic and metal crystals to cell membranes and biopolymers, *Chem. Rev.*, 88:221.
4. Williams, R. E. (1968). Space-filling polyhedron: its relation to aggregates of soap bubbles, plant cells and metal crystallites, *Science*, 161:276.

2
Physicochemical Behavior of the Components of Wheat Flour

I. INTRODUCTION

To understand the roles of the various components in wheat flour during dough mixing and baking, we have to know how the main components behave by themselves. This chapter deals with each group of the main components—their structure, interaction with water, and physical properties. With this as a basis, we can discuss in the next chapter how these components interact, and after that we are ready to tackle the dough structure.

II. LIPIDS

Lipids are usually divided in two groups: neutral lipids and polar lipids. This old classification, however, is misleading as many polar lipids are electrically neutral. A more natural classification based on our present knowledge is a functional description of their aqueous interaction. *Polar lipids* are thus the lipids that can interact with water and form aqueous phases, whereas *nonpolar lipids* do not give aqueous phases. This terminology will be used here.

Nonpolar lipids in cereals are dominated by triglycerides, and as they are liquid at room temperature we describe them as oils. They occur mainly in the embryo and as emulsion droplets (spherosomes) in the endosperm. Polar lipids originate from the cell membranes and are dominated by phospholipids and glyco- or galactolipids. These more complex lipid structures are shown in Fig. 1. Two of the most important phospholipids are indicated, and two other members that deserve mention are phosphatidylserine (PS) and phosphatidylinositol (PI), both negatively charged under physiological conditions.

In the living plant and even in a wheat flour there are both lipase and phospholipase activities. Lipase can digest a triglyceride molecule, with prefer-

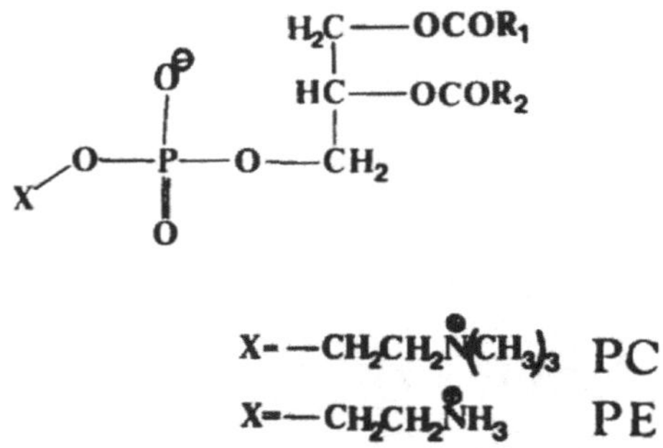

Fig. 1 The 1,2-diacyl-*sn*-glycerol unit of phospholipids and galactolipids. The groups in the 3-position giving phosphatidylcholine (PC) or phosphatidylethanolamine (PE) are shown below.

ence for the 1- and 3-positions, giving two free fatty acid (FFA) molecules and a 2-monoglyceride molecule. Phospholipase A (1- or 2-type) results in hydrolysis of one chain (in the 1- or 2-position, respectively). Thus a phospholipid with one chain is formed (and one FFA), and it is called lysophospholipid because of its ability to cause lysis of living cells such as erythrocytes. There are also two other phospholipases (C and D) which cleave the molecule on each side of the phosphorous atom.

Galactolipids, which comprise a common group of membrane lipids in plants, are also based on glycerol, with acyl chains in the 1- and 2-positions and sugar groups glucosidically linked at the 3-position. This type of lipid has a most important role; it forms the thylacoid membranes of the chloroplasts, where photosynthesis takes place. These membranes are about 50% diacylmonogalactosyl-glycerol (with linolenic acyls), 25% diacyldigalactocyl-glycerol, 15% sulfolipids, and 10% phospholipids. The two glycolipids play an important role in the physical properties of cereal flours. Lipids are not transported between the cells in higher plants; the cell is autonomous in this respect. Therefore the lipid composition can vary a lot from one tissue to another [1].

A. Chemistry of Wheat Lipids

Most studies of wheat lipids have been based on solvent extraction, using various solvents, but a serious complication is that nót all lipids can be recovered in this way. Morrison et al. [2] introduced acid hydrolysis and were able to obtain good quantitative data. The total lipid content of wheat flour is about 2.5% (w/w), and its composition is summarized in Table 1. Starch is about 1% lipids, mostly lysophosphatidylcholine in the form of an inclusion complex described later in this chapter.

Table 1 Lipid Composition of Wheat Flour

Lipid	Total (mg/100 g flour)	Nonstarch[a] (mg/100 g flour)	Starch[b] (mg/100 g flour)
Sterol ester (SE)	90	90	18
Triglyceride (TG)	709	674	35
Diglyceride (DG)	92	86	6
Free fatty acid (FFA)	129	110	19
Esterified monogalactosyl diglyceride (EMGDG) and monoglyceride (MG)	73	66	7
Esterified steryl glycoside (ESG)	77	71	6
Monogalactosyl diglyceride (MGDG)	93	87	6
Monogalactosyl monoglyceride (MGMG)	30	23	7
Digalactocyl diglyceride (DGDG)	226	214	12
Digalactosyl monoglyceride (DGMG)	83	58	25
N-Acyl phosphatidyl ethanolamine (APE)	72	72	—
N-Acyl lysophosphatidyl ethanolamine (ALPE)	34	34	—
Phosphatidyl ethanolamine (PE) and phosphatidyl glycerol (PG)	19	13	6
Lysophosphatidyl ethanolamine (LPE) and lysophosphatidyl glycerol (LPG)	69	10	59
Phosphatidyl choline (PC)	104	66	38
Lysophosphatidyl choline (LPC)	693	36	657
Phosphatidyl serine (PS), lysophosphatidyl serine (LPS), phosphatidyl myoinositol (PI), and lysophosphatidyl myoinositol (LPI)	35	11	24
Total lipids	2628	1703	925

[a]Nonstarch lipids are extracted by water-saturated butanol at 20°C.
[b]Starch lipids were extracted by water-saturated butanol at 90°C.
Source: Data compiled from [2].

Triglycerides occur as a storage reserve, emulsified into spherosomes in the aleurone layer and in the germ. The dominant role of the galactolipids among the membrane lipids reflects the significance of the photosynthesis; they originate from chloroplasts and in turn are transformed into amyloplasts where the starch is synthesized. One unusual lipid in foods is *N*-acylphosphatidylethanolamine.

The fatty acid pattern is dominated by linoleic acid (about 60%) followed by palmitic acid (about 20%) and oleic acid (about 15%). During storage, free fatty acids and lysophospholipids increase owing to lipase activity. Furthermore, the enzymatic oxidation of fatty acids may produce hydroxy fatty acids, with negative sensory effects. Thus lipoxygenase gives a hydroperoxy acid from linoleic acid, and this can form quite stable monohydroxy or trihydroxy fatty acids [3].

B. Significance of Wheat Lipids in Baking

Most reports in the literature on the effect of wheat lipids on breadmaking ignore the physical state of the lipids. The addition of lipid fractions to a wheat flour dough after lipid removal by extraction can give totally different effects depending on their physical state. The same lipid fraction can be delivered in various forms: as an oil, as an emulsion, or as a liquid-crystalline phase. The effect on breadmaking is directly related to this state, which will be described in the next section. First a summary of a few relevant studies on effects will be given.

One way to study the effects of lipids has been to first remove them by extraction and then add fractions of them back in different proportions. Aside from a change in physical structure, the most important objection to this kind of study is that not all of the lipids can be removed. When the most effective extraction agent (water-saturated butanol) is used, the baking performance is lost. This may be due to protein conformational effects. A more probable explanation in our view is that the lipids extracted last, which are the most polar ones, are of crucial importance to baking performance.

The loaf volume of a bread baked from a "defatted" (delipidized) wheat flour has been reported by MacRitchie and Gras [3] to be the same as the loaf volume of the bread baked from the corresponding nondelipidized flour (Fig. 2). Their classical work was well done, but at that time the effects of the different states of lipids were not known, and they added their lipids in dry form. MacRitchie and Gras found that extracted wheat lipids added back to the delipidized flour first give a decrease in loaf volume, then a minimum in the volume curve at a certain level of added lipid. When the same amount is added as was extracted, the loaf volume of the nondelipidized flour is reached. Increasing the level of added lipids increases the loaf volume even more. Contradictory results in studies of lipids added to delipidized flour are frequently reported in the literature. This may be due to the effects of the solvents on the protein functional properties and lack of control of the physical state of the lipids added.

It might also be noted that the improving effect calculated on the basis of the weight of an additive is much greater for lipids than for proteins [4], as illustrated in Fig. 3. The increase in loaf volume per percent of protein was 6.5 mL, and per

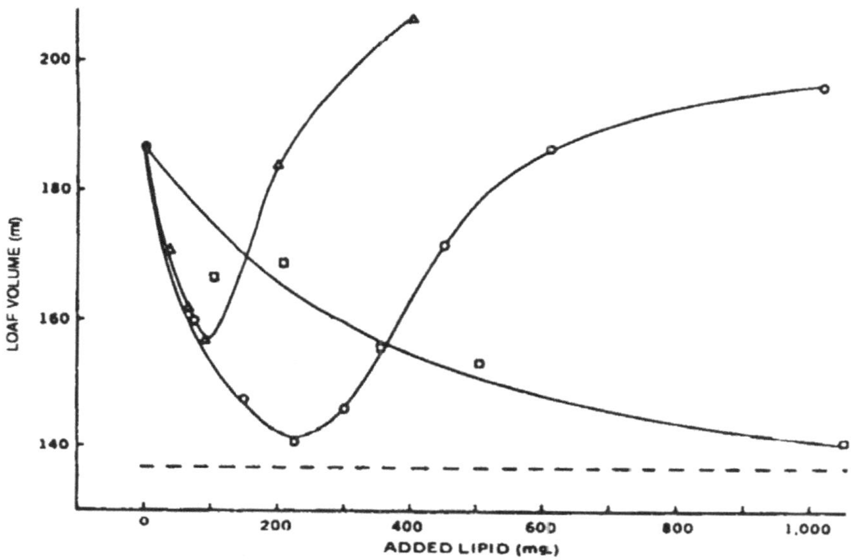

Fig. 2 The relation between loaf volume and lipid content of a wheat flour when extracted lipids are added back to the delipidized flour. (○) Total extracted lipids; (□) extracted nonpolar lipids; (△) extracted polar lipids. (Reprinted with permission from [3].)

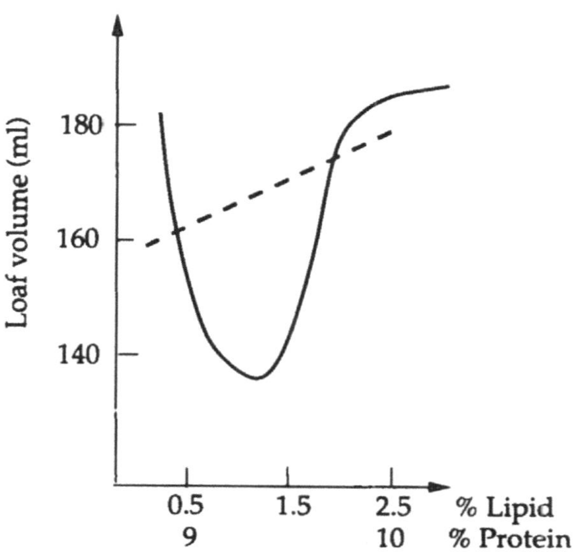

Fig. 3 The improving effect of lipids (solid line) compared with the improving effect of proteins (broken line). (Figure redrawn from [4].)

percent of lipid, 50 mL (in the increasing part of the curve in Fig. 3). But the beneficial effects are not the same for all types of lipids. The nonpolar lipids are even detrimental for loaf volume, as will be discussed below.

Although loaf volume is greatly influenced by the addition of wheat lipids, this is not the case with mixing behavior or rheological properties [4, 5]. The mixogram is identical for delipidized and nondelipidized flours [6]. The mixing process still has an influence on the lipids, which can be described with the concept of lipid binding (cf. [7]). That "lipid binding" is a concept illustrating extractability without any real relation to molecular binding will be discussed in Chapter 3. It has recently been shown that both "free" and "bound" lipids contribute to loaf volume [6] and that there is no relation between baking performance and the proportion of so-called bound lipids [8].

Toward the end of the proofing, the lipid content of the dough does not affect dough volume [4]. It might, then, be concluded that lipids are important primarily during the oven spring. This will be further discussed in Chapter 7.

1. Polar Wheat Lipids

The effects of adding polar wheat lipids on the loaf volume are illustrated in Fig. 2. The minimum in loaf volume is observed at a much lower level of addition and the volume increases faster than when the total lipid fraction is added. The effect of adding polar wheat lipids has been compared with the effect of added shortening, and it was found that the same improving effect (measured as loaf volume) was obtained with 0.5 g of polar lipid as with 3 g of shortening per 100 g of flour [9].

2. Nonpolar Wheat Lipids

When nonpolar wheat lipids are added to a delipidized wheat flour, a decrease in loaf volume is observed (Fig. 2). For a constant lipid level there is a linear decrease in loaf volume when going from 0% nonpolar lipids to 100% [6]. The loaf volume decreased from 220 mL to 160 mL at a total lipid content of 1.34% (calculated on flour basis). The proportion of nonpolar lipids in wheat nonstarch lipids is in the range of 49.1–63.1% [10].

The detrimental effects of nonpolar lipids in baking have been ascribed to the presence of free fatty acids [4]. The reasons fats or shortenings are beneficial in baking are discussed in Chapter 6. Although the nonpolar fraction is rich in triglycerides, it contains many other components, in addition to exhibiting a different melting point.

3. Differences Between Varieties

The curves in Fig. 2 showing the relation between loaf volume and amount of added total wheat lipids is not the same for all wheat flours [4, 11]. A good baking flour and a poor baking flour are compared in Fig. 4, and several

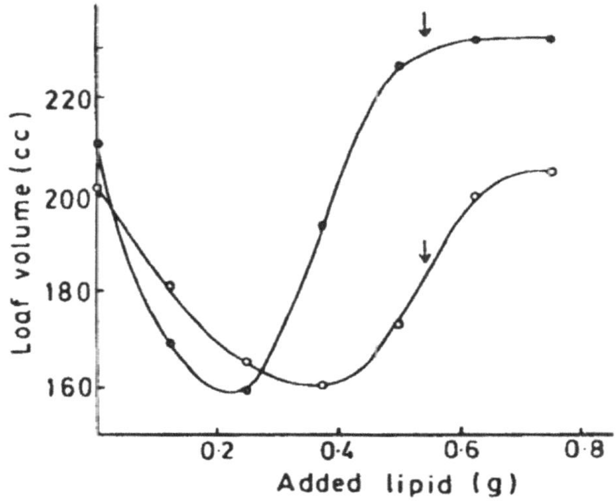

Fig. 4 A comparison of the loaf volume–lipid content relation for a good wheat flour (●) and a poor wheat flour (○). (Reprinted with permission from [4].)

interesting results are shown in this figure. The difference in loaf volume for the delipidized flours is much smaller than the difference for the flours with their native lipid content. The delipidized good flour had a loaf volume of 210 mL, and the value for the delipidized poor flour was 200 mL. With the native lipid content the good flour had a loaf volume of 230 mL, and the poor flour, 185 mL. The minimum in the curve occurred at a much lower lipid content for the good flour than for the poor flour (0.2 g of added lipid compared with 0.4 g). And, finally, more lipid was required for the poor flour to exceed the loaf volume of the delipidized flour than in the case of the good flour (0.4 g for the good flour, 0.6 g for the poor). These results indicate that for the potential loaf volume, which is governed by the proteins present, to be achieved, the lipids must play their role properly.

Pomeranz et al. [9] showed that when wheat lipids extracted from six different varieties were added to the same flour, only very small differences in volume were observed. Addition of the total lipid extract resulted in loaf volume in the range 873–905 mL, the polar lipids gave 910–938 mL, and the nonpolar lipids gave 818–868 mL. Also in other investigations it has been found that exchange of lipids between varieties does not influence the loaf volume [6, 8, 12]. However, the effect of added shortening seems to be related to the wheat variety [6, 9].

4. Correlation Between Baking Performance and Lipid Content and Composition

Many studies exist in which lipid content and composition (the lipids extracted in different ways) and baking performance are correlated (cf. [13],[14]). The best parameter for predicting baking performance seems to be the polar/nonpolar ratio of the so-called free lipids [10, 13, 15, 16]. To improve the correlations the loaf volumes are normalized to the same protein content. Seasonal variations seem to greatly influence the relation between loaf volume and lipid content [6, 17].

In recent work on reconstitution using lipid fractions in breadmaking, the effect of different types of lipids on bread volume was quantified [10]. At a constant lipid content [1.34% (w/w) nonstarch lipids], increasing proportions of nonpolar lipids gave a strong reduction in loaf volume (about 30%). This finding is also in agreement with earlier results on lipids in baking reported by MacRitchie [3]. McCormack et al. [10] examined flour from 15 cultivars and obtained the following equation from linear regression analysis of the results.

$$\text{Loaf volume (untreated)} = 11.8 + 0.95 \times \text{loaf volume (delipidized)} \quad (1)$$

For American and Canadian wheat, the glycolipids are claimed to be useful for predicting loaf volume, but this is not the case for wheat from the United Kingdom or New Zealand [17]. As shown below, it is quite obvious from the phase properties that it is not the glycolipids per se that are important, but the ratio of those with one sugar group to those with two.

C. Aqueous Systems of Wheat Lipids

The functional role of wheat lipids in baking relative to their aqueous interactions will be described below. We can imagine that cell membrane lipids (which belong to the polar lipid group) will interact with nonpolar lipids (such as triglyceride oil) during mixing of the dough to form aggregates of intermediate compositions.

There is a simple way to illustrate the aqueous interaction of all these possible intermediate compositions, and that is in the form of a ternary phase diagram (see Fig. 5). The position within the diagram defines the composition, and the boundaries of the different phases formed with water can then be illustrated as we draw a map. Such a phase diagram represents the thermodynamic equilibrium between phases at one temperature.

In the determination of ternary phase diagrams the lipids have to be extracted in a first step. The most effective extraction solvent we know is water-saturated *n*-butanol (WSB) (cf.[2]). About 90% (w/w) of the total wheat lipids can be obtained in this way. WSB is introduced as a two-phase mixture, 80:20 *n*-butanol

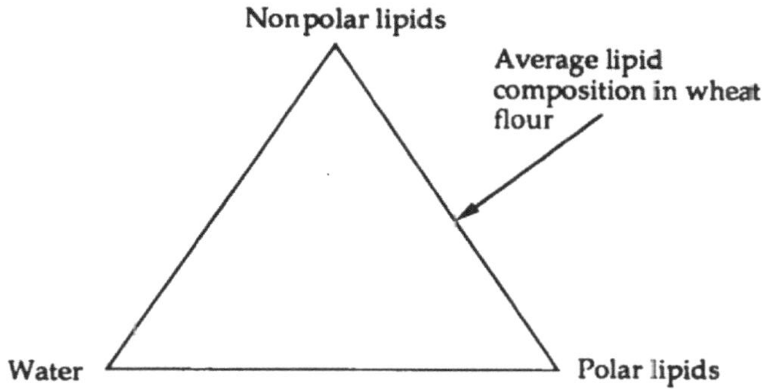

Fig. 5 A ternary composition diagram for polar lipids, water, and nonpolar lipids, illustrating the principles behind phase diagrams.

water (v/v), and as soon as a small amount of lipid is dissolved it is transformed into one phase. This phase is, in fact, a microemulsion consisting of water aggregates in a continuous butanol matrix. The lipids are then separated on a column, and the polar and nonpolar components are mixed with water to illustrate all possible combinations in the phase diagram [4]. The polar lipids are the lipids that interact with water to give aqueous phases. The nonpolar lipids are simply the rest.

Lipids from a strong spring wheat (Amy) give the phase behavior shown in Fig. 6 [18]. The addition of water at all proportions of polar to nonpolar lipids first gives a reversed hexagonal phase, then a lamellar liquid crystalline phase, and finally an L2 phase (a reversed micellar solution). In the proportions of polar to nonpolar lipids present in wheat flour and at higher proportions of nonpolar lipids there is an additional phase consisting of the oil (mainly triglycerides). This also means that in an excess of water there are in fact four phases, a situation that is in disagreement with the phase rule. This is, however, not surprising as there are more than three components (there are numerous lipid classes and species within each group of polar and nonpolar lipids). It is more surprising that the behavior of such a complex system can be illustrated by a ternary phase diagram.

The lamellar liquid crystalline phase exists over a wide composition range as shown in Fig. 6. The swelling, measured by X-ray diffraction of the total lipids (the dashed line), stops at about 45% (w/w) water, with a thickness of the water layer + lipid bilayer of about 75 Å. The polar fraction takes up more water, and the largest spacing is greater than 80 Å. From these X-ray data it is possible to calculate the thickness of the polar lipid bilayer, which is 49 Å. This is in good agreement with the spacing we can observe in a dried gluten powder preparation, and with the addition of water this spacing shifts continuously toward about 58 Å

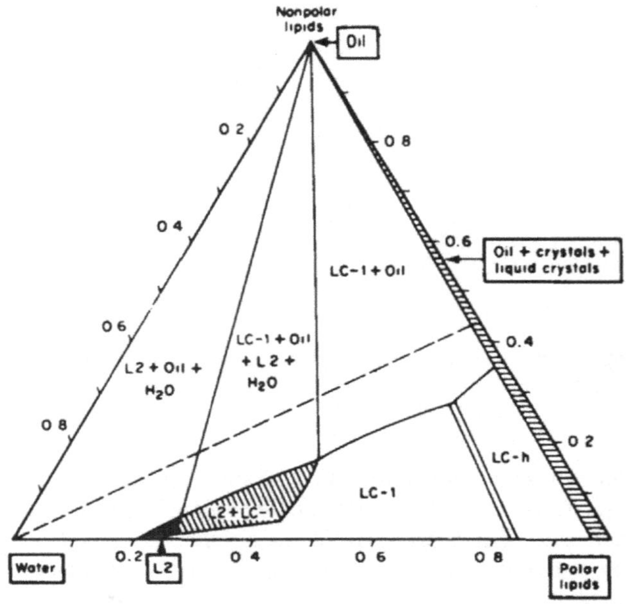

Fig. 6 Ternary phase diagram of wheat lipids at 25°C. The hexagonal liquid crystalline (LC-H), lamellar liquid crystalline (LC-L), L2, and nonpolar oil phases are the one-phase regions. The broken line corresponds to wheat flour. Regions where 2, 3, or 4 phases occur together are also indicated. (Reprinted with permission from [18].)

at total hydration. This is good evidence that the same lamellar liquid crystalline phase of wheat lipids exists in the dough.

In order to extract the starch lipids with WSB, it is necessary to gelatinize the granules. Some lipids, however, can be removed even at room temperature, probably from the surface zone of the granule. The corresponding phase diagram of lipids extracted from gluten illustrates this difference [19]. The gluten lipid–water phase diagram is shown in Fig. 7. The main difference is the extension of the lamellar liquid crystalline phase toward the water corner. This phase can accommodate more nonpolar lipids than that of the lipids from wheat flour (Fig. 6). The main features, however, are the same.

The water phase used in baking contains salt, and it was therefore of interest to study lipid interaction also with salt solution. The corresponding phase diagram is shown in Fig. 8. The main effect of the presence of salt is a smaller lamellar liquid crystalline region (less solubilization of nonpolar lipids) and less water in the L2 phase. Another remarkable difference is that almost no hexagonal phase is formed (it was observed only as traces dispersed in the lamellar phase, which could be recovered by centrifugation)[19].

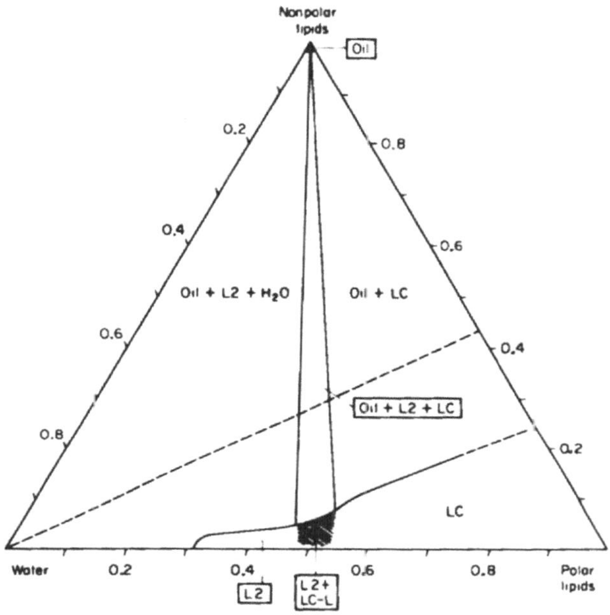

Fig. 7 Ternary phase diagram of the aqueous system of lipids from gluten separated into polar and nonpolar fractions. The same symbols are used as in Fig. 6. (Reprinted with permission from [19].)

The acqueous interaction of lipids from a wheat variety with poor baking performance as well as of lipids from other cereals is shown in Fig. 9. The phase behavior has been illustrated as a binary system where we neglect the presence of the nonpolar lipid–oil phase [20]. If we first consider the two wheat varieties, there is no region where the lamellar liquid crystalline phase exists alone in lipids from Maris Huntsman whereas the lipids from Amy show a large region of the lamellar phase. It has been demonstrated in different ways that baking performance is improved by the presence of the lamellar liquid crystalline phase (cf. [21] and [22]. Furthermore, it is possible to detect lipid aggregates consisting of lipid bilayers just like liposomes in gluten gels [23], which are prepared by simply removing starch from a wheat flour dough.

The lipids from rye and triticale (Fig. 9) do not exhibit any region with the lamellar liquid crystalline phase. Also, other studies in our department have shown good correlation between baking performance and the ability of the lipids to form the lamellar liquid crystalline phase. This is, of course, only one of many requirements, as will be further discussed in Chapter 6.

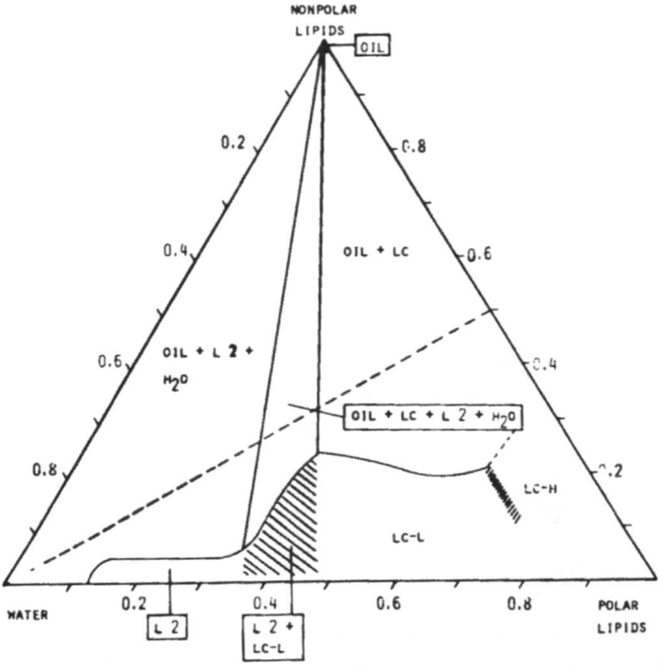

Fig. 8 Phase diagram of wheat flour lipids in an aqueous salt solution corresponding to that in dough (0.26 M NaCl + 0.07 M KCl + 0.016 M $CaCO_3$ + 0.004 M $CaCl_2$). The same symbols are used as in Fig. 6. (Reprinted with permission from [9].)

There have been many studies on wheat lipids in breadmaking. During recent years they have focused on relations between the polar/nonpolar lipids ratio and baking performance (cf. [10]). All studies show that a higher relative level of polar lipids improves baking performance. These results can be directly explained by reference to the lipid phase behavior demonstrated here. Lipids favoring the lamellar liquid crystalline phase (L_α) are positive, and those destabilizing this phase are negative for baking performance.

Phospholipids favor the existence of the L_α phase [24]. To determine the effect of glycolipids with regard to formation of the L_α phase we examined the ternary phase diagram of diacylmonogalactosyl-glycerol (DMG)/diacyldigalactosyl-glycerol (DDG)/water. The result is shown in Fig. 10. It can be seen that at a critical ratio of DMG to DDG, the $L\alpha$ phase is transformed into a reversed hexagonal phase. This result shows that if the amount of DMG is too high in relation to DDG and phospholipids, the wheat lipids will exhibit a poor contribution to baking performance.

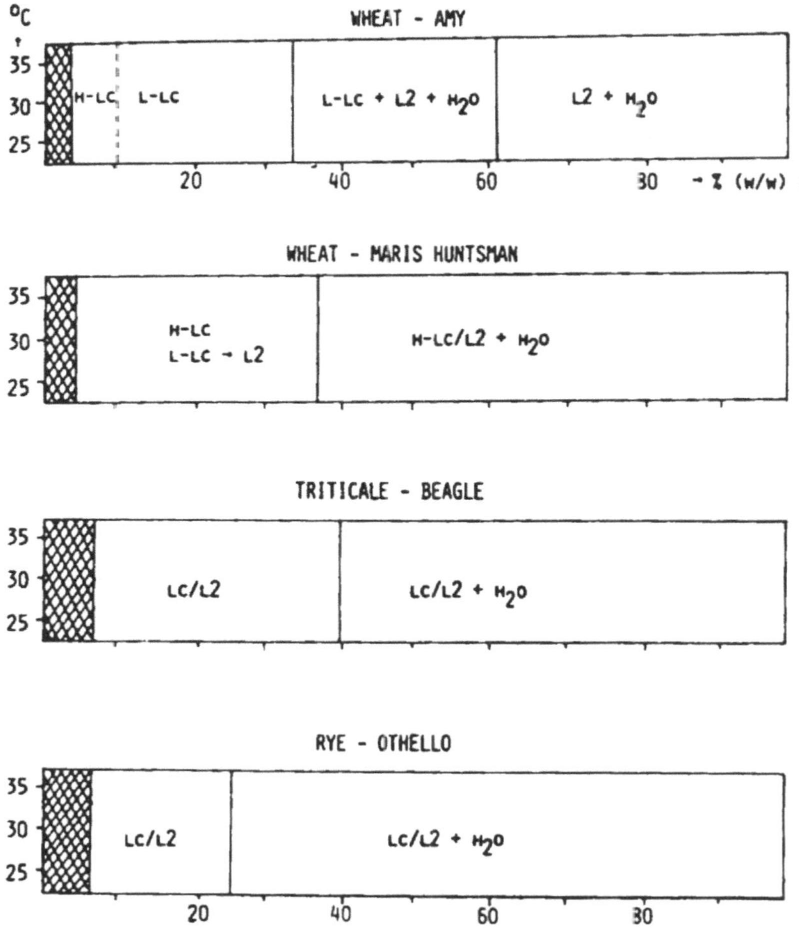

Fig. 9 Phase behavior of lipids from the winter wheat Maris Huntsman, from triticale (rye/wheat) Beagle, and from rye Othello compared to lipids from the wheat Amy (Fig. 6). The same symbols are used as in Fig. 6. (Reprinted with permission from [20], p. 205, by courtesy of Marcel Dekker Inc.)

D. Monolayers of Wheat Lipids

There are numerous studies on the lipid monolayer behavior at the air/water interface based on spreading from a solvent in the case of a so-called insoluble monolayer (cf. [25]). Almost nothing, however, has been reported on spreading from liquid crystalline phases, which is the physical state of wheat lipids during

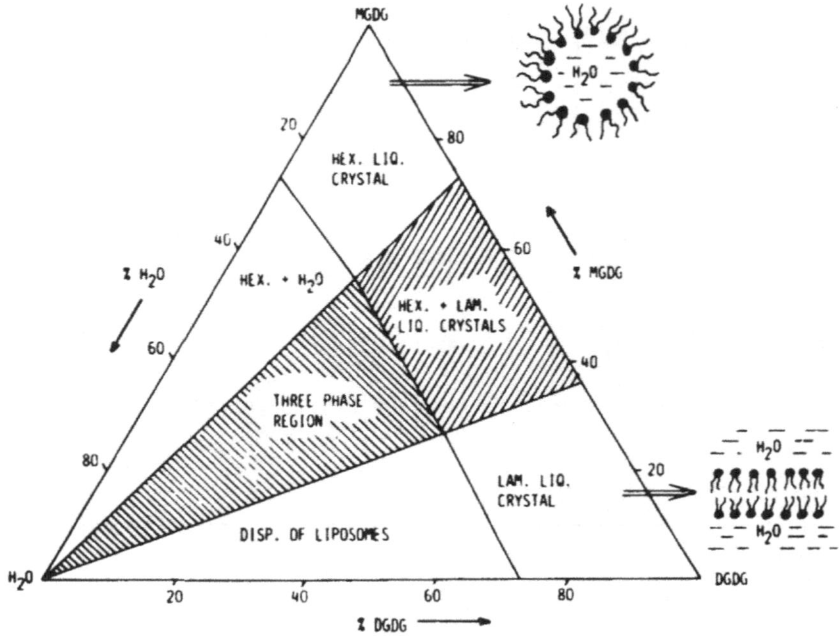

Fig. 10 Ternary phase diagram for DMG/DDG/water, (From [24].) Alternative symbols used in the literature are MGDG for DMG and DGDG for DDG.

dough mixing. We have done some simple model experiments in our laboratory [26] to try to simulate the circumstances during dough mixing.

The polar lipids are dominated by galactolipids and phospholipids, and the hydrocarbon chain melts below room temperature in an excess of water. To have a simple well-defined lipid, dimyristoyl phosphatidylcholine (DMPC) was chosen, with a chain melting point of 23°C. The experiments were done at 25°C, and the conditions using the Wilhelmy type of balance were the same as those discussed in connection with protein monolayers. If some crystals are dispersed in bulk, we should expect the formation of an adsorbed monolayer. However, no film pressure could be recorded even after many hours. This is due to the extremely low solubility of DMPC in water, about 10^{-11} mol/L, and the transport to the surface has to take place via the monomers. If the crystals are deposited on the surface (floating due to surface tension) the lamellar liquid crystalline phase will be formed in situ. Spreading will thus take place from the same phase that wheat lipids form in equilibrium with excess water (see above). After about half an hour a few crystals with a mass of about 0.1 mg will exhibit a film pressure of about 2–3 mN/m. The pressure will gradually increase toward the values exhibited by the pressure–area isotherm.

Alternatively, we can increase the film pressure using the surface balance (Fig. 1, Chap. 1) if we start compression by the barriers, and a pressure–area curve without any sign of phase transition or the occurrence of any linear region corresponding to solid-condensed monolayer phases is obtained. The isotherm is shown in Fig. 11, and it is characteristic for all lipids forming the lamellar liquid crystalline phase in an excess of water. The collapse point occurs at about 40 mN/m, and the cross-sectional area per chain is about 30 Å^2, which is in good agreement with the value observed in the lamellar liquid crystalline phase [26].

III. PROTEINS

The protein content of wheat is in the range 7–20% [27,28]. Since Finney and Barmore published their papers [12,27] on the relation between loaf volume and protein content an enormous amount of work has been done to understand this relation. In the agronomics field this has meant increasing the use of fertilizers to increase the protein content of the wheat grain, and in northern Europe high-protein wheat from North America is imported to support the home-grown wheat. The relation between loaf volume and protein content is linear, at least for protein contents in the range of 9–18% (see Fig. 12) [12,27]. If the regression

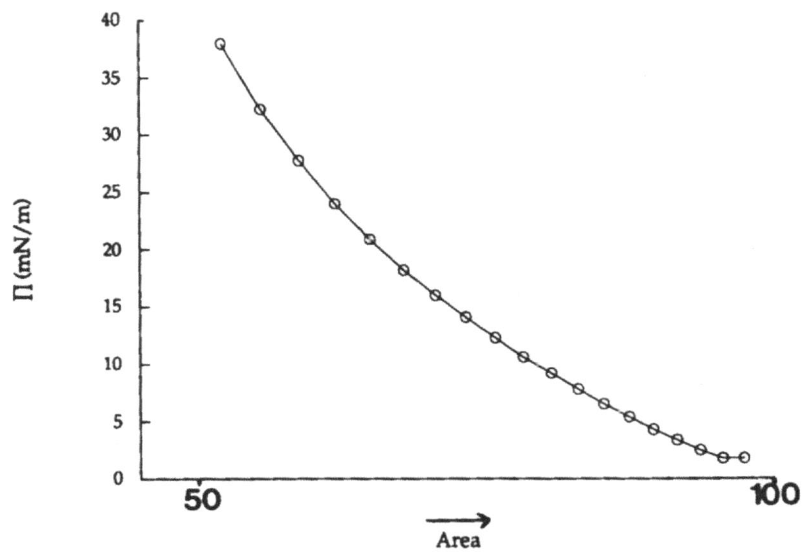

Fig. 11 Pressure–area isotherm of DMPC at 25°C, spread from crystals on the water surface. Estimated molecular areas are given in square angstroms per molecule.

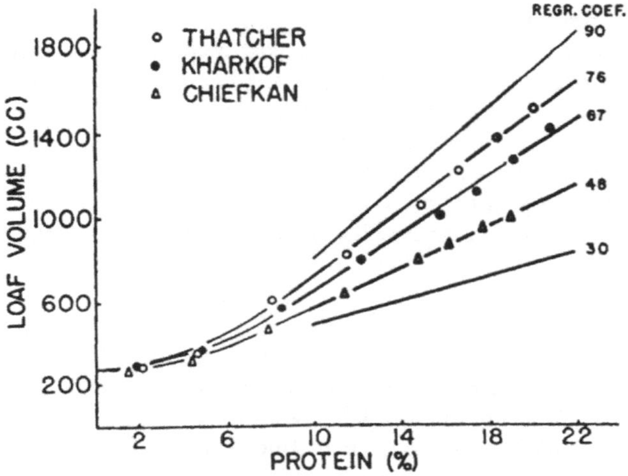

Fig. 12 Loaf volume as a function of the protein content of the flour for three varieties of wheat. (Reprinted with permission from [12].)

lines in Fig. 12 are extrapolated to the low protein levels they will cross for all varieties at about 6–8% protein. This protein content might be regarded as the amount necessary for a continuous protein phase in the dough.

About 90% of the variation in loaf volume for a single wheat variety was ascribed to the protein content [27]. However, when the relation between loaf volume and protein content is investigated for different wheat varieties, the correlation is not as good. For 71 Australian wheat varieties the correlation coefficient was only 0.39 [29]. The importance of the protein in baking is thus a question not only of quantity, but also of quality. The results in Fig. 12 show that it is possible to improve the baking performance of a wheat variety by increasing the protein content, but still we do not know why this is so. The uniqueness of wheat proteins among cereal proteins in viscoelastic properties and gas-holding capacity is striking (see e.g., [30],[31]). Some aspects of these quality parameters of the wheat proteins will be discussed in this section.

A. Composition of Wheat Proteins

Our understanding of the role of wheat proteins in baking is still incomplete, and two reasons for that are undoubtedly the complexity of their composition and their physical properties. The first step in the study of protein composition is often the fractional extraction procedure used by Osborne [32]. With the original procedure, four protein fractions are obtained, but this procedure is often modified to give five classes of proteins: albumins (soluble in water), globulins (soluble in salt solutions), gliadins (soluble in aqueous ethanol), glutenins (soluble, or rather dispersible, in dilute acid or alkali), and an insoluble residue. The percentage of the total protein content in each class is shown in Table 2.

As the fractionation is based on extraction procedures, there are no clear-cut limits between the classes. The composition of each class will depend on the exact conditions during the extraction, such as temperature and solvent/flour ratio [38–40]. For complete extraction of the wheat proteins, many different solvent systems have been tried. If surfactants [e.g., sodium dodecyl sulfate (SDS), or cetyltrimethyl ammonium bromide (CTAB)] or soaps are included, 90–95% of the proteins are claimed to be extracted [41–43]. It has been suggested that the extractability in certain solvents reflects the baking performance of the wheat flour; for example, less soap is needed to extract the proteins from a poor variety [41]. On the other hand, the extractability in SDS has been found to be related to the protein content but not to protein quality [44].

Another way to extract the wheat protein is illustrated in Fig. 13. The definition of gluten is also given in this figure. Wheat flour and water are mixed into a dough, which is then washed in excess water (or dilute salt solutions).

Table 2 Percentage of Total Protein Content in the Five Wheat Protein Classes

Albumin (%)	Globulin (%)	Gliadin (%)	Glutenin (%)	Residue (%)	Ref.
14.7	7.0	32.6	45.7	—[a]	33
15	3	33	16	33	30
30.7[b]	18.2[c]	16.0	24.3	10.8	34
24.7–30.9[d]		37.7–43.9	27.1–33.9	—[a]	35
7–10	6–10	40–45	40–45	—[a]	36
24[d]		33	43	—[a]	37

[a]Not included in the fractionation procedure.
[b]Also contains gliadins.
[c]Contains both albumins and globulins.
[d]Shows the sum of albumins and globulins.

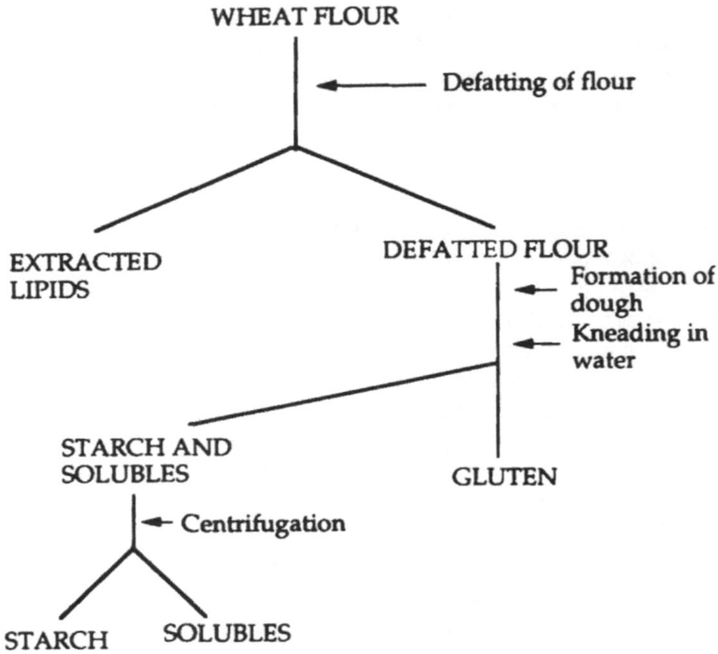

Fig. 13 Separation of flour components and the definition of gluten. (Adapted from [38].)

Gluten is formed as a cohesive mass, and starch and soluble components are washed away in the washing liquid. From this definition it is evident that "gluten" is not an unambiguous term. Gluten is what we get when we wash a wheat flour dough in water to remove soluble components and starch—a mixture of proteins, lipids, and carbohydrates. We will use the term *gluten gel* to describe gluten prepared as illustrated in Fig. 13. Some data on the composition of gluten are given in Table 3. In this chapter the albumins and globulins according to the Osborne fractionation procedure and solubles according to Fig. 13 will be referred to as *soluble proteins*. The gliadins and glutenins are often described as the *gluten proteins*. The definition in Fig. 13, however, implies that gluten is formed during dough mixing, and consequently gluten is not found in the wheat endosperm. A better term to describe gliadins and glutenins found in the endosperm could then be *wheat storage protein* (WSP). However, the term gluten is used in many of the cited papers, and we have not changed the nomenclature used by the authors. Therefore, "gluten," "gluten proteins," and "wheat storage proteins" are used to describe the same protein fraction.

Table 3 The Composition of Commercially Produced Gluten

Component	Example 1 (45)	Example 2 (46)	Example 3 (47)	Example 4 (48)
Water (%)	7.5	6.4		
Protein (%)	77.4[a]	77.5	77.0	77.6
Lipid (%)	5.8	5.7	1.15[b]	7.11[c]
Starch (%)		9.4		10.7
Pentosans (%)			1.42[d]	0.90[c]
Ash (%)	0.69	0.73	0.90	0.83
Number of samples, N	30	32	61	27

[a]Dry matter basis.
[b]Free lipids.
[c]$N = 8$.
[d]$N = 26$.

1. Soluble Proteins

a. Definition and Classification

The separation of the soluble proteins can be performed according to the gluten-washing procedure described in Fig. 13. After centrifugation to settle the starch, the nongluten proteins can be obtained from the supernate. This supernate contains not only proteins but also ions, peptides, amino acids, water-soluble pentosans, and glycoprotein [38]. If the supernate is dialyzed to remove low molecular weight components, the globulins will precipitate. Albumins might be recovered after heat treatment, and remaining in the supernate are water-soluble pentosans and glycoproteins. The water-soluble fraction obtained as described does not contain only protein, and consequently it has been described by several terms including solubles, water-solubles, nongluten proteins, metabolic proteins, and cytoplasmatic proteins (see, e.g., [49]).

Some literature data on the proportion of albumins and globulins of the total protein content are given in Table 2. Albumins are thus more prevalent than globulins. As albumins are water-soluble and there are always salts present in the flour, it is not possible to extract albumins directly from flour without obtaining globulins as well. Thus, a dialysis step is often included in the separation procedure. Globulins will then precipitate whereas albumins will remain in solution. An alternative procedure to extract the soluble proteins is to extract the wheat flour with a solution of NaCl, usually at a concentration of 0.04–0.5 M.

Dialysis against water as described above causes the globulins to precipitate, and after centrifugation the supernate is freeze-dried to recover the albumins. In this way the albumins will not be heat-denaturated.

As albumins and globulins are defined according to their solubility there is no clear distinction between them, and each fraction can easily be contaminated by gliadin protein (see Fig. 14). Even glutenins might be extracted [50]. As already discussed, the proportions of different classes of proteins depend on the flour/solvent ratio and on the solvent (water or salt, and concentration of the salt solution) [39,40]. Some examples of the extraction of different "soluble" proteins are given in Table 4. The amount of salt-soluble proteins is reduced if the flour is delipidized before extraction [40]. Even the physiological state of the tissue might affect the extractability [40]. Albumin-like proteins can also be entrapped in the glutenin structure or occur as subunits of glutenin. It is usually stated that the level of nongluten proteins is fairly constant and is not typical of the variety. The percentage of nongluten proteins will decrease if the total amount of protein is increased, for example, because of increased nitrogen fertilization. This is because an increase in nitrogen will cause an increase in storage protein but not in metabolic proteins.

b. Chemical Composition

The group of soluble proteins is very heterogeneous in composition (see Table 5). The group contains enzymes, both metabolic enzymes from the developing grain and hydrolytic enzymes necessary for the germination of the seed; and also enzyme inhibitors. Some proteins (phytohemagglutinins) have the ability to clot red blood cells. Other proteins are storage globulins in protein bodies, similar to

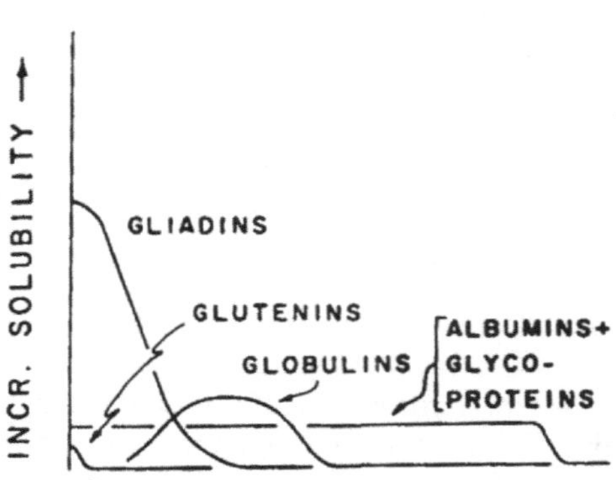

Fig. 14 Protein classes extracted at different salt concentrations. (Reprinted with permission from [50].)

Table 4 Soluble Proteins Extracted from Wheat

Protein[a]	Extraction method[b]	Percent of total protein extracted	Protein content in fraction (%)	Ref.
WSF	Distilled water	15.5–20.3		51
Calcium-binding albumin	Remaining in solution after precipitation with ammonium sulfate and binding with EDTA	1.5		52
Lipid–protein complex	Petroleum ether		30–60	53
Globulin	6% NaCl	7	12.8	53
"S"-Protein	Ammonium sulfate precipitation of HAc-soluble proteins in AUC solvent	5		54
Ligolin	In supernate after ammonium sulfate precipitation of acetic acid solubles in AUC	10[c]		55 34
Water-solubles	Gluten preparation	8.5	27	50
Albumins, globulins	0.04 M NaCl		18.2	34

[a]WSF = water-soluble fraction.
[b]AUC = acetic acid–urea–cetyltrimethylammonium bromide.
[c]Percent of gluten.

the globulins in legumes. Another group of "soluble" proteins are described in the literature as lipoproteins. One example is the so-called CM proteins, named for their extractability in chloroform-methanol.

Although the nongluten proteins are a complex mixture of many different proteins, they have some properties in common that differentiate them from the gluten (or storage) proteins. The soluble proteins have a higher mobility than gliadins in electrophoresis at pH 3, and they also have the highest mobility in sodium dodecyl sulfate polyacrylamide gel electrophoresis (SDS-PAGE) [30]. Most of the soluble proteins seem to have a molecular weight below 40,000 [57]. The isoelectric point is in the pH range 5–8. The pattern obtained in isoelectric focusing (IEF) differs between varieties.

The heterogeneity of the soluble proteins is easily detected with electrophoresis or high-performance liquid chromatography (HPLC). In gel electrophoresis

Table 5 Proteins Included in the "Solubles" Group, Their Biological Functionality, and Their Role in Baking

---

Enzymes

Examples: α-amylase, β-amylase, protease, pullulanase, lipase, phosphatases (e.g., phytase), lipoxygenase, polyphenoloxidase, oxidoreductase (e.g., peroxidase and catalase).

Biological functionality: catalyze chemical reactions, the nature of which depends on the enzyme in question.

Role in baking: Hydrolyze starch to fermentable sugars (the amylases), might affect the rheological properties of the doughs (amylases and proteases), might affect SS/SH exchange (lipoxygenase), increase the level of free fatty acids (lipase), can cause darkening of the dough (polyphenol oxidase), catalyze the conversion of ascorbic acid to dehydroascorbic acid (ascorbic acid oxidase).

Enzyme Inhibitors

Examples[a]: chymosin inhibitors, bifunctional α-amylase/subtilisin inhibitors, bifunctional α-amylase/trypsin inhibitors, embryo trypsin inhibitors.

Biological functionality: inhibit hydrolytic enzymes, might affect resistance against insects.

Role in baking: not known.

Lipoproteins

Examples: purothionins, CM proteins, ligolin, "S" protein.

Biological functionality: constitute membranes.

Role in baking: thought to cause "lipid binding" during mixing and to facilitate protein–lipid interactions during baking.

Lectins

Examples: wheat germ agglutinin.

Biological functionality: imparts disease resistance.

Role in baking: not known.

Storage protein

Examples: globulins

Biological functionality: storage protein

Role in baking: not known.

---

[a]According to [40].
Source: Compiled from several sources, including [30], [40], [49], [56], [57].

at pH 8.9, 21 bands were discovered and at pH 3.2, 12 bands. HPLC has revealed 50–100 peaks [30]. Two-dimensional electrophoresis has been reported to show about 160 albumin components and 140 globulin components. The amino acid composition is characterized by high levels of lysine and low levels of glutamine compared with the gluten proteins. Many of these proteins are located in the embryo and in the aleurone layer, but they might also be found in the endosperm. α-Amylases in the developing kernel are located in the pericarp, whereas the α-amylases in the germinating seed are located in the aleurone [49]. The β-amylases are located in the endosperm. The purothionins are found only in the endosperm [40].

c. Description of Individual Components

Some of the nongluten proteins are known in more detail, and a few examples will be discussed here. Information on extraction procedures is given in Table 4, and a summary of the soluble proteins, their biological role, and their role in baking is given in Table 5.

A *calcium-binding albumin* has been described [52], and it was suggested that this protein, through the Ca^{2+}, will also bind phospholipids and added surfactants. Protein-lipid interactions will be discussed in Chapter 3. The molecular weight of the protein was estimated to be 16,500. The amino acid composition was characterized by low levels of sulfur-containing amino acids but high levels of lysine compared to the albumin fraction as a whole. The protein represents about 0.2% of the flour.

Some of the globulins seem to be very similar to the globulins in legumes and could therefore be described as *storage globulins*. They are not, however, similar to the ordinary storage proteins in wheat and are not included in the wheat storage protein fraction. They have molecular weights in the range 150,000–160,000, and after reduction they give monomers of molecular weights in the range 20,000–25,000 [30]. The amino acid composition is similar to that of the globulins in legumes. These globulins are the main storage protein in the embryo [58].

Most of the albumins and globulins are *enzymes*, but the only enzyme present in a large amount in sound wheat is β-amylase. This enzyme exists in a "free" form and a "bound" form, the latter being bound to glutenins through S—S bonds (cf. [57]). The bound β-amylase can be released by the addition of cysteine. It can also aggregate with itself. In several aspects this enzyme is similar to the storage proteins. The molecular weight of the monomer is about 64,000 [40]. β-amylase has a high mole percent of the amino acid lysine, 4.0 mol % and more in high-lysine varieties [40]. In sound wheat the level of α-amylase is very low, whereas its activity is a problem in sprouted wheat. Several isomers of α-amylase are present [49].

The proteases are usually not easy to extract. They might be sensitive to thiol-blocking agents, and some of them are associated with gluten [30,49].

Proteins that are *enzyme inhibitors* are also present. The α-amylase inhibitors have been reported to have molecular weights of 12,000, 24,000, and 60,000. They are thermostable but are usually inactivated during bread baking [57]. The trypsin inhibitors constitute about 5–10% of the water-soluble proteins [40,49]. The cysteine content is high, whereas the content of glutamic acid and glutamine is low. The molecular weight of the trypsin inhibitors has been reported at 12,000–13,000 [57].

Many of the protease inhibitors are bifunctional and also inhibit α-amylases [40]. There is considerable homology in the amino acid sequences of different α-amylase inhibitors as well as between amylase inhibitors and other inhibitors [40].

A whole range of *protein–lipid complexes* or *lipoproteins* have been described in the literature. Such complexes can be obtained from the petroleum ether–extracted oil, or they can be extracted by 6% sodium chloride (NaCl). Hydrochloric acid (HCl) disrupts the complex [53]. Protein–lipid complexes extracted in this way were further divided in two fractions, one containing 60% protein, 35% carbohydrate, and almost no lipids, and the other containing 30% protein, 7.5% carbohydrate, and 62.5% lipid. It was shown that the lipids were not covalently bound and were polar. Also, when 6% NaCl was used to extract solubles, a globulin was obtained that contained 13.8% lipid [53]. As in the case of "ligolin" discussed below, it is quite possible that this is not a true complex, but only lipids that happen to have the same solubilization behavior as the actual protein. The effect of hydrochloric acid might support this, as a change in pH alone could not be expected to dissociate the complex.

There are low molecular weight (LMW) thiol peptides present, which might facilitate thiol–disulfide (SH/SS) exchange reactions. One example of the cysteine-rich proteins is gluthathione; another is the purothionin group [56]. Purothionins are extracted with petroleum ether, salt solutions, or acids. There are two proteins, α- and β-purothionin, both consisting of 45 amino acids and having a molecular weight around 5000. Viscosity measurements indicate that these proteins are compact, spherical globular proteins. The α-helix content was estimated to be about 40% [49]. These proteins are toxic against certain bacteria, yeast, and insects. The mechanism of action is their ability to modify the permeability of membranes [57] The lipid part of this complex consists of phospholipids and glycolipids [49]. The strongest association is obtained between the protein and the phospholipids [57]. Although the purothionins are extracted as a protein–lipid complex, they are not believed to be involved in "lipid binding" during dough mixing [55].

CM proteins are of low molecular weight (below 25,000), and the amino acid sequence is similar to that of the amylase inhibitors. As mentioned above, they are extracted with chloroform–methanol and are difficult to dissolve in water. The complex is dissociated by acids, and the protein is then soluble in water [49].

Among the protein–lipid complexes extracted from wheat flours or doughs, ligolin, described by Frazier et al. [55], has attracted special attention in connection with baking. This protein associates with triglycerides in a molar ratio of 1:1 when doughs are mixed in nitrogen. The protein has a low molecular weight (<10,000) and is thought to be responsible for most of the "lipid binding" during dough mixing. Ligolin will be further discussed in Chapter 3. The amino acid composition of the protein is characterized by more asparagine and less glutamine than in glutenin, as well as less proline and more cysteine, although the cysteine content is not as high as in purothionin. The protein–lipid complex has a great tendency to aggregate, and the complex can be dissociated by SDS. It was suggested that the interaction between protein and lipid in the complex is hydrophobic. Ligolin also has a strong tendendy to associate with glutenin.

Another protein–lipid complex, "S"-protein, soluble in 0.1M NaCl, was described by Zawistowska et al. [54]. This protein constituted about 5% of the total amount of protein in the flour and was associated with 20% of the polar lipids. The composition did not differ between varieties and was not affected by defatting of the flour. This protein was high in cysteine, glutamic acid, and proline The lipid content was 22–25%, mostly polar lipids. The SDS-PAGE molecular weight was 14,000–16,000. "S"-protein is globulin-like. It has been suggested that "S"-protein and ligolin are the same protein [7].

As a final remark on these lipoproteins, we think that they often are membrane proteins. Due to the hydrophobic character of the major part of the proteins, the separation conditions become complex, and membrane lipids will behave like true lipoproteins when exposed to lipophilic solvents.

The *agglutinins* or *lectins* are found in the embryo and are known as *wheat germ agglutinin* (WGA). They are glycoproteins that bind to the surface of the cell and can cause cells to clot. The molecular weight has been reported to be 23,500 for one agglutinin and 17,800 for another. Glucose is the main carbohydrate in the former protein [30].

The glycoproteins found in the water extract of a wheat flour are included among the pentosans. They can form gels in the presence of oxidants, and they will be discussed further later in this chapter.

d. Significance in Baking

It seems to be a common opinion that the composition of albumins and globulins does not vary between wheat varieties, and there is no correlation between the amount of albumins or globulins and baking performance [30,38]. It is thus not possible to use the albumins and/or globulins to identify varieties or to discriminate between varieties differing in baking performance. However, Lásztity [57] described albumins and globulins that were reduced when treated with 2-mercaptoethanol (2-ME), and these proteins were found in varieties that give good baking performance. There are also results that indicate that more soluble protein might be extracted from a good variety of wheat (Monopol) than from a poor one (Maris Huntsman) [51]. The albumin content differs between hexaploid and tetraploid wheats, so it may be possible to use the albumin composition to identify bread wheat in durum.

MacRitchie [38] showed in reconstitution and exchange experiments that the soluble fraction is not responsible for the difference in baking performance between varieties. If the water-soluble fraction (WSF) is omitted in reconstitution experiments the loaf volume will be decreased [12,50,59]. The differences in volume are observed during fermentation [12]. The dialysate obtained when albumins and globulins are separated might improve the baking performance. The reason for this improvement is that the dialysate functions as a yeast nutrient [38,59]. The albumins and globulins in the soluble fraction seem not to influence the loaf volume [59].

In micro-baking tests, it was found in our laboratory [51] that there was an optimal level of addition of water-soluble fraction from the wheat variety Kadett. The maximum increase in loaf volume was 15%. The water-soluble fraction (WSF) from Maris Huntsman did not increase the loaf volume in a similar experiment. It was concluded that gliadins present in the WSF might contribute to the increase in loaf volume. Hoseney et al. [50] found no correlation between the source of water-solubles and loaf volume. Frazier et al. [55] showed that the distribution of protein in different solubility classes might change during mixing of the dough and that it might also depend on the atmosphere (air or nitrogen) during dough mixing. However, the level of soluble proteins seemed to be fairly constant and not related to the mixing intensity or atmosphere. It was given as around 14–15 mg/g dough. Marais and D'Appolonia [60] found that mixing behavior might be affected by the addition of albumins and that the bromate requirements are to some extent related to the water-solubles [61].

As many of the soluble proteins are enzymes, they might affect the baking performance in a profound way. For example, if the enzyme activity is excessive, as in malted flour, the result is a bread of inferior quality, and excessive proteolytic as well as amolytic activity should be avoided. The activity of another enzyme, lipoxygenase, will be discussed in relation to oxidation and reduction reactions in doughs (Chapter 6). As mentioned before, the only enzyme present in a large amount in sound wheat is β-amylase. Other enzymes that might be of importance are dehydroascorbic acid reductase or ascorbic acid oxidase, which catalyzes the conversion of ascorbic acid to dehydroascorbic acid, and polyphenol oxidase, which might cause darkening of whole meal doughs [30].

The proteins rich in cysteine might influence the SS/SH exchange among proteins. The addition of purothionin to a wheat flour did not improve the baking performance [30]. However, a sulfur-rich peptide, gluthathione, is known to affect the rheological behavior of the dough [56].

Water-soluble proteins are concentrated around the starch granules. These are probably enzymes associated with starch granule synthesis. Some proteins extracted from starch granules have protease activity [62]. Soluble proteins have been claimed to play an important role in the hardness of the endosperm [62,63]. A low molecular weight protein has been found to be present in soft wheats and absent in hard wheats [64]. The role of such an interaction between starch and protein will be discussed in relation to milling (Chapters 3 and 5).

The lipoproteins described by Hoseney et al. [53] were not necessary for the improving effects of polar lipids in baking. Proteins very similar to purothionins have also been isolated from barley [55].

Glycoproteins, or pentosans, are claimed to be necessary for good baking performance [65]. The structrure of the pentosans is not known in detail (see Section V). Some of the pentosans are not even glycoproteins as the carbohydrate is not covalently bound to the protein. It has also been claimed that the pentosans are not important for baking, as the enzyme pentosanase does not affect the loaf volume [30]. MacRitchie [38], on the other hand, claimed that the water-soluble pentosans and the glycoproteins contribute to the extensibility of gluten and to the gas-holding capacity of the dough.

2. Gluten Proteins—Wheat Storage Proteins

Whereas the metabolic proteins are found among the soluble proteins, the gluten proteins constitute the main storage protein in the wheat endosperm. This has important implications for the composition of the gluten proteins. They are rich in glutamine and proline but poor in lysine and arginine. With time many mutations have occurred in the storage protein complex without detrimental effects on their function (which is to be digested once the seed germinates). The known number of individual, only slightly different, proteins has increased tremendously. Gluten could perhaps be viewed as a protein mixture with a continuous molecular weight distribution from about 30,000 to perhaps 20 million. Nevertheless, attempts are made to divide this complex mixture into less complex subgroups in order to understand their behavior better. The most common separation is into gliadins and glutenins (see Table 2), and these two groups will now be described in more detail.

a. Gliadins

Gliadins are defined as the wheat proteins soluble in aqueous ethanol in the classic Osborne extraction procedure [32]. As discussed earlier, this is not an unambiguous definition; the amount and type of protein extracted will depend on the conditions during the extraction procedure. An alternative nomenclature has been proposed by Shewry et al. [66] in which the gliadins are defined as monomeric proteins. The disulfide bonds present are thus all intramolecular. The gliadin conformations are stabilized by hydrogen bonding and hydrophobic interactions. They constitute about 30–35% of the total protein when they are extracted in 70% ethanol (see Table 2). The gliadin group is heterogeneous, and the mobility in aluminum lactate starch gel electrophoresis has been used to divide the gliadins in four groups: the α-, β, γ-, and ω-gliadins [67]. α-Gliadins are the fastest moving, and ω-gliadins the slowest.

The proportion of the total protein content has been estimated for different gliadins from the electrophoretic patterns. The ω-gliadins constitute about 8–13% of the protein, whereas the sum of the α-, β-, and γ-gliadins constitutes about 34–48% [68]. An average value of 4.6 g gliadin/100 g flour has been calculated for 70 wheat varieties [37]. The ω-gliadins are found in the lowest amount, followed by α-gliadins, β-gliadins, and γ-gliadins, respectively, in increasing amounts [37]. The amount of one individual gliadin component is in the range 0.06–3.00 g/100 g storage protein.

An SE (size-exclusion)-HPLC chromatogram of gliadins is shown in Fig. 15. The molecular weight of most gliadins is in the range 30,000–40,000, but the ω-gliadins have a molecular weight around 60,000–80,000 [69,70]. It has been claimed that there are also high molecular weight (HMW) gliadins with a molecular weight in the range 100,000–200,000 [69]. These HMW gliadins can be reduced to gliadins of lower molecular weight and would then be defined

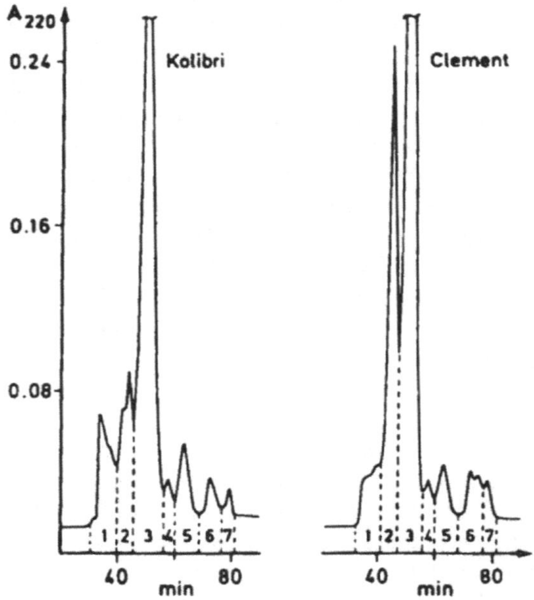

Fig. 15 Size-exclusion (SE) HPLC of gliadins from the wheat varieties Kolibri and Clement. (Reprinted with permission from [70].)

as low molecular weight (LMW) glutenins [66,71]. Conformational similarities between HMW gliadins (or aggregated gliadins) and LMW glutenins have been shown [72]. It should be borne in mind that molecular weights obtained from SDS-PAGE are overestimations [73,74].

The amino acid composition of the gliadins is characterized by a high level of glutamine and proline. A typical composition is shown in Table 6 [70]. The content of sulfur-containing amino acids differs between ω-gliadins and the other gliadins. The cysteine content in ω-gliadins was found to be 0–8 residues per 100,000 g of protein, whereas it was 17–22 residues in α-, β-, and γ-gliadins [75]. Tatham and Shewry [76] suggested that the gliadins could be classified according to their content of sulfur-containing amino acids. The ω-gliadins then belong to the sulfur-poor prolamins, whereas α-, β-, and γ-gliadins belong to the sulfur-rich prolamins. The content of ionic amino acid residues is very low,

Table 6 The Amino Acid Composition[a] of Gliadins

Amino acid	ω	α, β, and γ
Asx	0.8– 1.4	2.1– 3.5
Thr	0.6– 3.5	1.6– 2.5
Ser	4.5– 6.4	5.2– 7.3
Glx	40.0–44.4	34.8–41.0
Pro	25.0–31.0	15.4–18.8
Gly	1.3– 2.7	2.0– 3.5
Ala	0.4– 1.8	2.7– 3.6
Val	0.5– 1.7	4.2– 5.3
Met	0.0– 0.2	0.4– 1.6
Ile	2.0– 3.8	3.4– 4.9
Leu	3.9– 5.3	6.5– 8.3
Tyr	0.6– 1.6	0.5– 1.9
Phe	6.6– 7.6	3.5– 5.9
His	0.4– 1.1	1.3– 2.6
Lys	0.3– 0.9	0.2– 0.8
Arg	0.8– 2.0	1.3– 2.4

[a]Mole % (without Cys and Trp).
Source: Adapted from [70].

whereas the content of hydrophobic residues is high. This results in a very low water solubility of the proteins. It has been suggested that the gluten proteins are among the lowest charged proteins existing [49]. Among the gliadins the ω-gliadins have the lowest charge of all, with only six basic groups and about the same amount of free carboxyl groups as calculated for a protein with a molecular weight of 65,000 [49]. Reversed phase (RP) HPLC has also been used to characterize the gliadins. The gliadins are eluted in the order ω-, α-, γ-gliadins, the ω-gliadins being the most hydrophilic and the γ-gliadins the most hydrophobic [70].

The amino acid sequences of some gliadins are known, and more are being solved with time. An example is given in Fig. 16, where the hydrophobicity of the amino acid residues is also indicated. A very common N-terminal sequence found in α-, β-, and γ-gliadins is Val-Arg-Val-Pro-Glu-Leu-Glu-Pro [57]. The sulfur-rich gluten proteins have their sulfur amino acid residues in their C-terminal part [66].

The hydrophobicity of gliadins has been calculated from the amino acid compositions to be higher than that of the glutenins [77]. Recent experimental efforts to determine a kind of surface hydrophobicity, however, seem to indicate that glutenins have the higher surface hydrophobicity [78,79]. Among the gliadins, γ-gliadins exhibit the highest surface hydrophobicity [79]. The number of hydrophobic sites was not affected by folding of the gliadin molecule, and it was concluded that hydrophobic interactions are not of great importance for the gliadin structure [73]. However, when the gliadins aggregated, the number of hydrophobic bonding sites decreased considerably [73].

The high level of proline will certainly affect the secondary structure of gliadins in that the proline residues disrupt α-helices. The secondary structure has been estimated for gliadins by CD (circular dichroism spectroscopy), and some results are presented in Table 7. The ω-gliadins differ from the other gliadins in that they do not contain α-helix or β-sheet structures. Instead, the ω-gliadins contain β-turns. This structure is stabilized by hydrophobic interactions between aromatic amino acid residues and by hydrogen bonding involving glutamine residues [76]. As the cysteine content is very low, disulfide bonds are not important for the conformation of the ω-gliadins. The α-helix structure of the other gliadins is very stable; for example, it is stable against heat [76]. Although the disulfide bonds may help in stabilizing the structure, the main stabilizing forces are attributed to the very high number of hydrogen bonds [73].

The gliadin composition, analyzed electrophoretically or by HPLC, is characteristic of the wheat variety. It does not change over the years, but improved electrophoretic techniques might affect the pattern so that a greater number of bands are detected [81]. The gliadin pattern is not affected by growth conditions, by total protein content, or by sprouting. Severe sulfur deficiency, however, might have an influence [57]. This method thus affords a possible means of

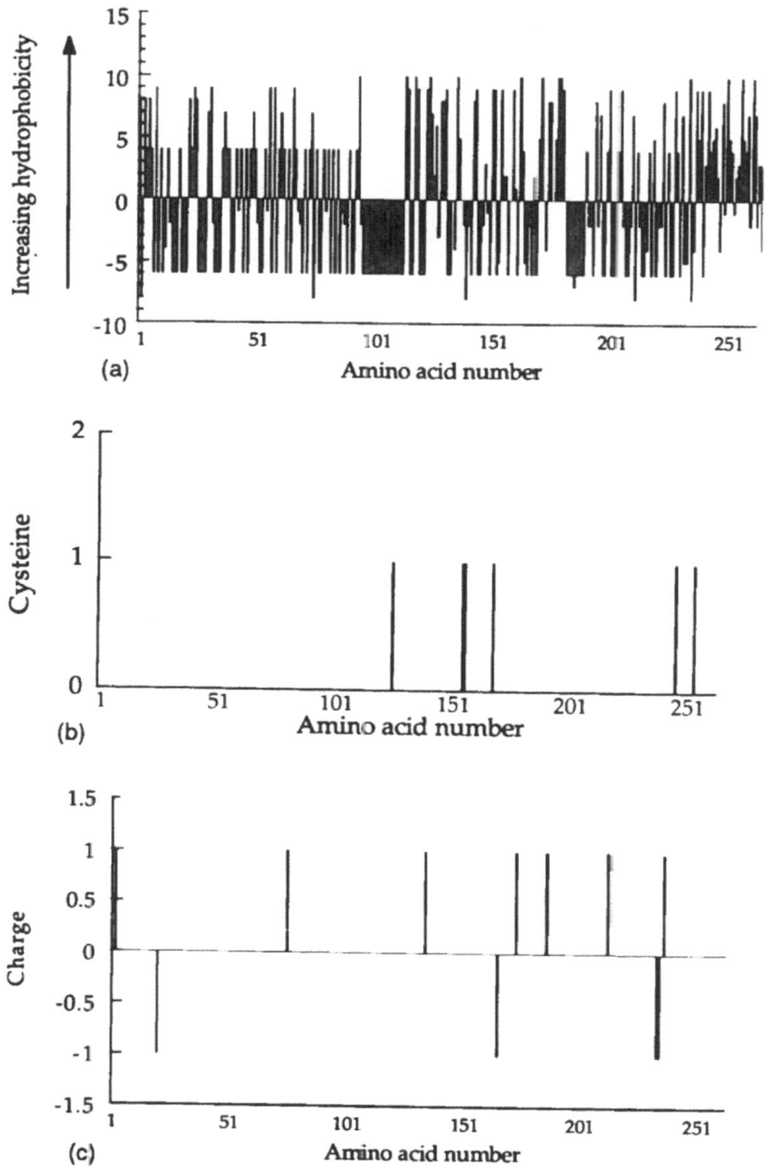

Fig. 16 (a) The hydrophobicity of amino acids in the primary sequence of A-gliadin, (b) the position of cysteine, and (c) the position of charged amino acids. (Data from [30].)

Table 7 Percentage of Secondary Structure of Some Gliadins

Gliadin	α-Helix (%)	β-Structure (%)	β-Turns (%)		Unordered structure (%)
A-Gliadin [80]	27	23	29		21
α-Gliadin [76]	36–37	11–12		52–53[a]	
Aggregated gliadins [72]	34–37	18–24		44–48[a]	

[a]Sum of β-turns and unordered structure.

identification of wheat varieties. The number of gliadins is high—in one single variety there are 20–25 different gliadins—but the total number of different gliadins is even higher. In an investigation of the electrophoretic patterns of 70 different wheat varieties, 60 gliadin bands were identified. The greatest variation was found among the ω-gliadins, with 24 bands [37]. In another investigation, 49 different gliadins were found among 71 wheat varieties [29]. By RP-HPLC more than 30 gliadins are revealed [82].

When two-dimensional electrophoresis is used, the number is even higher. Moreover, amino acid sequence analysis seems to indicate that there might be differences in some of the neutral amino acids, and in that case the total number of gliadins might be even higher, because such delicate differences are very difficult to detect with the physicochemical methods used. As will be discussed later, the glutenin HMW subunits are found in pairs. Some gliadins are also found together, but the appearance of certain gliadins seems not to be linked to the presence of certain HMW glutenin subunits. Furthermore, an ethanol extract from flour contains not only protein, but also 10% nonprotein material, of which 0.6% has been identified as carbohydrate and 9.4% as lipid [57]. This has led many researchers to describe the gliadins as a protein–lipid complex, a possibility that will be further discussed in Chapter 3.

b. Glutenins

According to the classification of Osborne, glutenins are the proteins remaining after the albumins, globulins, and gliadins have been extracted. They constitute 40–50% of the total protein present in wheat flour (see Table 2). An average value of 5.9 g/100g flour was determined for 70 wheat varieties [37]. The amount of glutenins increases when the protein content of the wheat increases [83]. Th glutenins are soluble, or rather dispersible, in dilute acid or alkali, in denaturants such as urea, and in surfactants [82]. One reason for the low solubility of glutenins is their high molecular weight. Molecular weights up to 20 million have been reported [82,84–86]. According to the nomenclature suggested by Shewry et al. [66], the glutenins belong to the polymeric prolamins.

The molecular weight distribution of native, nonreduced glutenin is difficult to measure because of the insolubility of the proteins. A few attempts have been made, for example by gel chromatography, but the results show little detail.

Transmission electron microscopy (TEM) has shown that the glutenins are aggregates built up of spherical particles [86]. The molecular weight of such an aggregate is around 8×10^5, which corresponds to a particle diameter of 10 nm. If the glutenins are reduced, the subunits can be studied by, for example, electrophoresis or HPLC. An example of a size-exclusion HPLC chromatogram is given in Fig. 17. The glutenin subunits are usually divided into two groups, LMW subunits and HMW subunits, but some researchers add a group of middle molecular weight (MMW) subunits [71]. The HMW subunits have a molecular weight in the range 90,000–150,000, whereas the LMW subunits have molecular weights in the range 30,000–51,000 [87]. Ng and Bushuk [74] determined the molecular weight by SDS-PAGE and reported it to be in the range 91,200–172,000 for the HMW subunits, whereas Shewry et al. [88] reported 95,000–145,000. However, SDS-PAGE overestimates the molecular weights, and when this overestimation is compensated for, the molecular weights are found to be in the range 68,100–104,000 [74]. Reversed-phase HPLC is also used to study

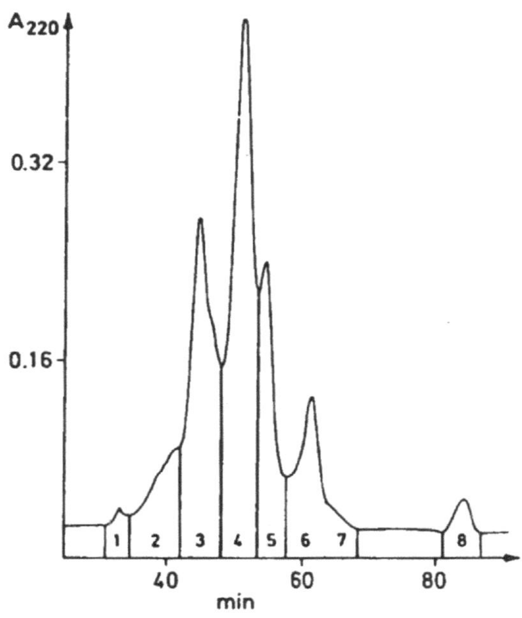

Fig. 17 SE-HPLC of reduced glutenin from the wheat variety Kolibri. (Reprinted with permission from [71].)

glutenin subunits, and three groups of glutenin subunits differing in hydrophobicity are obtained. The most hydrophobic group contains the LMW glutenin, whereas the MMW glutenins belong to the least hydrophobic group. The HMW glutenins are thus of intermediate hydrophobicity [71].

An example of the amino acid composition of glutenins is given in Table 8 [70]. The amino acid sequence for a few glutenin subunits is known, and an example is given in Fig. 18. The sequence is composed of three domains: the N-terminal domain consisting of about 100 amino acid residues, the C-terminal domain consisting of about 50 residues, and a central, repetitive domain of about 600–800 residues. The N- and C-terminal domains contain almost all cysteines and ionic amino acid residues, but these domains are low in proline [88,89]. The central domain contains high proportions of glutamine, glycine, and proline, and a very common sequence is Pro-Gly-Gln-Gly-Gln-Gln [89]. The different HMW

Table 8 The Amino Acid Composition[a] of Glutenins

Amino acid	LMW glutenins	MMW glutenins	HMW glutenins
Asx	2.3–3.0	2.0–2.1	2.1–3.5
Thr	2.4–3.2	1.5–1.8	3.3–3.7
Ser	7.5–10.4	5.3–6.4	8.2–9.7
Glx	33.1–37.3	44.0–45.1	29.2–34.9
Pro	12.6–13.9	15.4–15.7	10.9–14.6
Gly	4.7–7.1	6.4–7.4	13.5–18.4
Ala	3.1–4.1	2.0–2.1	3.3–3.9
Val	4.5–5.0	1.3–1.5	2.6–3.1
Met	1.0–1.5	0.2–0.4	0.3–0.4
Ile	3.6–4.4	3.5–4.0	2.0–2.4
Leu	6.8–7.6	3.4–4.1	4.2–5.2
Tyr	0.8–2.0	0.7–1.2	2.7–3.7
Phe	3.7–4.1	6.5–7.5	1.6–3.2
His	1.5–1.9	1.2–1.4	1.0–1.4
Lys	0.8–1.4	1.2–1.3	1.5–1.8
Arg	2.3–2.8	1.4–1.5	2.2–3.2

[a]Mole % (without Cys and Trp). LMW, MMW, HMW = low, medium, and high molecular weight, respectively.
Source: Adapted from [70]

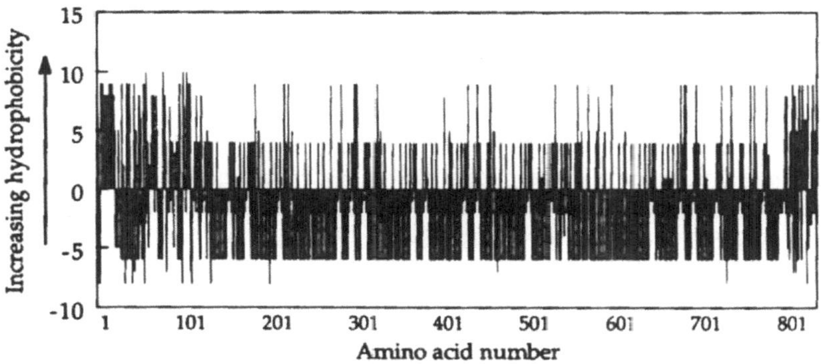

Fig. 18 The hydrophobicity of amino acids in the primary sequence of glutenin. (Data from [30].)

subunits have several amino acids in common in their N-terminal sequence [88]. However, the amino acid composition of the total subunit differs somewhat between the different subunits; for example, one HMW subunit contains three or four cysteines per mole whereas another subunit contains seven [88].

The conformation of the HMW glutenin has been studied by CD and is similar to that of the ω-gliadins. The conformation is characterized by a large proportion of β-turns in the central domain. The terminal domains contain some α-helix structure. The existence of β-turns has been suggested as the reason for the elasticity of glutenins [90].

The HMW subunits constitute only about 1% of the dry matter content of the endosperm [91]. The total number of different HMW subunits is around 20; in an investigation of 71 wheat varieties, 20 different HMW glutenin subunits were found [29], whereas in another investigation 17 different HMW subunits were found in 70 wheat varieties [92]. Ng and Bushuk [74] detected 24 different HMW subunits with SDS-PAGE. A single variety usually contains three to five different subunits [87,91]. The glutenin subunits are linked together by disulfide bonds to the huge polymeric glutenin molecules. This is consistent with the location of the disulfide bonds in the terminal domains of each subunit, as is evident from the cysteine residues located there.

The exact number of interchain S—S bonds per chain is not known, although it has been suggested that each intramolecular bond between glutenin subunits is composed of one S—S bridge, not two [93]. It might be impossible to determine because of the difficulty of discriminating between inter- and intramolecular S—S [94]. If there is only one S—S bond per chain, linear concatenations might be formed, as suggested by Ewart [95]. There might also be more than one S—S bond per chain, resulting in other types of structures. Kasarda [89] suggested that the subunits are linked into sheeted structures, which contributes liquid crystalline behavior to the dough.

The exact subunit composition of the glutenin molecules and the sequence of subunits in the molecule are not known. However, there are some relations between molecular weight and subunit composition in that the glutenins of low molecular weight contain a higher proportion of LMW subunits [30,56]. If only traces of 2-mercaptoethanol are used in SDS-PAGE, bands are obtained in the molecular weight region above that of the HMW subunits [96]. It is further observed that these oligomers are always composed of the same HMW subunits. All glutenin subunits are built into aggregates; no monomeric glutenins exist [89]. The smallest native glutenin polymers are dimers of LMW subunits with a molecular weight of 60,000–80,000 [68]. Some workers assume that the glutenin molecules are polymerized in the kernel during synthesis [95], whereas others claim that polymerization occurs when the glutenin subunits are exposed to air during milling and further during dough mixing (or gluten preparation) [97]. An oxidative strengthening of gluten occurs only with strong varieties; that is, only strong varieties can provide subunits for further polymerization.

c. Significance in Baking

The importance of a certain class of proteins for the baking performance of wheat flour could be shown in several ways. The correlations between baking performance (e.g., measured as loaf volume) and the content of the protein class in question (or one or several gliadins or glutenins) in several flours could be established. Baking experiments might be performed with the proteins of that class added to wheat flour at increasing levels. Finally, the protein might be interchanged between flours that differ in baking performance. These experiments are fairly easy to design—in theory. They also might be (and have been) performed practically, unfortunately not always with convincing results. The difficulties in these experiments lie in the definition, analysis, and preparation of the different classes of proteins. Moreover, each class is a mixture of many different proteins, as described earlier in this chapter.

The amino acid composition of the gluten proteins might also be related to the baking performance. MacRitchie [98] found a relation between glutamine content and baking performance. An investigation of the amino acid composition of peptides obtained from glutenin revealed that the presence of an HMW peptide rich in glycine (and glutamine and proline) could be correlated to good baking performance [99].

The gliadins seem not to be crucial to the baking performance. When the gliadin fraction is interchanged between wheat flours of different baking performances, the effect on the result is very minor compared with that of the interchange of glutenins [8, 100]. The correlation between loaf volume and the proportion of glutenin (determined by SE-HPLC) is much better than the correlation between the proportion of gliadins and loaf volume [35]. No relation between gliadin pattern and mixing behavior has been found [101].

No single gliadin component has been found to correlate with baking quality or other quality attributes (e.g., Chopin values, Zeleny data, dough strength) [37], although groups of gliadins are related to endosperm hardness, dough strength, Chopin values, or the Zeleny test [29,37,81,102]. The glutenin composition seems to be of much greater importance.

Glutenins influence the baking performance of a wheat flour in at least three ways: through the gliadin/glutenin ratio, through the molecular weight distribution of glutenins, and through the presence of certain HMW glutenin subunits [30,56,69,84,100,103]. It has long been recognized that the glutenins control the mixing requirements [59], but they may also influence loaf volume.

Although there has been much speculation as to the importance of the gliadin/glutenin ratio, there are no definite results [56]. The implication of the gliadin/glutenin ratio will be further discussed in relation to the gluten structure later in this chapter.

Molecular weight distribution differs among wheat varieties, and a strong wheat with medium-long mixing time contains more of the high molecular weight material [84]. Moreover, these wheat varieties also contain more residual protein.

MacRitchie [104,105] fractionated gluten proteins by successive extraction procedures, using water, urea, and sodium hydroxide [104] or dilute HCl [105]. The fractions obtained in the HCl fractionation procedure have been investigated by SDS-PAGE, and it was found that when the fraction number increases (which is the same as a decrease in pH) the proportion of glutenin increases. However, in the last fractions obtained, the protein content was low, and occasionally decreased proportions of glutenin were observed in these fractions [68]. The protein is more easily extracted from a poor flour than from a good flour. In this way, up to 10 fractions are obtained, fractions that differ in, among other things, molecular weight distribution. The proportion of high molecular weight protein increases in the successive fractions as the HMW proteins have a lower solubility than LMW proteins. These fractions have been added to wheat flours to evaluate their influence on baking performance. It was found that fractions rich in LMW proteins decrease the mixograph developing time as well as the loaf volume in test baking [105]. The fractions with a high proportion of HMW proteins, on the other hand, increased the mixograph developing time as well as the loaf volume in test baking. The LMW fractions are rich in gliadins according to SDS-PAGE, whereas the HMW fractions are rich in glutenins. In other investigations, a modified AUC (acetic acid–urea–cetyltrimethylammonium bromide) solvent was used to solubilize gluten, and gel filtration of this gluten resulted in one fraction (F1) in the void and one fraction (F2) that was separated. The F1/F2 ratio was found to correlate with the baking performance; the higher the F1/F2 ratio, the better the baking performance [43].

As in the case of the gliadins, relationships between certain glutenin subunits and the quality of the flour have been sought. Such a relationship seems most promising in the case of HMW glutenin subunits. These subunits are of greater importance for dough strength and dough stickiness than the LMW glutenin subunits [106]. Each subunit has been assigned a number, and the presence of certain combinations of these subunits are related to different quality aspects [107]. The HMW subunits 5+10 are said to be present in varieties of good baking performance, whereas 2+12 are present in varieties of poor performance [29,30,91,102,108]. The presence of subunits 5+10 is also associated with a high sedimentation volume in the SDS sedimentation test, whereas the presence of subunits 2+12 is associated with a low sedimentation volume [107]. For varieties containing both 5+10 and 2+12, the sedimentation volume will be intermediate.

On the basis of these results, quality scores are assigned to each of the HMW subunits, and by adding the scores for the different HMW subunits present, a quality score, called the Glu-1 score, is obtained for each wheat variety [91]. The highest score is obtained for the subunits 5+10, and for British wheats the highest score calculated was 10 and the lowest 3. The Glu-1 score is positively correlated with baking performance in the case of bread, and 47–60% of the variation in breadmaking qualities could be accounted for by the variation in HMW subunits [91]. The quality score is negatively correlated with the baking performance in the case of biscuits [91], which is in accordance with what we should expect. The correlation between subunits 5+10 and baking quality seems not to be so strong in all countries; for Australian wheats a somewhat weaker relation was observed, perhaps related to the presence of subunits 17+18 [29]. In other wheats, where high Glu-1 scores were obtained (9 or 10), the variation in baking performance was still great; loaf volumes were in the range of 750–1050 mL [109].

The HMW glutenins seem not to influence loaf volume during fermentation or proofing [110]. In one experiment the volume of the dough before entrance into the oven was 30 mL, irrespective of the HMW subunit composition. With varieties lacking HMW glutenin subunits, the volume increased to 32 mL during the oven step, whereas 48 mL was obtained when HMW subunits (5+1, 1, 17+18) were present [110]. It could also be noted that the variety without HMW subunits, described above, had a greater protein content, 13.5% compared with 8.3% for the variety with the complete combination of HMW subunits.

It is quite evident that the HMW glutenin subunits are important to baking performance. However, it is not known if the presence of a specific HMW subunit is directly responsible for the quality observed or if the presence of the subunit coincides with other properties that are the direct cause. It has been argued that it is the *amount* of glutenin and not its *composition* that is most important [89]. The presence of subunits 5+10 could also be related to a higher proportion of glutenin, and a high amount of glutenin is beneficial [29,35,92].

The loaf volume is related directly to the proportion of HMW subunits determined by SDS-PAGE, a further indication that the amount might be more important than the type of glutenin [110].

B. Monolayer Properties

A fundamental property of wheat flour–water mixtures is the ability to encapsulate gas cells and ultimately form a stable foam structure. In order to understand the mechanisms involved in stabilization of the gas/water interface we must start with the monolayers formed at this interface. The formation of wheat protein monolayers and their properties will be considered below.

Protein monolayers can behave in two ways during dough mixing, and their effect on interfacial tension can be followed by two corresponding methods. Using the surface balance technique described in Chapter 1 it is possible to follow the spreading of "insoluble" monolayers, either directly from the solid state or via an organic solvent in which the protein is dissolved. During such experiments all protein molecules remain at the monolayer, even though they may have some solubility in water. The other method is to follow adsorption of the protein from water to form a monolayer at the air/water interface. This type of measurement can also be performed in a surface balance, but it is simpler to use the drop-volume technique, which makes use of the fact that when a hanging drop falls its surface tension balances the weight.

1. Surface Tension Measurements in the Presence of Wheat Storage Proteins

We will start with the spreading of a protein monolayer from solid powder put on the surface of water. The reduction in the surface tension of water as discussed in Chapter 1 gives the surface film pressure (π) according to the equation

$$\pi = \gamma_0 - \gamma \tag{2}$$

where γ_0 is the surface tension of water and γ is the surface tension in the presence of the surface-active material (such as the wheat proteins). It is thus possible to follow the spreading behavior of wheat proteins by measuring the changes in surface tension of the water [111]. In Fig. 19 such a π–t (time) isotherm is illustrated. The surface film pressure was measured by the Wilhelmy plate method in a surface balance. The isotherms presented in Fig. 19 illustrate that different wheat protein fractions have different spreading behaviors. The HMW fraction spreads very slowly, whereas an LMW fraction spreads more quickly. The spreading behavior is changed if the proteins are heat-treated. A laboratory-prepared gluten that was heated to 70°C spread more slowly and to a lower π value than the control sample, which was freeze-dried without any previous heat treatment [112].

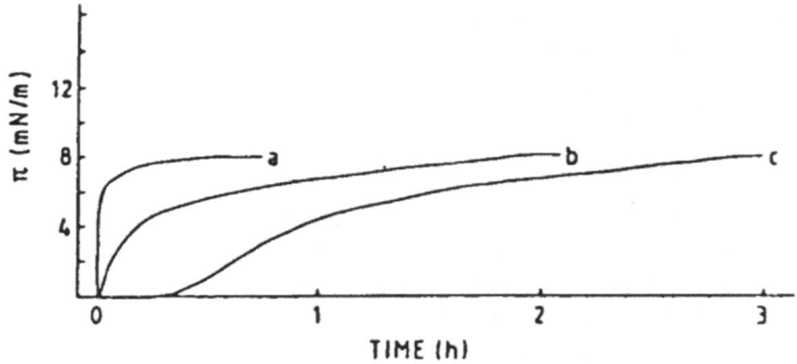

Fig. 19 Pressure–time (π–t) isotherms obtained when different wheat protein fractions are spread at the air/water interface. (a) A low molecular weight fraction; (b) a medium molecular weight fraction; (c) a high molecular weight fraction. (Reprinted with permission from [111].)

The proteins form a monomolecular film at the air/water interface. After the film has formed and has equilibrated for some time, the film can be compressed, and the π–A isotherm (A = area) recorded. An example of such an isotherm is given in Fig. 20. When the protein film is compressed, the π value increases; that is, the surface tension of water decreases further. The behavior of gluten films at different substrates (aqueous substrates as well as nonpolar) was investigated by Tschoegl and Alexander [113]. They spread gluten dispersions (0.4 mg/mL in anhydrous acid chloroethanol) on aqueous substrates that were adjusted to different pH values and different ionic strengths. The straight portions of the π–A isotherms were extrapolated to zero pressure, and the "area of close

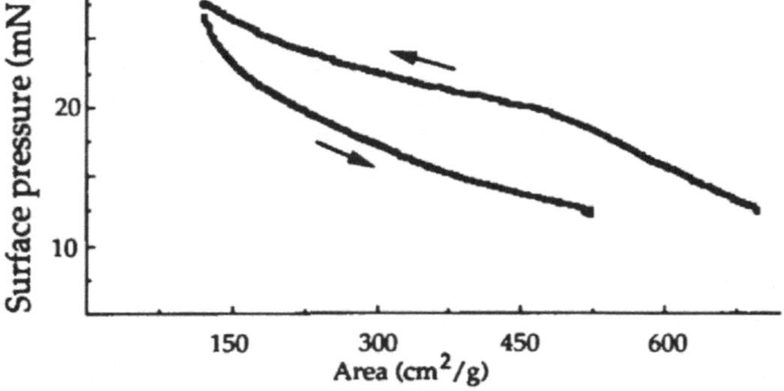

Fig. 20 Surface pressure–area (π–A) isotherm obtained for wheat gluten. The arrows indicate compression (above) and expansion (below). (Adapted from [112].)

packing" was calculated. It was found to be 1.33 m^2/mg at the air/water interface and 1.54 m^2/mg at the oil/water interface. They also found that the gluten films were remarkably stable. Other protein monolayers collapse at π values around 20–25 mN/m, but in a gluten film there was no sign of collapse. In fact, it is possible to compress and expand a wheat protein film repeatedly without losing the molecules into the substrate [114]. It has been shown that after each compression/expansion cycle the molecular weight of HMW proteins (corresponding to glutenins) increases, whereas the molecular weight of gliadins is unchanged (Fig. 21).

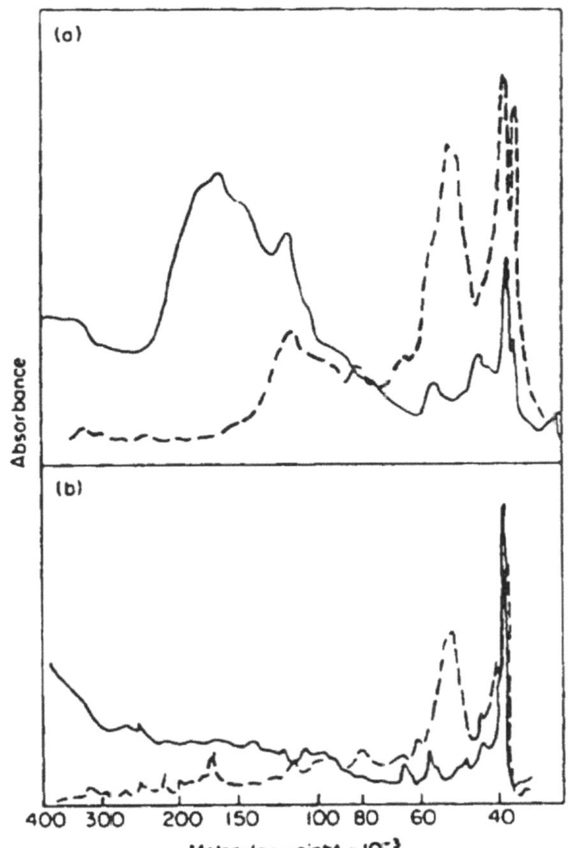

Fig. 21 SDS-PAGE of a high molecular weight wheat storage protein fraction (a) before and (b) after compression at the air/water interface. The SDS-PAGE analysis was performed with (broken line) and without (solid line) 2-mercaptoethanol. (Reprinted with permission from [114].)

The surface behavior of more "pure" wheat protein fractions has also been studied. Carlson [21] studied A-gliadin, and Tao et al. [115] studied glutenins. In these investigations the π-A isotherms were recorded during one compression. The A-gliadin showed signs of collapse at π values above 25 mN/m; that is, it behaved like most other proteins (Fig. 22). The glutenins, on the other hand, showed no tendency toward collapse, and a π value of about 25 mN/m was reached at an area of about 0.15 m^2/mg. No inflections were obtained in the π-A isotherms for glutenins, as we have seen in the case for gluten or flour (compare Fig. 20). Tao et al. [115] compared glutenins from wheat varieties differing considerably in baking performance. The π–A isotherms of these glutenins were nevertheless very similar. This might not be surprising but rather to be expected. The glutenins might all be very similar in their surface behavior, whereas the differences between varieties might be found in the complete protein mixture.

The surface concentrations of some wheat proteins are compared at two different π values in Table 9. The glutenins gave values in the range 2.0–3.2 mg/m^2 at π = 10 mN/m and 3.6–6.3 mg/m^2 at π = 20 mN/m. The corresponding values for A-gliadin were much lower, 1.6 mg/m^2 at π = 10 mN/m and 2.8 mg/m^2 at π = 20 mN/m. These results indicate that a much lower amount of gliadin is required to achieve a certain decrease in surface tension. The data reflect the thickness of the condensed monolayers. In order to form a foam, a large decrease in interfacial tension is beneficial. Then the LMW gluten proteins

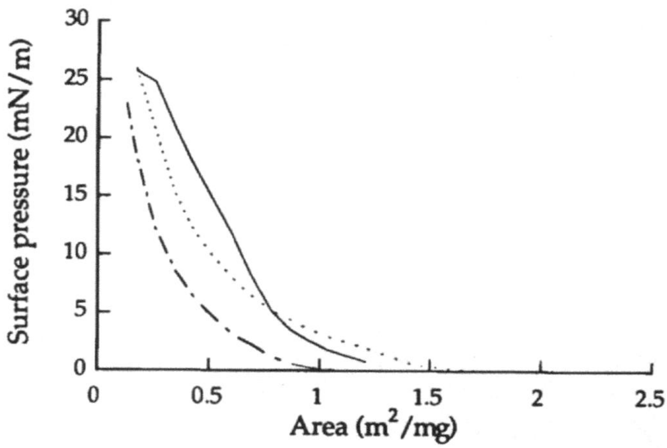

Fig. 22 Surface pressure–area (π–A) isotherms of A-gliadin and glutenins. The A-gliadin (solid line) was spread from a water solution (pH 5) on 0.44 M NaCl, and the glutenins (dotted and dot-dash lines) were spread from acetic acid on 0.05 M NaCl. (Redrawn from [21] and [115].)

Table 9 Surface Concentration of Wheat Protein Fractions at Two π-Values

	π = 10 mN/m		π = 20 mN/m	
Wheat variety[a]	Area (m^2/mg)	Surface concentration (mg/m^2)	Area (m^2/mg)	Surface concentration (mg/m^2)
RR68	0.35	2.9	0.17	5.9
CS	0.31	3.2	0.16	6.3
Cnn 1	0.44	2.3	0.23	4.3
Cnn 3	0.51	2.0	0.28	3.6
Cnn 4	0.4	2.5	0.23	4.3
A-Gliadin	0.64	1.6	0.36	2.8

[a]RR68 = Red River 68 (strong mixing character, poor to good baking quality); CS = Chinese Spring (weak mixing character, poor baking properties); Cnn = Cheyenne (medium to strong mixing character, good baking properties); Cnn1 = highest molecular weight; Cnn 4 = lowest molecular weight.
Source: From [115].

might be of importance, together with the soluble proteins. However, to stabilize a foam, the interfacial film should have suitable mechanical properties. In the dough these mechanical properties could be provided by the HMW glutenins.

The surface behavior has been shown to vary among commercial gluten samples [116]. Two extreme behaviors were found: some gluten samples spread at a high rate to a high π value, whereas other samples spread at a much lower rate to a low π value.

The study in our laboratory of A-gliadin [21] will be considered here in more detail, as it makes possible an interpretation of the structure. The A-gliadin study was performed by spreading the protein from a solvent. The equilibrium spreading pressure (ESP) value of the protein was also recorded and found to be 25 mN/m. That is identical to the collapse value of the monolayer spread from solvents, which indicates that the film structure is the same. One first-order phase transition was observed. The collapse occurred at 0.30 m^2/mg, and the other transition was observed at 0.536 m^2/mg. The transition was proposed to resemble the transition in lipid monolayers from liquid to crystalline hydrocarbon chains. The interpretation proposed is shown in Fig. 23. At the phase transition the area was about 2600 Å^2 per molecule, and at collapse the area was 1.567 Å^2 per molecule, corresponding to a thickness of 26.3 Å.

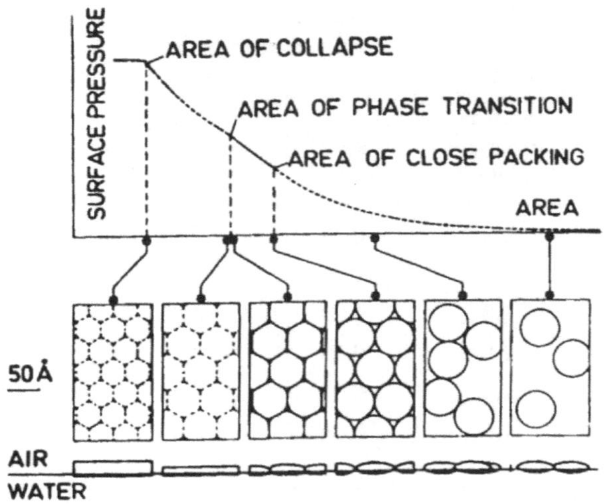

Fig. 23 A schematic picture of the geometry of the A-gliadin monolayer. A π–A isotherm is shown together with three characteristic areas: collapse, phase transition, and close packing. The packing of the molecules is illustrated beneath the isotherm in both plane and vertical sections. (From [21].)

This proposed A-gliadin structure at the air/water interface differs from the general behavior of proteins, which are thought to unfold at the interface. The structure is consistent with the bulk structure of the gluten proteins.

Surface tension measurement by the drop volume technique was used to characterize the surface activity of water-soluble fractions from four different flours [51]. The soluble fractions showed surface activity, but there was no difference in behavior between WSFs from different varieties. The surface activity was about the same as in the case of caseinate (Fig. 24).

2. Wave Damping of Wheat Storage Proteins

Damping of the Faraday waves on a water surface was described in Chapter 1. Surface-active molecules of the lipid type increase the wave damping, as we can observe on the surface of the sea ("sea slicks" are usually due to the presence of a lipid film). In collaboration with Henderson [118], we have recently studied the effect of gliadin and glutenin monolayers on wave damping. Their effect was found to be much stronger than those of lipid monolayers. Damping rates up to 15 times the value of a clean water surface were observed. From these studies the following conclusions could be drawn.

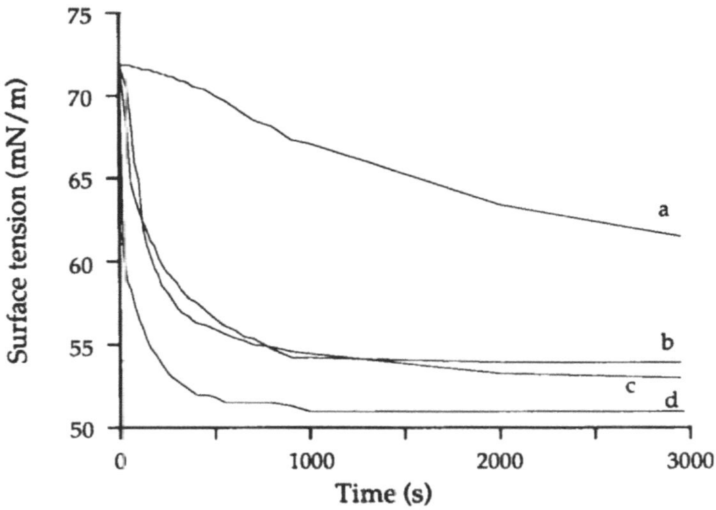

Fig. 24 The surface tension of soluble proteins in distilled water; (a) soy protein, (b) whey protein, (c) caseinate, and (d) water soluble fraction from wheat. (Data from [51] and [117].)

1. The presence of a small concentration of a wheat storage protein (WSP) monolayer increases wave damping rates by a factor of 10 compared to the damping effect of a simple surfactant.
2. Cross-linking among molecules may be responsible for the large observed damping rates.
3. For WSP concentrations above saturation, cross-linking may be responsible for an observed maximum in damping rate.
4. Cross-linking among molecules did not affect the wave rates as a function of time.
5. The WSP monolayer wets glass more effectively with increasing concentration.
6. Wheat storage protein monolayers spread from the solid phase (the glutenin fraction) affect wave damping the same as WSP monolayers spread from the ethanol solution (the gliadin fraction).
7. The surface dilatational viscosity of WSP monolayers is not responsible for the high damping rate.
8. The dilatational viscosity of the WSP monolayer depends on the frequency of surface straining, which probably reflects kinetic effects on cross-linking reactions.

A general impression from monolayer studies of the storage proteins is that they exhibit properties in two dimensions similar to those of the gluten gel. Thus monolayers show quite remarkable mechanical properties with the ability to easily cross-link and form larger molecules. We can probably find polymers that in bulk can show a mechanical behavior similar to that of the gluten gel, but we know of no substance with similar interfacial properties.

C. Foam-Stabilizing Effects

A foam is an agglomerate of polyhedral gas cells separated by a thin liquid film (see Fig. 13 in Chapter 1). If a gas is dispersed in a liquid, we first get spherical bubbles forming a so-called gas emulsion. When they are close-packed, a foam is obtained. Each cell in a foam is closed, and if the walls open up we obtain a porous structure (see Chapter 1). The stability of a foam is related to the drainage, which makes the walls thinner and leads to collapse of the foam lamellae. Drainage is caused by gravity and also by the pressure in the so-called plateau borders (where the lamellae meet). The pressure in the plateau border is smaller than that in the central part of the lamellae, as the lamellae are almost planar whereas the plateau border is curved. Thus the liquid within each lamella will flow toward the plateau border, and then there will be a flow in the connected systems of plateau borders. There are, however, forces acting against this flow, owing to the repulsion between the surface monolayers including each electric double layer.

If we consider just a vertical film, which can be a part of a foam, equilibrium occurs when the hydrostatic pressure is balanced by the disjoining pressure. The gravitation field defines the hydrostatic pressure, and the disjoining pressure is due to repulsive forces between the opposing film surfaces. In addition to this there is an obvious effect of viscosity on the drainage of the foam.

Surface viscosity, elasticity, and the Marangoni effect are the most important factors in determining the stability of a foam, the resistance to collapse of the walls between adjacent gas cells. The Marangoni effect is due to differences in surface tension between different points at the surface of the foam lamellae. We can understand the Marangoni effect in the following way. Assume that a foam lamella starts to burst. When the "wound" in the film is formed by local thinning of the lamella, the surface area will increase, and because of the low diffusion rate of the surface-active molecules in the outer monolayers, the surface tension there will be higher than that of the surrounding part of the lamella. This spot with greater surface tension will therefore gain monolayer molecules from the surroundings, which, owing to viscosity, will also drag liquid from the inner part. Thus the "wound" will be closed.

In wheat systems the foam can be stabilized by polar lipids, by water-soluble proteins, and finally by the storage proteins. The different roles of these stabilizers of the gas cell interface from dough mixing to the final bread are described in Chapters 6 and 7.

Foam stabilization by the water-soluble fraction was studied by Lindahl [51] in our laboratory. A remarkable observation was that this fraction was as efficient as egg white proteins in foaming performance. The effect of the storage proteins in an aqueous dispersion on foam stability has been reported in many studies by Mita and coworkers [119]. The lifetime of a gluten foam depends on pH and ionic strength, and the stability is highest close to the isoelectric point and at high salt concentrations [119]. Furthermore, the foam stability is higher for a delipidized gluten than for a nondelipidized sample. The gluten foam is more stable than a foam of egg albumin, gelatin, or peptone.

The monolayer properties reported earlier, particularly the mechanical properties of the film, are certainly a very significant factor in this foam stabilization.

D. Rheological Properties

The viscoelastic behavior of the glutenins is often quoted as one factor contributing to the unique baking performance of wheat, and one that differentiates wheat from other cereals. Rheological measurements (physical dough testing) are often used for quality control and to increase our understanding of the interactions taking place during dough formation [104]. In this section a short description of fundamental rheological methods will be given, followed by examples of the rheological behavior of wheat protein fractions and of gluten. The rheological behavior of wheat flour doughs is described in Chapter 6.

1. Fundamental Rheological Measurements for the Study of Wheat Proteins

a. Viscosity Measurements

In viscometric measurements the relation between the shear rate ($\dot{\gamma}$) and the shear stress (σ) is studied. The test sample is exposed to increasing or decreasing shear rates, and the response of the material determines if it is a Newtonian liquid or if it is shear-thinning (pseudoplastic) or shear-thickening (dilitant). If the shear stress is linearly proportional to the shear rate according to the equation

$$\sigma = \eta \times \dot{\gamma} \tag{3}$$

where the proportionality constant η is the vicosity, the material is said to behave as a Newtonian liquid. However, for a material like gluten the relation between shear stress and shear rate in not of the same kind as Equation (3). Instead, the viscosity depends on the shear rate, and it is described as an apparent viscosity (η_{app}). Therefore, to characterize the material, it is not enough to measure the viscosity at a single shear rate; instead, a flow curve (i.e., shear stress versus shear rate) must be registered. Such a flow curve might be described with a power law of the type

$$\sigma = K \times \dot{\gamma}^n \tag{4}$$

where K and n are constants, and n indicates the deviation from Newtonian behavior. For a Newtonian liquid, $n = 1$. For a shear-thinning material, $n < 1$; and in the case of gluten, n depends on the concentration. The flow curves obtained for gluten dispersions in 0.01 M acetic acid at pH 4.5 show that the behavior is Newtonian up to a concentration of 4% [120]. At a concentration of 16% the value of n is 0.81. Also time-dependent changes of the gluten samples were observed. When a sample was sheared at a fixed $\dot{\gamma}$, the shear stress decreased with time. Wall and Beckwith [121] found that gliadin solutions in 70% ethanol showed Newtonian behavior even at the highest concentrations studied (10%). Glutenin, on the other hand, showed shear-thinning behavior even at a concentration of 4% (Fig. 25). Furthermore, at the same concentration and shear rate, the shear stress was always higher in the case of glutenin. For solutions, or dispersions, of gluten proteins it is thus clear that a gluten dispersion changes from Newtonian to shear-thinning behavior at some concentration, and that glutenin shows shear-thinning behavior at a lower concentration than gliadin.

Viscometric measurements might also be used to characterize the shape and size of macromolecules. The viscosity of a solution is then measured at different concentrations, and the specific viscosity, η_{sp}, is calculated:

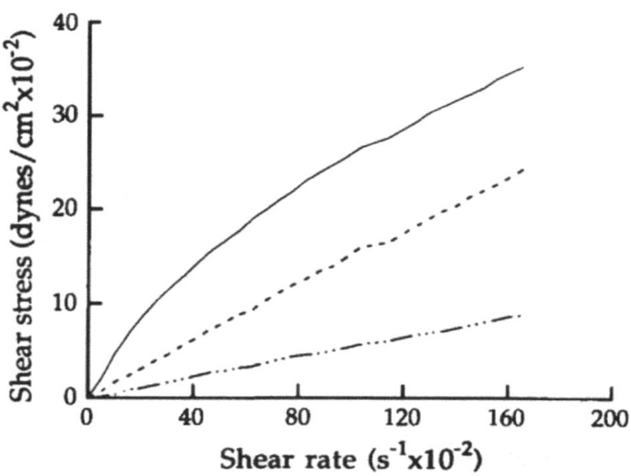

Fig. 25 Flow curves of gliadin and glutenin solutions; 4% (w/v) gliadin (dashed and dotted line), 10% (w/v) gliadin (dotted line), and 4% (w/v) glutenin (solid line). (Redrawn and recalculated from [121].)

$$\eta_{sp} = (\eta - \eta_s)/\eta_s \quad (5)$$

where η is the viscosity of the solution and η_s is the viscosity of the solvent. The specific viscosity is plotted against the concentration, and the regression line is extrapolated to zero concentration. The viscosity at zero concentration is defined as the intrinsic viscosity, $[\eta]$. Some values of $[\eta]$ are shown in Table 10. The results indicate great differences between gliadins and glutenins. For globular proteins, $[\eta]$ is in the range 3–4 mL/g, and thus both gliadins and glutenins deviate from the behavior of a globular protein. A-Gliadin might approach this value at certain conditions, and it has been suggested that A-gliadin is a globular protein but that under most conditions it is not present in a monomeric state [126]. The presence of aggregates causes the value of the intrinsic viscosity to be increased. From the intrinsic viscosity values obtained for glutenin, Field et al. [127] calculated the length and diameter of an HMW glutenin subunit as 492–500 Å and 17.5–17.9 Å, respectively.

A relation between $[\eta]$ for glutenins and baking performance has been found. Loaf volume increases with increasing $[\eta]$ values of the glutenin fraction [128].

b. Creep and Creep-Recovery Measurements

Creep and creep-recovery measurements can be performed in shear as well as in extension. The test material is subjected to a defined stress, and the deformation

Table 10 Intrinsic Viscosity [n] (in m L/g) of Gluten Protein Fractions

Buffer system	Gluten	Gliadin	Glutenin
Sodium lactate buffer, pH 3.1, 25°C,			
0.003 M	57	21	114
0.03 M	25	16	43
Aluminum lactate buffer, 0.0167 M		16	88
0.001 M CH_3COOH + 0.01 M NaCl (pH 4.0)		4.0	
5.9 M GuHCl (pH 5.0)		18.5	
3 M urea–0.11 M KCl, 25°C			
pH 3.3–3.5	27.6		
pH 4.1–4.3		18	58
pH 10.1–10.3	31	19	82
3 M urea + KCl, pH 4.8–5.5			
I = 0.0035[a], 20°C		24	124
I = 0.5		17	51

[a]The ionic strength was 0.0025 for the gliadin preparation.
Source: Data compiled from [122–126].

due to this stress is measured. After some time the stress is released, and the recovery is followed. For a completely elastic material the recovery would be 100%. When a gluten gel is exposed to a constant stress, the material will deform; that is, it will creep [129,130]. The recovery after release of load will be 100% only if the stress is below a certain value, the yield stress σ_y. Gluten behaves as an elastic body below σ_y. When the stress is above σ_y, the recovery will not be complete, and thus the material will also show viscous flow (Fig. 26).

c. Stress-Relaxation Measurements

Stress-relaxation measurements could be described as being the reverse of creep and recovery experiments. A well-defined deformation (strain γ) is applied to the test material, and the stress that the test material responds with is measured. As we are dealing with viscoelastic materials, this stress relaxes with time. An example for gluten is shown in Fig. 27. It is important that the loading period be short compared to the relaxation time of the material; otherwise the stress has

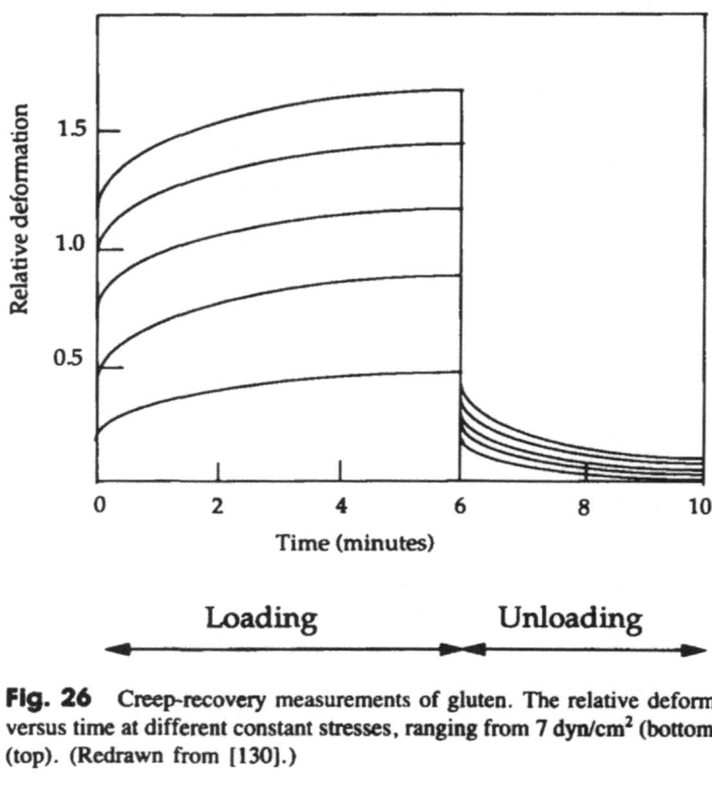

Fig. 26 Creep-recovery measurements of gluten. The relative deformation is plotted versus time at different constant stresses, ranging from 7 dyn/cm^2 (bottom) to 50 dyn/cm^2 (top). (Redrawn from [130].)

time to relax during the loading. To characterize the stress-relaxation behavior, the relaxation modulus G, or shear modulus, can be calculated. This modulus is defined in the following way:

$$G(t) = \sigma(t)/\gamma \tag{6}$$

$G(t)$ is the relaxation modulus at time t, and γ is the strain. The stress might approach zero after infinite time, and then the material could be regarded as a liquid. If the stress approaches a finite value after a considerable time, the material could be regarded as solidlike. It is evident from the stress-relaxation curve in Fig. 27 that gluten could be described as a liquid; only 3% of the initial stress remained after 10^4s [131,132].

The stress-relaxation behavior of gluten and dough has been analyzed according to a theory of cooperative flow [131,132]. The relaxation flow behavior of gluten was characterized by two relaxation processes. The primary flow process, occurring at short times (0.1–10 s), is characterized by a coordination number of 4. This corresponds to a non-close-packed structure formed either by oriented fibrillar structures or by a particle structure. The second flow process occurs at longer times (10–10^4 s) with a coordination number of 2, corresponding to a lamellar structure. This structure is illustrated in Fig. 28.

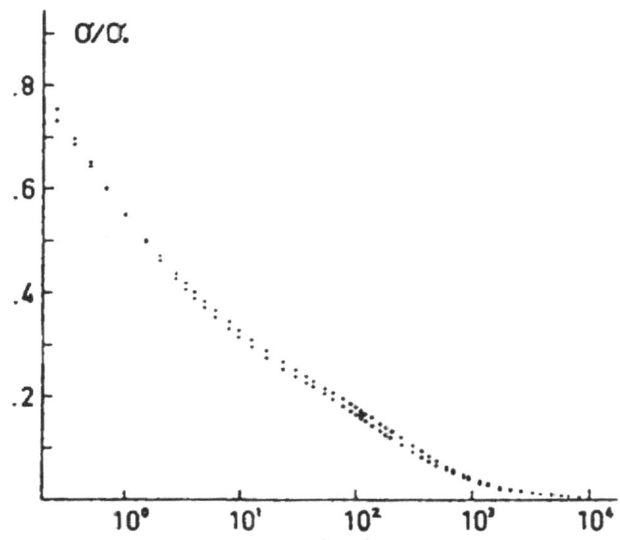

Fig. 27 Stress-relaxation curves for gluten at the shear strain of 0.2 (σ_0 = 172 N/m^2) and 1.0 (σ_0 = 264 N/m^2). (Reprinted with permission from [131].)

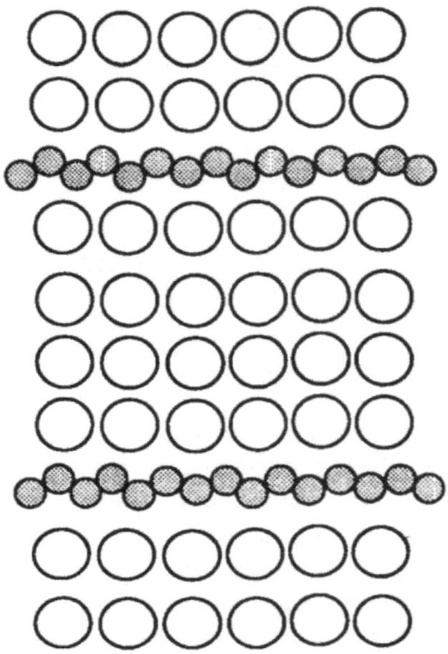

Fig. 28 The structure of flow units in gluten according to stress-relaxation measurements. The large circles represent cross sections of non-close-packed fibrils, and small circles represent an interface between homogeneous domains creating a lamellar superstructure. (Adapted from [132].)

There are several molecular causes for the creep and the stress-relaxation observed for polymeric materials [133]. Chain scission and bond interchange both result in creep and stress relaxation; viscous flow caused by slipping of linear chains might be a third cause. Physical cross-links are in a constant motion, and this might cause reversible relaxation.

d. Small-Amplitude Oscillatory Measurements

In small-amplitude oscillatory measurements (dynamic measurements), a deformation varying sinusoidally with time is applied to the test sample. The resulting stress will also vary sinusoidally with time, but, depending on the structure of the material, the stress may be completely out of phase with the strain (i.e., phase shift $\delta = 90°$) or completely in phase with the strain ($\delta = 0°$), or the phase shift may be somewhere in between ($0° < \delta < 90°$). If $\delta = 0$ the response is purely elastic, whereas a phase shift of 90° means that the material is purely viscous. For a viscoelastic material such as gluten and dough, the phase shift is thus somewhere in between, and for doughs δ is usually in the range of 25–30° [111,134].

In oscillatory measurements, a storage modulus, G', and a loss modulus, G'', are obtained together with the phase shift. Usually the measurements are performed so that these moduli are measured as a function of frequency; such measurements are illustrated in Fig. 29.

The storage modulus of gluten is influenced by the water content, but it is not possible to prepare gels with a water content of more than 65% [136]. The values of G' and δ are strongly dependent on the wheat variety; for gluten gels from four varieties, G' was in the range 5–10 kPa and tan δ in the range 0.4–0.7 [136].

Low molecular weight substances present in gluten influence the rheological behavior; if gluten is washed in extra water during the preparation, the G' values obtained are greater [137]. These substances might very well be low molecular weight proteins of high sulfur content [30].

For the oscillatory measurements to be of real value they should be performed in the linear viscoelastic region. This means that G' and G'' should not depend on

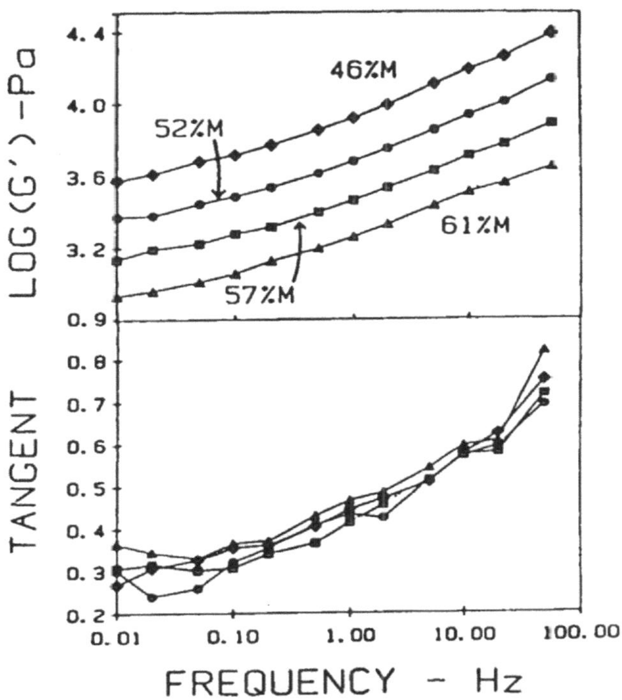

Fig. 29 Small-amplitude oscillatory measurements on commercial gluten mixed with different levels of water (%M in the figure). The storage modulus G' and the tangent (G''/G') are shown as a function of frequency. (Reprinted with permission from [135].)

the strain value. It is a common experience that colloidal systems are linear only at very low strain values (see, e.g., [138]). This is also the case with gluten and dough, and in some studies a linear region is claimed to be found [134,139], whereas others have concluded that the dough is nonlinear over all strains [140–142]. A linear region is more easily found in gluten than in dough [140].

e. Surface Rheological Behavior

The fundamental rheological methods described above are usually employed to characterize the bulk behavior of materials. However, it is also possible to study the rheological behavior of surfaces. Such measurements are of special relevance for the investigation of the properties of wheat proteins because of their proposed role as film-forming substances in the dough.

The surface rheological properties of gluten films were first investigated by Tschoegl and Alexander [143] and more recently by Kokelaar and Prins [144]. A gluten film exhibits viscoelastic properties, and these properties change with time, even after that the surface pressure has attained a stable value [144]. The viscosity of the interfacial protein film is highest at pH values close to the isoelectric point [143]. It is also noticed that the phase angle δ is much lower in the gluten film than in the dough. For the dough, δ values of about 28–30° have been observed [111], whereas δ for the gluten film is about half these values. The values obtained for the film thus indicate that the gluten film is almost completely elastic.

2. The Origin of the Viscoelastic Behavior of the Wheat Proteins

The viscoelastic behavior of gluten (and dough as well) is well documented from measurements using different fundamental rheological methods, as described above. What is it, then, that causes the viscoelastic behavior of gluten? It is often assumed that the glutenins contribute the viscoelastic behavior to the dough [145]. Linear concatenations could be expected to cause entanglement coupling, at least when the molecular weight has reached a certain value [146]. *Elasticity* will then emerge from the formation of a three-dimensional network formed because of the entanglement coupling. It is not necessary to postulate a permanently covalently linked network. The role of the disulfide bonds may be to increase the size of the glutenin molecule to enable entanglement coupling. The elasticity might also emerge from the presence of β-spirals, as described by Tatham et al. [90]. If the molecules are stretched, for example, during dough mixing, they will strive to return to the initial, more energetically favored state, which could be the β-turn [90]. The *viscous flow* in gluten and dough is related to the breaking of noncovalent bonds such as hydrogen bonds and to the change of hydrophobic interactions.

Also, an LMW protein like A-gliadin may exhibit viscoelastic behavior [147,148]. When a solution of A-gliadin at a concentration above 5% is sheared, it transforms into a gel (Fig. 30). At a fixed shear rate $\dot{\gamma}$, the stress σ increases with time after an initial lag period. Then σ is constant, followed by a decrease. It was concluded that the increase in σ was due to the formation of a three-dimensional network; that is, a gel was formed. This gel was found to be stable if left unperturbed for 24 h, but with continuous shearing the gel eventually broke down. The formation, stability, and breakdown were all dependent on $\dot{\gamma}$. On a colloidal level the gel was described as being composed of aggregates of A-gliadins forming fibrils that were elongated during shear. These elongated fibrils formed a network. The fibrillar phase was described as a liquid crystalline phase.

E. The Gluten Gel Structure

The molecular structure of the protein components in gluten has already been described, and in this section the structure of the aqueous gel on a colloidal and molecular level will be considered. Before describing the structure of the gluten gel, we will discuss the importance of the gliadin/glutenin ratio for determining the properties of gluten.

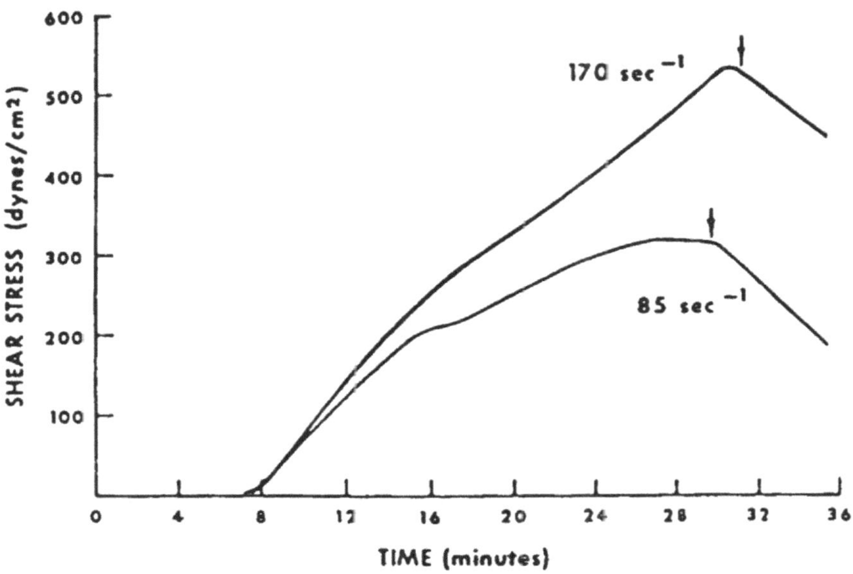

Fig. 30 The change in shear stress with time of shearing for a solution (5.7%) of A-gliadin. Two different shear rates, 85 s^{-1} and 170 s^{-1}, were used. The measurements were carried out at 20°C. The arrows indicate the point of gel breakdown in each case. (Reprinted with permission from [147].)

1. Gliadin/Glutenin Ratio

As was previously discussed, the gliadin/glutenin ratio has been suggested as an important parameter for baking performance. However, owing to the difficulties of preparing well-defined gliadin and glutenin fractions, a relation between the gliadin/glutenin ratio and baking performance has not been strictly proved [56].

For a good variety the gliadin/glutenin ratio was reported to be 46.5/43.5 and for a poor variety the ratio was 54.5/45.5 [149]. These values were based on extraction procedures. Determinations based on SDS-PAGE gave the glutenin proportion in the range of 41.3–47.1% for a good variety and 29.5–42.0% for a poor variety [68]. Although the absolute values might depend on the method of evaluation, the reported values indicate that a high proportion of glutenins is beneficial for baking performance.

It is possible to affect the rheological behavior of a flour by adding gluten or high molecular weight fractions of gluten. The best result, however, is obtained with an addition that contains both gluten and the HMW fraction; that is, there must be a balance between the various proteins [146]. Mixing behavior is very sensitive to the gliadin/glutenin ratio. If the proportion of glutenin is increased, the mixing requirements increase as well, and a longer time is needed to obtain the peak in, for example, a mixogram. When the proportion of gliadins is increased, the reverse, that is, a shorter mixograph development time is observed [100]. If the gliadin/glutenin ratio is changed toward a higher proportion of gliadins, the loaf volume will decrease. For a wheat flour of good baking performance there is a broad range of gliadin/glutenin ratios over which the loaf volume is unaffected [100].

To explain the significance of the gliadin/glutenin ratio it has been suggested that the gliadins act as a plasticizer for the glutenin. Gliadins help to solubilize or disperse the glutenins [128]. The rheological behavior of gluten and its dependence on the gliadin/glutenin ratio has also been interpreted in this way [103]. The gliadin interrupts the glutenin–glutenin interactions and thus weakens the gluten [86]. However, too much glutenin would result in a very stiff gel, which could prevent the expansion of the gas cells [44].

2. Some Earlier Models of the Gluten Structure

The existence of a lamellar structure was proposed by Grosskreutz [50] on the basis of diffraction spacings measured by the X-ray diffraction technique. Gluten shows a *d*-spacing of about 48 Å and gliadin a *d*-spacing of 42 Å [150]. At least the *d*-spacing in gliadin is independent of the wheat variety. Another aspect of the *d*-spacing in gluten is that it increases with hydration, and a *d*-spacing of 55 Å is observed for a fully hydrated gluten gel [151]. As the *d*-spacings disappeared when gluten was delipidized, it was suggested that gluten is a lipoprotein complex [150]. It has since been shown, however, that spacings are due to the lipids exhibiting a liquid crystalline phase in the gluten gel [151]. A lamellar structure of gluten was also discussed by Kasarda [89], who suggested that the glutenins are polymerized into lamellar structures, thus imparting some kind of liquid crystalline behavior to gluten. Furthermore, the rheological behavior of gluten in stress-relaxation measurements has been interpreted as due to a lamellar suprastructure imposed on a non-close-packed fibrillar structure [132].

Fibrils of hydrated protein are observed in the light microscope when wheat endosperm is hydrated [152]. When these fibrils are observed in the scanning electron microscope (SEM), more and thinner fibrils are visible, and it is evident that the fibrils are drawn out from continuous sheets of protein. They seem to result from a breakdown of the sheeted structure [153]. These fibrils have diameters of 100–600 Å, and it was suggested that they might be composed of smaller fibrils such as α-gliadin fibrils.

Gluten, washed out from a dough, can contain sheets of proteins as described above. If such structures are stressed (for example, by stretching), the protein sheets have been reported to form a network structure, although the networks are still arranged in layers [154]. If gliadin is washed away from these gluten samples, the same type of structure is obtained, although the sheets may now be transparent.

Transmission electron microscopy (TEM), used to characterize the gluten samples at a higher magnification than in light microscopy (LM) or SEM, shows a uniformly packed protein structure [151]. This structure is not influenced by lipids, as delipidized gluten shows the same structure. Moreover, the structure does not change after heat treatment at 95°C. Also, different storage protein preparations show the same TEM texture. The structure has been described as a globular aggregate with a hydrophobic core and a hydrophilic surface zone forming a continuum [151]. If such gluten gels are treated with reducing agents, threads (formed from the protein aggregates) may be observed. The molecular weight of these threads was estimated at several million (2.2–6.1 million according to Amend and Belitz [155]. At a low degree of extension they may be the basic units in the gluten structure obtained by mixing. Gluten gel may also be transformed into sheets or threads (fibrils) by mechanical treatment, depending on the type and degree of the treatment. The mechanical treatment in dough mixing, which involves protein being squeezed between starch granules, may give similar effects.

In the protein framework, osmophilic globular shapes of 15–300 nm diameter have been observed [155]. It has been conclusively shown that the structure of gluten is determined by the protein and is not affected by the presence or absence of lipids [151–153, 156].

3. The Gluten Gel Structure—A New Model

After this introductory summary of earlier reports on the gluten gel structure we will try to give a comprehensive description of the structural behavior thought to be most significant. Finally a proposed model of the gluten gel will be shown.

The pronounced film-forming properties of the gluten gel should be expected to be related to some kind of ordered structure. In fact, it is possible to observe birefringence under the polarizing microscope when a thin film is stretched, and this birefringence is reduced to zero with time. The stretching direction will also be equal to one of the extinction directions. The occurrence of birefringence is in itself proof of a periodically ordered structure, which in this case can be classified as a liquid crystal. The stretching induces a cooperative orientation of all molecules so that the observed region appears as a single crystal. If, however, the oriented gluten film is formed by just pressing down the cover slip, the anisotropy is radially symmetric from the point of maximum deformation. These deformations are further discussed at the end of this section. The time involved in this structural change from anisotropic to isotropic is of the same order of magnitude as the stress-relaxation time according to rheological measurements (see the preceding section). Results from X-ray scattering, electron microscopy, atomic force microscopy, and these polarizing microscopy observations are all consistent with the gluten gel structure shown in Fig. 31.

A unique feature of wheat storage proteins is their ability to swell in water to a well-defined phase that can coexist in equilibrium with an outside water phase. In this respect they behave like lipids (Section II). Lipids that form such phases are also characterized by partial disorder, and therefore by an ability to accommodate various related molecular species in a single aqueous phase. Although the gel appears to be homogeneous on the colloidal level we still have to ask whether the gluten gel really is one phase in the same way as a lipid–water phase is. The following simple experiments were done to enable us to answer this question.

We separated WSP fractions according to the method of MacRitchie based on dilute HCl [105], and the SE-HPLC chromatograms of two fractions used are shown in Fig. 32. One is richer in LMW proteins, and the other in HMW proteins.

The limit of swelling in water was first determined by formation of the gel in excess water and separation of the gel in the ultracentrifuge (200,000*g*). The gliadin-rich LMW fractions had a water content of 56% (w/w), and the HMW fraction was 54% (w/w) water. Thus, with respect to water content and consequently volume, there are no indications of LMW fractions and HMW fractions separating into one continuous and one dispersed structure of the gel. On the contrary, the swelling behavior of the two fractions appears to be identical, and this must reflect a close structural relationship.

We also colored the gel of one protein fraction with erythrocin and shook it (without stirring) into the uncolored gel of the other fraction. Light miscroscopy showed that the composite gel of the two fractions was completely homogeneous.

Circular dichroism studies have indicated that glutenin subunits are rich in

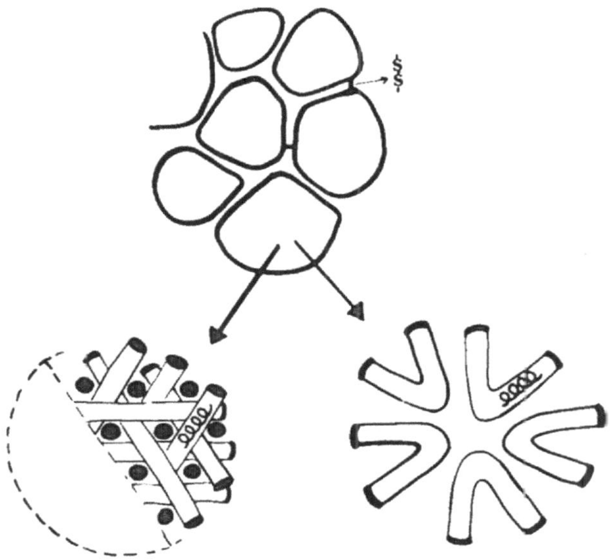

Fig. 31 Schematic model of the proposed gluten gel structure. The hydrophobic core of the globular molecules is shown, with the β-spirals indicated. In the peptide aqueous continuum there are disulfide bridges, and two such bridges are indicated with lines joining adjacent units. Small units of LMW fractions are not indicated in the figure; they contain less of the peptide chain in β-spiral conformation. Only one direction of β-spirals within each strutural unit is shown. Possible arrangements of the rod-shaped molecules in the micellar units are indicated below. The size of the micelles will be about the same as the length of the β-spirals, which is about 600 Å in the HMW fraction [157]. Either the stiff conformation of rods can be arranged in three directions to form a micelle (as shown to the left by a construction of interpenetrating rods), or some of them can be bent to fill space (as shown to the right by a model showing the polar ends and the axes of the β-spirals).

β-turns [72], and repetitive β-turns have been suggested to result in the formation of a β-spiral [90]. This structure is assumed to be responsible for the elasticity of elastin. It is in this connection worth noting that a tunneling electron microscopic study with very high resolution was recently reported [157]. A regular β-spiral conformation of the HMW fraction was observed (see Fig. 33). The samples were prepared from trifluoroethanol by evaporation on graphite, and the forces determining the conformation must be expected to be quite different from those in a water environment. Still it is possible that the same conformation can occur in a gluten gel, and it is also interesting to note that a similar conformation was seen in elastin, a protein with elastic properties like the gluten protein.

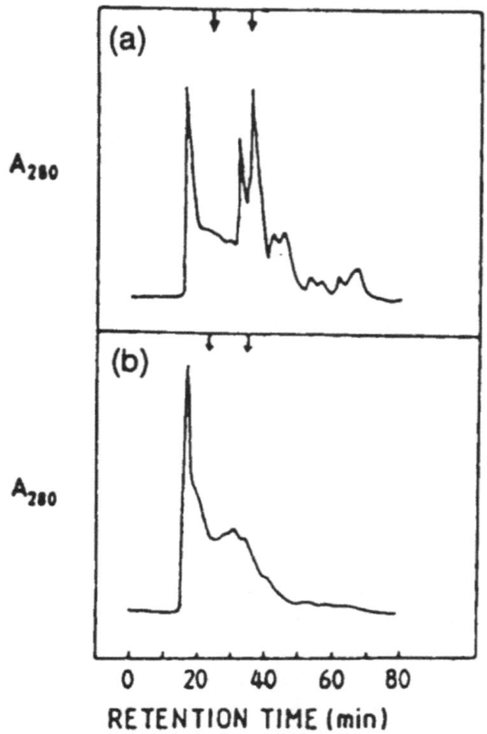

Fig. 32 Size-exclusion HPLC characteristics of two fractions of wheat storage proteins fractionated by hydrochloric acid. Fraction (a) is regarded as a low molecular weight fraction and (b) as a high molecular weight fraction. (Adapted from [111].)

Elastin is characterized by repeats of a hexapeptide, a pentapeptide, and a tetrapeptide [90]. All contain the dipeptide Pro-Gly, which favors β-turns. This is a very common sequence in the gluten proteins. The β-spiral conformation has been analyzed as a model of the elastin structure, and the repetition unit along the spiral is considered to be 9.45 Å with 13.5 residues per turn. The β-spiral rods in Fig. 33 have a lateral packing distance of 31 Å, and the helical periodicity along the rod is 14.9 or 15.8 Å (different domains). The periodicity seems to vary with the amino acid sequence. Schofield [44], in a review on flour proteins, also pointed out the analogy with elastin and the β-spiral content in connection with dicsussions of the gluten structure. We consider this background and the recent electron microscopic results as convincing evidence for a core with β-spirals in the gluten gel structural units. Stretching of such a molecule (or gel structural unit, which also might be an association complex in the case of small molecules)

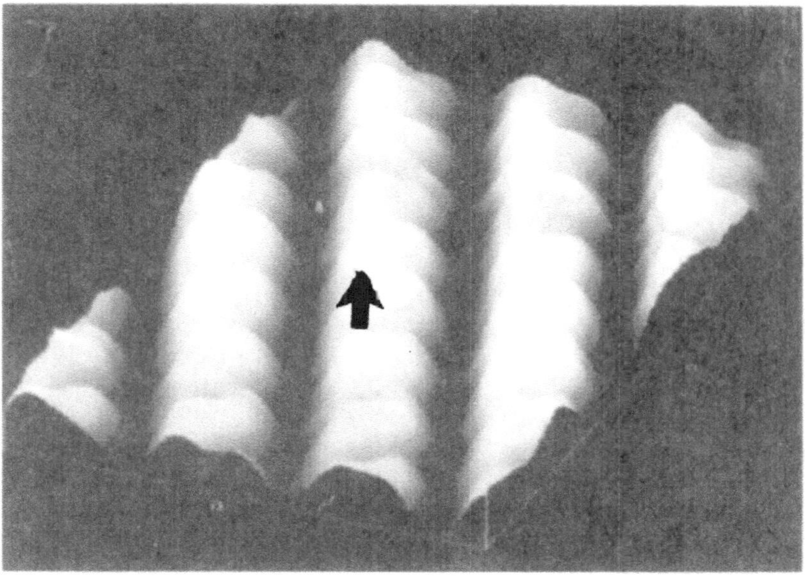

Fig. 33 Tunneling electron microscopic image of high molecular weight glutenin fraction. (Adapted from [157].)

will give an orientation and then disrupt the β-spiral as shown in Fig. 34, and the spiral conformation will be adopted again when the stress is removed.

The thermal effects on the gel are also relevant to the gel structure. A few minutes at 95° will induce loss of adhesiveness of the gels from the protein fractions described above. Particles in excess water cannot be forced mechanically to fuse, and the deformability is lost almost completely. According to the definitions of gel structures given in Chapter 1 we should regard this change as a gel → coagel transition. It takes place as a consequence of the chemical changes of the gluten proteins. We prefer not to describe the changes as a denaturation (which normally is a cooperative process taking place at one specific temperature). As described, for example, in the extensive review on wheat protein by Schofield [44], the degree of disulfide bridging increases rapidly with temperature, as is evident from molecular weight determinations. The thermal effects on the storage proteins are discussed further in Chapter 7.

The combined X-ray and electron microscopic study of gluten gels and delipidized gluten gels showed unambiguously that lipids self-assemble into aggregates; liposome-like aggregates occur frequently [151]. Futhermore, the freeze-fractured and freeze-etched EM micrographs indicated that the structure is very similar to that of the L2 phase of lipids, with an aqueous zone of a specific

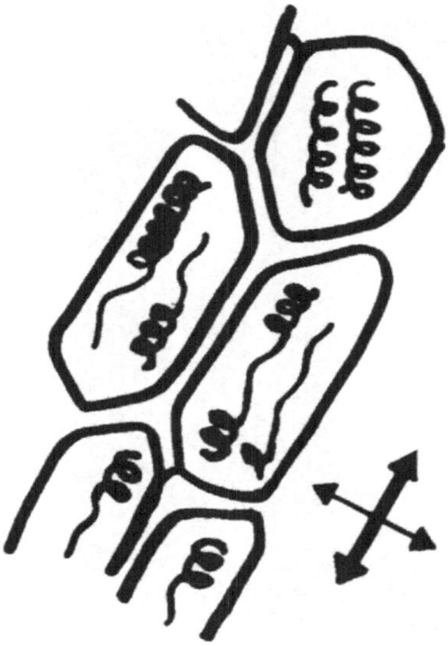

Fig. 34 Illustration of a linear deformation of a gluten gel according to the proposed structure, and the effect as indicated by the behavior in the polarizing microscope. The directions of the analyzer/polarizer cross are indicated, and one of these directions is the deformation direction. The rupture of β-spirals is also illustrated.

thickness forming a continuum between globular aggregates. In the case of the gluten gel, this L2 type of structure will consist of the globular units of the protein molecules of varying size, separated by such a water zone. The core of the protein units is hydrophobic, and the hydrophobic force (see Chapter 1) is the driving force behind this structure. A sample of the total gluten gel (including the lipids), a delipidized gluten, and the gel formed by salt-fractionated storage proteins all showed the same type of structure; a characteristic EM picture is shown in Fig. 35.

Perhaps the most important result of this study was the observation that the EM appearance of the gel was the same even after heat treatment at 95° for 30 min. As we know that the globular units are cross-linked to such a degree that the characteristic cohesiveness of the gel is lost (thus it is a coagel), it is clear that disulfide bridging has taken place. Certainly this occurs in the aqueous continuum between the molecules, but probably also intramolecularly, making the globular structural units (see Fig. 31) less deformable. This is in accordance with general knowledge of the location of intermolecular S—S bonds (usually at the ends of the peptide chains of the HMW units) and of intramolecular S—S linkages (within LMW-type molecules). This structural model is indicated in Fig. 31.

The basic conclusion supporting this model is that the structure must be

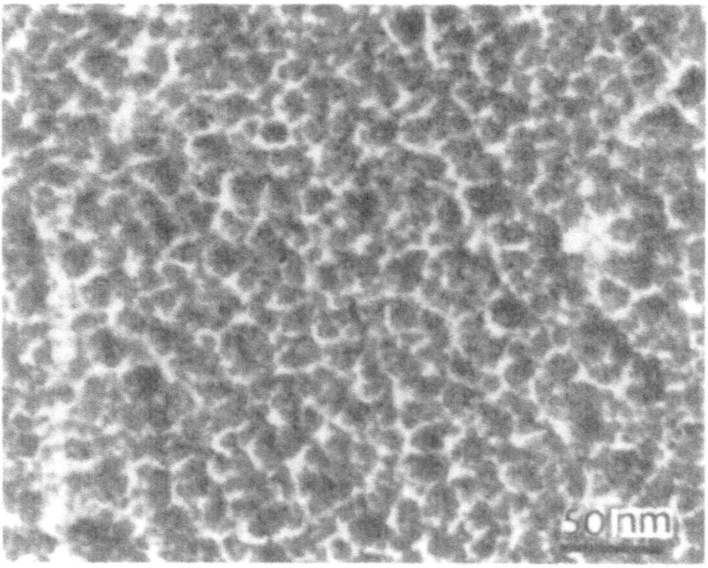

Fig. 35 Electron micrograph of a gel formed from salt-fractionated storage proteins. The sample was freeze-fractured and freeze-etched (rotary shadowing). The dark regions show individual protein molecule units separated by the aqueous zone. (Reprinted with permission from [151].)

analogous to general amphiphile–water phases like those formed by lipids. Furthermore, the protein molecules from the smallest to the largest are arranged similarly, as they exhibit the same swelling. The amount of β-spiral and the possibility to form intermolecular S—S bonds increases with molecular weight and contributes more to elasticity, but this is just a gradual change.

The mechanical properties of such a structure will be mainly determined by two features:

1. The aqueous zone between the molecules, which tends to keep a constant thickness (like any L2 phase).
2. The deformability of the hydrophobic core, which is related to the possibility of reorientation and the extensible character of the β-spiral. The reorientation is, of course, successively reduced by disulfide bridging.

If we imagine two simple types of deformations of this phase, a linear deformation would result in elongation of the molecules in one direction (see Fig. 34), whereas compression in one direction of a three-dimensional sample should give a morphological change from globular units to disklike units. These structural changes, as illustrated in Fig. 34, are in agreement with the birefringence observed under the polarizing microscope. The optical characteristics were discussed earlier in this section. The radially symmetric deformation, as can be seen in the polarizing microscope, will have the same type of deformation as that shown in Fig. 34, with the difference that the elongated units are radially directed, starting from the point of maximum compression. After relaxation the structure shown in Fig. 34 is obtained again. There is also a marked tendency to form one-dimensional structures. The LMW fraction defined in Fig. 32, for example, can be drawn to strong threads thinner than a human hair without rupture. The elasticity that exists during this stretching disppears within a few seconds.

The LMW fraction according to Fig. 32 gives, after ultracentrifugation, a completely transparent phase, that, owing to its homogeneous character, is a true protein–water phase in the strict meaning. Again it is pointed out that no other protein has been described to have such swelling behavior, whereas this is standard behavior for another group of amphiphiles, the lipids. The structural analogy proposed here is therefore justified. If a triglyceride oil is added to this phase, there is no interaction. By stirring, the oil can be emulsified into the phase, however, and the strong tendency to form spherical interfaces, illustrated in Fig. 36, demonstrates a significant interfacial tension. Water, on the other hand, spontaneously forms a rippled interface when exposed toward a gluten phase swollen to its maximum (see Fig. 36). This behavior is consistent with a water-continuous gluten phase structure.

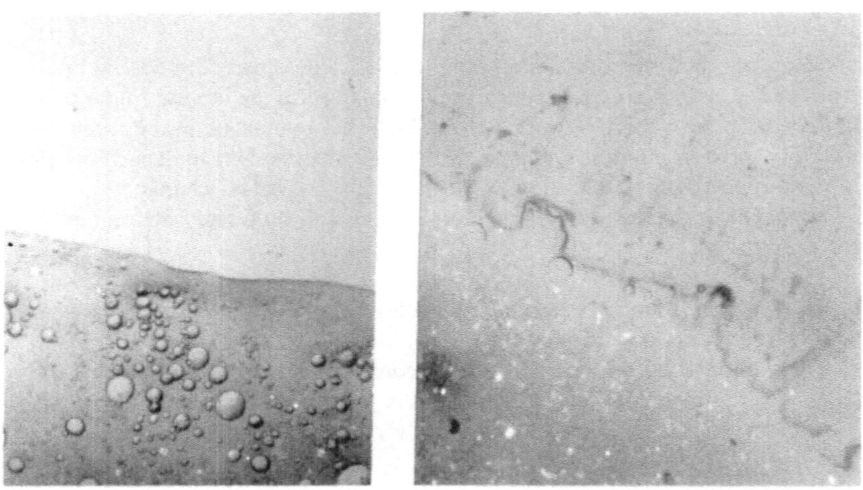

Fig. 36 A gluten gel containing oil droplets as seen in the light microscope. To the right the interface toward outside water can be seen.

The X-ray small-angle scattering curves of the two fractions at full water content are given in Fig. 37. These scattering curves contain information on electron density variations throughout the sample. When there are identical structural units in an aqueous continuum, the size and shape of these units can be obtained. The plot of log(I) versus S^2 is then linear, and the slope gives the radius of gyration. The two scattering curves in Fig. 37 do not give this linearity, showing that the units are polydisperse in size. It is evident, however, that the HMW fraction has structural units of larger average size than the LMW fraction (according to the steeper falloff in intensity). A very noticeable feature is the lack of diffraction effects at this high concentration, which demonstrates that the particles (or scattering units) have a very wide size distribution range. These scattering features are also exhibited by the gel formed by the total gluten proteins. Our conclusion from the X-ray data is that the scattering curves are consistent with an L2 type of structure with individual molecules forming the structural units separated by a water layer.

We believe that the structural model of the gluten gel presented above is in agreement with available chemical/analytical data and explains, at least qualitatively, the physical properties of the gluten gel. The model is still highly simplified, leaving open many questions on molecular details. The most important is the relation between the gliadin and glutenin types of constituents. We

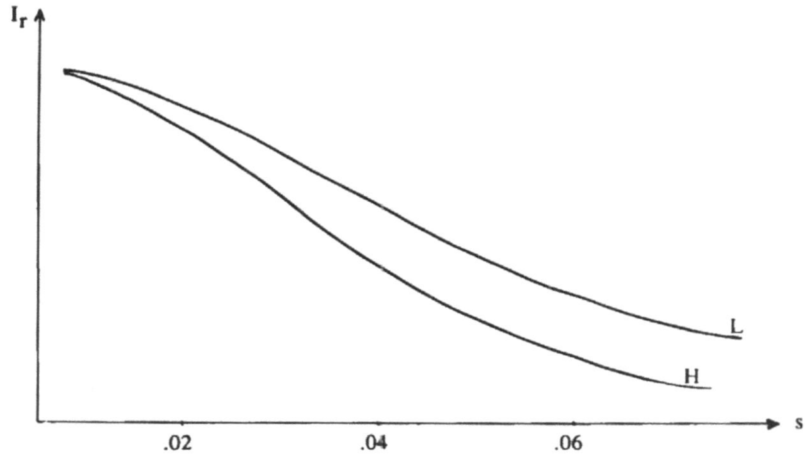

Fig. 37 Small-angle X-ray scattering curves of the fully swollen aqueous phase of (H) the HMW fraction and (L) the LMW fraction according to the SE-HPLC pattern in Fig. 32. The relative intensity (I_r) is given versus $S = 2 \sin v/\lambda$, where v is the scattering angle and λ is the X-ray wavelength. Identical exposure conditions were used for all samples in a Luzzati type of small-angle camera..

believe that they are arranged in a similar way, the gliadins, owing to their smaller size and lesser content of β-spirals, being partly space fillers and contributing less to elasticity.

There are two general alternatives to consider when we make the analogy between the gluten phase and structures of amphiphile–water phases exhibiting a limit of swelling. One is the structure discussed above with aggregates of hydrophobic cores in a continuous aqueous zone, reminiscent of a foam. The other is a bicontinuous structure, where the hydrophobic regions also are fused into a continuum. This latter alternative will form the same type of cubic structure as shown in Fig. 10 of Chapter 1. The hydrophobic rods of the β-spirals would then be arranged in lamellae with perpendicular orientation, their polar ends forming aqueous zones on each side.

There is, however, one argument against such a structure. A necessary condition for forming the curved lamellae of cubic phases (the cubic symmetry is just an average organization and need not show X-ray diffraction) is that the cross section in the center of the hydrophobic lamellae must be smaller than that nearer the polar ends. This shape corresponds to the wedge shape discussed in Chapter 1. This is not the case; the rod-shaped molecule has a constant average cross section. The β-spiral pitch might vary, however, and therefore the molecule could attain a wedge shape toward the center. Alternatively, the "hairpin" conformation shown in Fig. 31 (micellar model) would also allow the cubic structure. We therefore still consider this alternative to be possible.

In conclusion we want to emphasize once again that the gluten gel is an amphiphile–water phase similar to that occurring in lipid–water systems, including a liquid crystalline character (induced by stress). It is thus not a gel in the conventional sense, in which various amounts of bulk water are immobilized. We continue to use the term gluten "gel," however, as the phase has the viscoelastic and cohesive properties characteristic of a gel. The only adequate alternative we can think of is the term "L2 phase," which is correct according to the proposed model of the structure. It should also be noted that the existence of a swelling limit in itself has significant structural consequences. Thus protein networks, like those proposed in entanglement models, with large water compartments of various sizes, cannot exist. The well-defined single-phase character means that there must be a uniform water-layer thickness.

IV. STARCH

Starch constitutes the major component of a wheat flour, 74–90% on a dry basis, depending on the extraction rate. In baking experiments, starches from different sources (i.e., including non-wheat starches) were compared, and the result was clear. To get the same baking result from reconstituted flours as from a wheat flour, the starch in the reconstituted flour has to be wheat starch. Acceptable results were obtained with barley and rye starches, but all other starches behaved poorly [158,159].

Differences in baking behavior have been noted [160] between spring wheat and winter wheat starches, and when bread was baked from blends of gluten, starch, and water-solubles, the loaf volumes were in the range of 166–204 mL for 16 different wheat starches [161]. Wheat starch thus has properties that are crucial to the normal function of the wheat flour, and these properties seem to be one reason for the differences in baking performance between wheat varieties.

Whereas the role of lipids and proteins might be described as that of contributing gas-holding capacity and viscoelastic behavior to the dough, the role of starch is manifold. Starch provides the yeast with fermentable sugars, and it contributes to the structure of the crumb and to the structure and color of the crust. Some of these roles might be played by ingredients other than starch. For example, sugar might provide nutrition for the yeast but also add color to the crust because of carmelization. During storage of bread the starch component also causes deterimental changes. The role of starch has sometimes been described as that of an inert filler, although it was shown 30 years ago that another inert filler (glass beads) cannot be used to replace starch [162]. It should be pointed out that a solid filler will induce changes in the mechanical properties of the dough as well as in the bread. For example, the foam lamellae of the fermented dough (cf. Chapter 6) derive part of their stability from the oriented starch granules in the interstitial space. So even if the role of starch was only to be an inert filler, that might be a very important role. However, with the use of more sophisticated analytical methods, an awareness of an active role for starch has evolved. Rheological measurements of dough and gluten show that starch does not act only as an inert filler. Starch–protein interactions are important during all stages of the baking process. If this type of interaction is not optimized, the film surrounding the gas cells might rupture. Starch–protein interactions have also been suggested to be involved in the staling of the bread [163].

A. Chemical Composition of the Starch Granule

1. The Starch Granule

Starch is laid down in the shape of particles in special cells—amyloplasts—in the plant. These particles are called granules, and they are the means by which the wheat plant stores energy for the developing seed. The starch granule is a very efficient way not only to store the carbohydrate in a space-saving way, but also to make the energy easily accessible when the seed germinates. One starch granule is synthesized in each amyloplast, and the shape and size of a starch granule is typical of its botanical origin. Wheat starch granules are illustrated in Fig. 38.

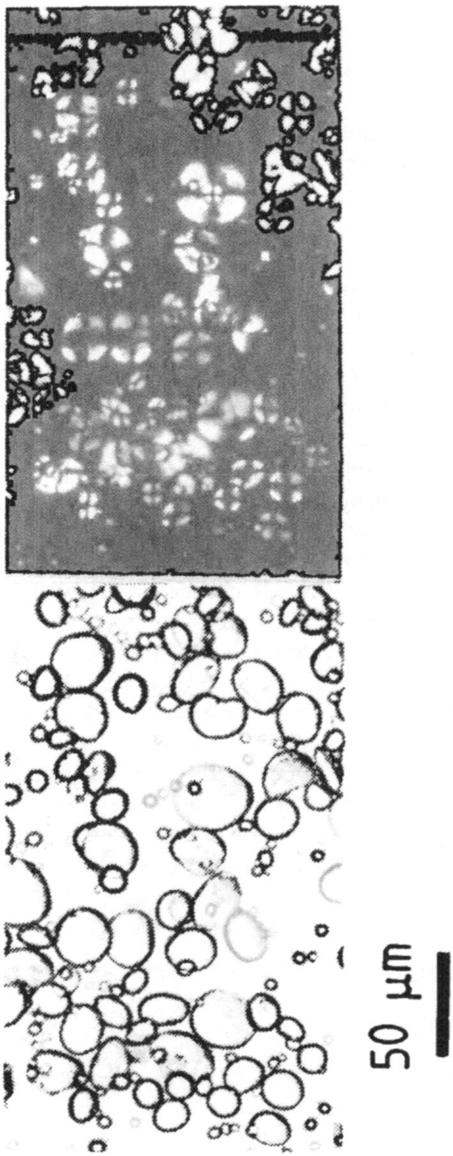

Fig. 38 Wheat starch granules observed in the light microscope. (Left) Ordinary light; (right) polarized light.

The size of the wheat starch granule is in the range 1–30 μm [164–166], the size distribution being bimodal (Fig. 39). Such a bimodal size distribution is characteristic of wheat starch, as well as of rye and barley starches. The small B-granules are spherical and have a diameter of less then 10 μm; a mean value of about 4 μm has been reported [166]. The large A-granules are lenticular and have a diameter greater than 10 μm, with a mean of 14.1 μm. The A-granules also have a typical equatorial groove or score [167]. B-Starch constitutes 23.9–28.2% of the total starch volume but about 90% of the starch granules in number [168,169].

It is possible to observe starch granules in the endosperm 4 days after anthesis. These first granules are A-granules, as the synthesis of B-granules does not start until about 2 weeks after anthesis [167,170]. This means that there is a change in the number of starch granules as well as in the particle size distribution during development and maturation [165]. The B-granules could be thought of as space fillers in between the large A-granules. They develop under conditions of tight packing and may therefore not always be spherical [167]. Now and then indentations are observed in the surface of A-granules, which have been interpreted as being due to the B-granules. As the A-granules are found at an earlier stage than the B-granules, it might be concluded that the A-granules are not just a bigger version of the B-granules. There are several differences between A- and B-granules in composition as well as in properties. Some examples are given in Tables 11 and 12 [160, 166, 171] and they will be discussed in this chapter.

The proportion of small granules is important for the baking performance of a wheat flour. When the starch is replaced with different wheat starches in gluten–starch–water-solubles blends, the loaf volume was in the range 166–204 mL, as discussed above. When a small-granule fraction was used, the volume

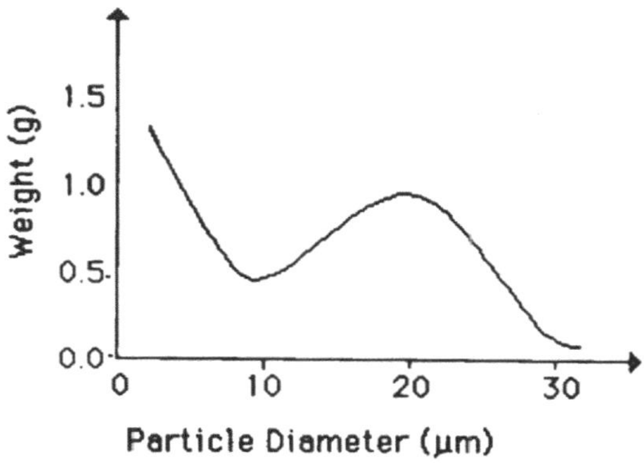

Fig. 39 Particle size distribution of wheat starch. (Adapted from [165].)

Table 11 Composition of A- and B-Granules in Wheat Starch

Component	A-Granules	B-Granules
Amylose[a] (%)	29.2 ± 1.0	27.4 ± 1.6
Amylose[b] (%)	27.3–31.0	26.3–31.5
Lipids (mg/100 g)		
FFA[a c]	84 ± 25	162 ± 50
LPL[a c]	845 ± 65	1062 ± 90
FFA[b c]	27–39	58–79
LPL[b c]	707–1014	1058–1398

[a]The mean value and standard deviation of 23 wheat varieties.
[b]The highest and lowest values for six wheat varieties.
[c]FFA = free fatty acid; LPL = lysophospholipids.
Source: Data compiled from [166], [171].

Table 12 Properties of A- and B-Granules in Wheat Starch

Property	A-Granules	B-Granules
Mean volume (μm^3)	1824 ± 221	56.6 ± 11.6
Mean diameter (μm)	14.1 ± 0.6	4.12 ± 0.28
Specific surface area[a] (m^2/g)	0.265 ± 0.011	0.788 ± 0.058
Gelatinization temperature		
BEPT[b], initiation (°C)	54.8	55.9
midpoint (°C)	58.4	61.5
final (°C)	62.0	64.8
DSC[a,c], T_o(°C)	56.8 ± 2.4	55.9 ± 1.9
T_m(°C)	62.3 ± 2.0	62.8 ± 1.7
ΔH (J/g)	9.9 ± 1.5	9.4 ± 1.2

[a]The mean value and standard deviation of 23 wheat varieties.
[b]Birefringence end-point temperature.
[c]Differential scanning calorimetry.
Source: Data compiled from [160], [166].

attained was only 147 mL, whereas it was 209 mL for a large-granule fraction [161]. Kulp [160] also noted a lower baking potential for small granules compared to regular starch. Baking experiments were performed with mixtures of A- and B-granules from different wheat starches (75% A-granules and 25% B-granules) [171]. With the same type of B-granules, and with A-granules from different wheat varieties, the loaf specific volume was in the range 3.43–3.97 cm^3/g. With the same type of A-granules, and B-granules from various sources, the specific volume ranged from 3.82 to 4.03 cm^3/g. When the proportion of B-granules was changed, the results in Fig. 40 were obtained. There is thus an optimal combination of large and small granules, and this combination seems to be close to the one found in the native flour.

The dimensions of the starch granule are in the upper range of what are usually defined as colloids (10^{-6}–10^{-9} m). Starch granules are far bigger than even the biggest glutenins, even those with molecular weights of around 20 million. Although they are large compared to other components in the dough, they are still small enough to create a large interface. The A-granules have a specific surface area of about 0.27 m^2/g, and the corresponding value for the B-granules is about 0.79 m^2/g (see Table 12). The specific surface area of the total starch is around 0.4m^2/g [166], resulting in a starch–gluten interface in the dough on the order of 0.2m^2/g dough. A substantial part of the gluten is thus required just to surround the starch granules.

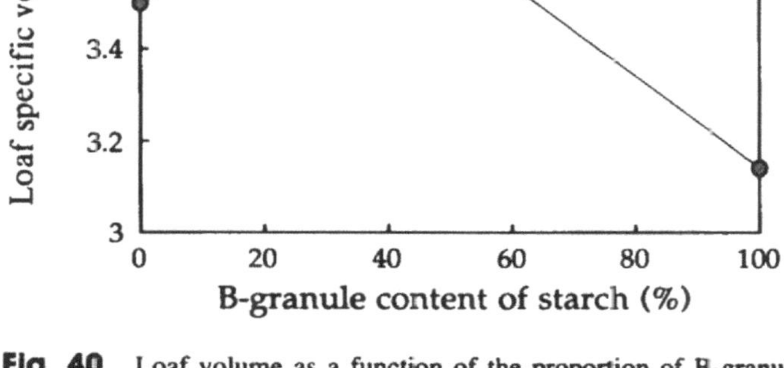

Fig. 40 Loaf volume as a function of the proportion of B-granules in wheat starch. (Adapted from [171].)

2. Amylose and Amylopectin

The starch graunle is almost 100% composed of the polysaccharides amylose and amylopectin; 98.5±0.5% of the starch granule is α-glucans [166]. Both amylose and amylopectin are glucose polymers linked by the α(1→4) linkage (Fig. 41). Amylopectin also contains a few percent (4–5%) of α(1→6) linkages, leading to a branched molecule [172, 173]. Both amylose and amylopectin have a wide distribution of molecular sizes. The proportion of amylose in wheat starch is about 29% [166]. The amylose content seems to be somewhat higher in the A-granules than in the B-granules (29.2% compared with 27.4% in bread wheats, see Table 11). In wheat starch, as in other starches, there exists a fraction that is neither amylose nor amylopectin. This intermediate material amounts to about 5–10% in cereal starches [174]. There is some confusion regarding the structure of this fraction—whether it should be described as an anomalous amylopectin or an anomalous amylose. In any case, its branching pattern differs from that of its "normal" counterpart [175].

Amylose is essentially linear, with an average degree of polymerization (DP) of about 1290 in wheat starch [176]. Amylose from cereal starches usually has a lower DP than amylose from other starches. In a recent study it was found that 1.6% of the glucose in amylose was in fact found in an α(1→6) branching point [173]. That amylose is not completely linear is observed in the fact that the β-amylolysis limit of wheat amylose is in the range 77–79% [177]. This means

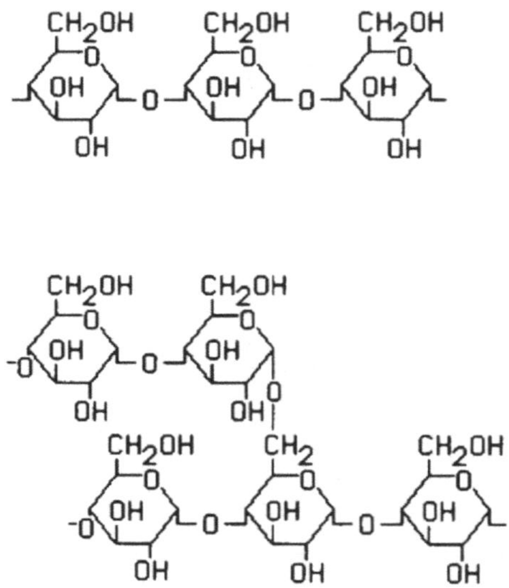

Fig. 41 The structures of amylose (top) and amylopectin (bottom).

that β-amylase, an enzyme that hydrolyzes $\alpha(1\rightarrow4)$ linkages but not $\alpha(1\rightarrow6)$ linkages, cannot completely hydrolyze amylose. The amylose molecule must then also contain $\alpha(1\rightarrow6)$ linkages. An average of 4.8 chains per molecule has been reported for wheat starch [176]. The solution behavior of amylose has been interpreted as being that of a random coil, with no marked helical content in aqueous solution in the absence of complexing agents [178–179]. Small-angle X-ray scattering data suggest that the amylose chain is highly disordered but that short-range helical structures might exist [180]. These helical structures are both irregular and labile, and an amylose molecule in solution might be illustrated as in Fig. 42. Although amylose is slightly branched, as described above, the branching does not influence its behavior in solution; amylose behaves as a linear polymer.

There are two features of amylose in solution that are of special interest in relation to baking. The first is the great tendency to form intramolecular hydrogen bonds, which means a strong tendency toward crystallization (also referred to as ***retrogradation***). An amylose solution is therefore, not very stable. For a 2.4% amylose solution, turbidity is observed only a few minutes after quenching to 32°C [181]. A limiting value was obtained after 1–2 h. The turbidity is interpreted as being the result of phase separation, as described in Chapter 1. The development of crystallinity can be followed by the X-ray powder technique, and it is found that crystallinity develops at a slower rate than turbidity. Moreover, the crystallinity develops in the polymer-rich phase. The second feature of interest with respect to amylose in solution is its ability to form helical inclusion complexes. When the proper ligand is present, amylose forms a helix with the

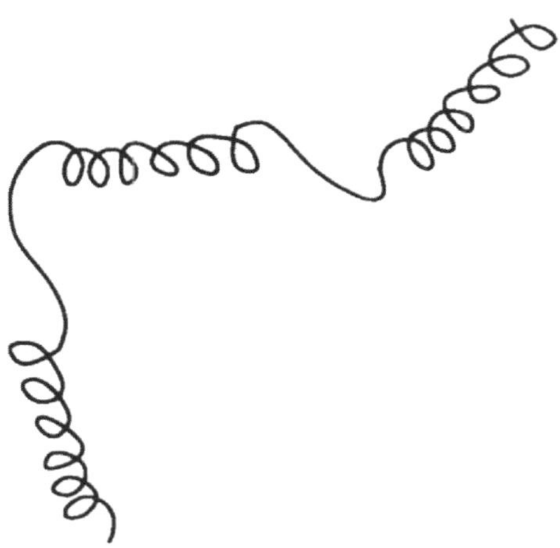

Fig. 42 Amylose in solution. (Adapted from [178].)

ligand in the cavity. The formation of a complex between amylose and iodine gives rise to the typical blue color, as this complex has an absorption maximum at $\lambda \approx 640$ nm [178]. Other ligands are the polar lipids, and the amylose–lipid complex will be discussed in section IV of Chapter 3.

Amylopectin is a much larger molecule than amylose; DP values of 10,000 and more are reported [182]. In fact, amylopectin is one of the biggest molecules known in nature. Different methods for the determination of the molecular weight of amylopectin give a wide range of results. This could be related both to the differences due to the source of amylopectin and to the use of different methods for the determination of the molecular weight. The ratio M_w/M_n is about 300, indicating a wide distribution of molecular sizes. The amylopectin molecule is highly planar in dimethylsulfoxide (DMSO) and behaves as an oblate ellipsoid [182, 183]. In water, aggregates are formed that are somewhat more spherical than the molecule in DMSO. The a and b axes of the ellipsoid molecule were determined in D_2O to be a = 1200 Å and b = 150 Å [183]. There is some variation in the molecular size of wheat starch amylopectin from one cultivar to another and from one growth location to another [182].

As already discussed, amylopectin is frequently branched. In wheat starch amylopectin, the average chain length (CL) has been reported to be 25 anhydroglucose units [184]. However, chain length in amylopectin shows a wide distribution, and in wheat starch a trimodal distribution is found (Fig. 43), or even a tetramodal distribution [185]. The shortest chains (F_2 in Fig. 43) have an average chain length of 17, and the second group of chains (F_1 in Fig. 43) has a CL value of 56. The longest chains have CL values of about 200 and more, and

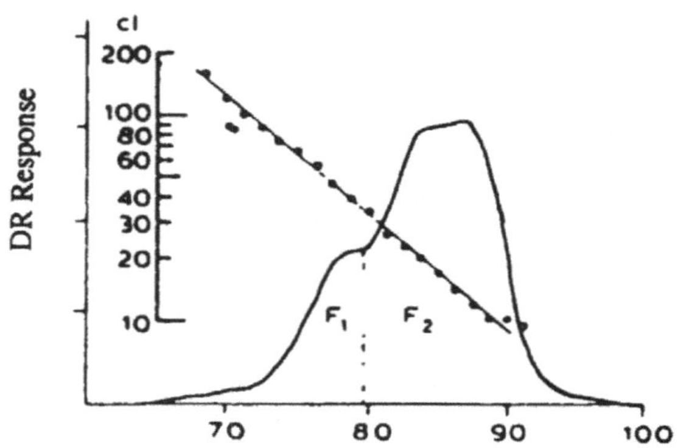

Fig. 43 The chain length distribution in amylopectin from wheat starch; DR = differential refractometer. (Reprinted with permission from [184].)

there are chains as short as 10 [184]. The CL value has been found to differ somewhat between wheat varieties, from 17 in durum to 20 in a soft wheat [177]. On the other hand, the branching pattern has been found to be the same for amylopectin from 10 different wheat varieties [185]. In order to elucidate the structure of the amylopectin molecule, results from enzymatic hydrolysis have been used [175]. The analysis differentiates between three types of chains: A-Chains are linked to the molecule through the $\alpha(1\rightarrow6)$ linkage, but the other linkages are all $\alpha(1\rightarrow4)$ linkages: B-chains are linked in the same way but may also carry other chains (A- and/or B-chains) through the α $(1\rightarrow6)$ linkages; and finally, there is one C-chain, which is the chain carrying the only reducing end group in the molecule. The ratio of A- to B-chains is important for the modeling, and it has been reported to be within the range of 1–1.5 [175]. One possible structure of an amylopectin molecule is illustrated with the cluster model in Fig. 44. Such a molecule may have a diameter of about 100 Å and a length of several thousand angstroms [187–189].

Amylopectin solutions do not show the strong retrogradation tendency as described for amylose. However, under the proper conditions, amylopectin will also crystallize. Turbidity is observed in solutions with a concentration above

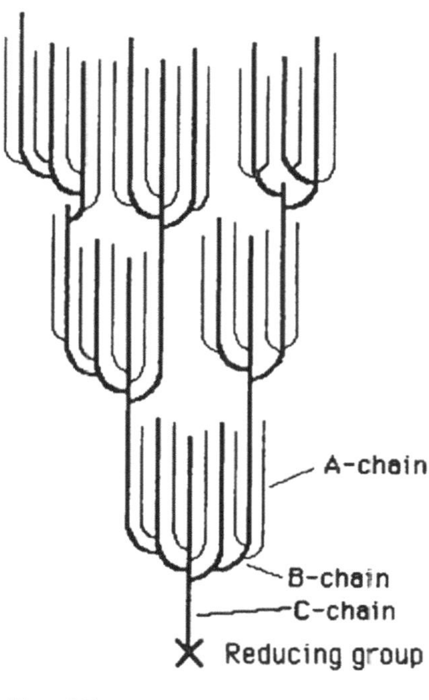

Fig. 44 The cluster model of amylopectin. (Reprinted from [186] with permission.)

10% and at temperatures below 5°C [190]. For a 20% amylopectin solution, the turbidity approached a limiting value after 4–5 days, in contrast to the results given above for amylose (1–2 h for a 2.4% solution). The possible crystal polymorphs of amylose and amylopectin are discussed later in this chapter.

The ability to form inclusion complexes in amylopectin is much less pronounced than in amylose. A red-brown complex is formed with iodine, the complex having an absorption maximum at $\lambda \approx 530$–540 nm [178]. However, as will be discussed in Chapter 3, the ability of the outer branches of the amylopectin molecule to complex with polar lipids has been investigated thoroughly in our laboratory recently.

An intriguing feature of amylose and amylopectin, when one tries to picture their distribution in the starch granule, is their incompatibility [191]. The incompatibility is observed when solutions of amylose and amylopectin are mixed. With time a two-phase system will develop, with an upper amylose-rich phase and a lower amylopectin-rich phase [19]. This result is also consistent with the starch structure (with amylose and amylopectin located separately in the starch granule). It has been suggested that the starch granule is initiated from a coacervate droplet formed by amylose, and the the hilum thus contains only amylose [188]. If during the biosynthesis amylose exists as a dispersed phase, it is not surprising that it will be incorporated in a disordered state within vacancies in the amylopectin lattice as described below.

3. Minor Components

In addition to the polysaccharides (amylose and amylopectin), wheat starch also contains components at levels of 1% and less. Becuse of the small amounts present in the starch granule, these components are usually referred to as "minor"—not because they are of minor importance for the properties of the starch. The minor components include polar lipids and proteins, substances that greatly affect the properties of the starch. These components will be described in this chapter, and their impact on starch behavior will be more thoroughly discussed in Chapter 3.

The minor components present in highest amounts are the polar lipids. They constitute about 1% of the wheat starch, but there are differences between A- and B-granules (see Table 11). The B-granules contain about 1220 mg lipids per 100 g starch, whereas the corresponding value for A-granules is about 930 mg/100 g starch [166]. The composition of the wheat starch lipids is given in Table 1. In wheat starch the lysophosphatidyl choline constitutes about 90% of the polar lipids present. The true starch lipids are difficult to extract; usually the starch has to be gelatinized (i.e., swelled with a combination of water and heat) for the lipids to be completely extracted. The best solvent is *n*-propanol–water [3:1 (v/v)] at a temperature of 90–100°C [192]. These lipids could be described as "internal lipids," and they are thought to be present as a helical amylose inclusion complex [193].

Another minor component in starch is protein. The protein content of purified starch can be estimated from the nonlipid nitrogen content, and the nitrogen content in isolated starch is in the range of 0.03–0.08% [161]. Half of this nitrogen comes from the lipids (lysolecithin [194]), and the protein content is then estimated to about 0.1–0.25%. Further purification might decrease the protein level to below 0.1% [166]. The protein present is of different types, depending on, among other things, the starch preparation procedure. SDS-PAGE has shown that proteins whose mobility is similar to that of the gliadins may be found at the starch granule surface after that the starch has been recovered from a gluten washing procedure [195]. Other proteins present might be enzymes, either remaining from the starch synthesis or those necessary for the hydrolysis of starch when the seed germinates. Two types of amylases are present: α-amylases and β-amylases. α-Amylse is an *endoenzyme*, meaning that it breaks α(1→4) bonds randomly (more or less). The size of the starch molecules will quickly be reduced owing to the action of α-amylase, and consequently the viscosity of a starch suspension rapidly decreases when α-amylase is present. This is the idea behind the *falling number* method, a method that is used to detect excess enzyme activity in grains. The α-amylase activity is normally low and will increase upon germination. High α-amylase activity is thus an indication of sprouting. β-Amylase is an exoenzyme, meaning that the α(1→4) glucosidic bonds are hydrolyzed from the nonreducing end during the release of maltose. However, when the enzyme meets an α(1→6) linkage, its action is stopped, and the β-amylase leaves β-limit dextrins as a result of its action. β-Amylase is also found in sound grain.

Starch proteins that are neither enzymes nor of the gliadin type are also present. One such protein is salt-extractable and has a molecular weight of approximately 30,000. The amino acid composition is quite different from those of the gluten proteins; the levels of glutamine and proline are very low, only 15% of the amino acid residues [196]. The protein shows no enzymatic activity. This protein accounts for 8% of the total protein associated with the starch. Because of the ease of extraction, it was thought to be located at the starch granule surface. The presence of another protein at the starch granule surface has been found to correlate with the endosperm hardness and thus with the milling characteristics of wheat [64]. This aspect of starch–protein interactions will be discussed further in Chapter 5. Other proteins can be extracted from starch but only after gelatinization [196]. These proteins are therefore described as internal, and SDS-PAGE shows that they comprise several components with molecular weights of 50,000–90,000. The proteins present at the starch granule surface can be extracted with a solution of 1% SDS containing 1% 2-ME [197]. After an extraction period of 24 h, no protein was present at the granule surface as judged from staining with protein-binding dyes. From 10 g of starch, 3.39 mg protein was extracted in this way. With five consecutive extractions, a total of 4.20 mg proteins was extracted. SDS-PAGE was used to study the extracted protein, and bands in the molecular weight range of 30,000–57,200 were observed.

When starch is separated from flour, as in the gluten washing procedure illustrated in Fig. 13, the protein content of the starch is usually higher than 0.1%. Moreover, if the starch is fractionated into primary starch (mostly A-granules) and secondary starch (mostly B-granules) it is found that the protein content differs between these starches, with 0.34–0.35% reported for the primary starch, whereas the secondary starch might contain as much as 4% protein [198]. The reason for these high values is, of course, that protein is attached to the starch granule surface during processing. Treatment with toluene or proteases decreases the protein content of the starch considerably [169, 199]. The proteins identified as gliadins, discussed earlier, may very well have adhered to the starch granule surface during the washing procedure. The protein fibrils formed when wheat flour is hydrated have starch granules adhering to them [152, 153].

4. The Starch Granule Surface

The arrangement of amylose and amylopectin at the starch granule surface is not known in great detail except that the starch molecules have the nonreducing sugar groups at the surface. In SEM micrographs the surface looks very smooth, but those pictures are usually shown with a low degree of magnification. The surface has been described as resembling a hairy billiard ball [200].

The amylose content increases with maturation [170]. As the starch granule grows by apposition, it is reasonable to suppose that the outer part of the granule will have a higher proportion of amylose. Furthermore, the amylose molecules increase in size with maturity. Also phosphorus, which comes almost completely from lipids, is enriched toward the periphery of the granule [192].

The starch granule surface is remarkably stable. Cereal starch granules are especially difficult to disintegrate, and fragments of the surface structure remain even after autoclaving [178]. When enzymes attack wheat starch granules they usually form pits and hydrolyze the starch granule from the inside [201, 202]. The large A-granules are attacked preferentially to the B-granules, and they are attacked at the central groove [201]. After some days of sprouting, the starch granule has holes all over the surface.

The starch granule surface contains other components besides amylose and amylopectin. The zeta potential of the B-granules depends on pH [203]. At pH values below 3.7 the zeta potential is positive, and above pH 3.7 it is negative. Extraction of the granules with different solvents changed the pH dependence, which was interpreted as showing that phospholipids are present at the surface of the wheat starch granules. The difference in extractability of lipids from wheat starch is also related to the presence of internal lipids and surface lipids [192]. An ESCA (electron spectroscopy for chemical applications) analysis showed that the following elements are present at the starch granule surface: carbon, oxygen, nitrogen, silicon, phosphorus, and sulfur [204]. Only carbon and oxygen are present in large amounts. The nitrogen present corresponds to amide-linked nitrogen, suggesting the presence of protein (quaternary nitrogen could not be distinguished because of the poor signal/noise ratio). The sulfur signal was found to correspond to a sulfide environment, and the phosphorus to phosphate. Silicon might arise from silicates or silica in starch but could also be a contamination. It might therefore be concluded that the starch granule surface contains protein as well as phospholipids [204].

These surface components will certainly play a role in the interactions between starch and other components. Treatment of the starch granule surface with a protease gave a complete loss of surface nitrogen and phosphorus [204]. It was therefore concluded that the lipids were bound to the surface by the protein. Extraction with methanol also caused a loss in phosphorus but an increase in nitrogen, perhaps by exposing protein farther away from the surface [204]. The presence of protein at the starch granule surface is also shown by the binding of protein-specific dyes to that surface [205]. The starch granule surface is thought of as being hydrophilic, and it has a poor oil-binding capacity [206]. It is possible to make it hydrophobic by treating it with chlorine or heat, which affect the protein at the starch granule surface [206, 207]. If the hydrophobic starch is treated with pepsin, its oil-binding capacity is lost [207]. This is further discussed in Chapter 3. In these experiments the oil-binding capacity was determined in the presence of water. If water is not present, starch is easily dispersed in oil also.

B. Organization of the Starch Granule

Native wheat starch, as well as other native starches, is birefringent when viewed in polarized light (see Fig. 38). Moreover, the native starch shows an X-ray diffraction pattern, and in the case of native wheat starch this pattern is referred to as the A-pattern. Gelatinized wheat starch might show the V-pattern due to the amylose–lipid complex, whereas retrograded starch shows the B-pattern. The birefringence indicates that the starch granule is anisotropic; that is, there is some kind of organization in the starch granule. The X-ray diffraction patterns show that the starch granule is partly crystalline.

The component in the crystalline domains is not amylose, as one would expect, but amylopectin. This is evident from the fact that waxy starches, almost free of amylose, give a good X-ray diffraction pattern. Furthermore, the crystallinity is retained even when amylose is leached from the granule [188].

The degree of crystallinity has been estimated in different ways, and values of 0–60% are reported [208]. Differences in the extent of crystalline organization of wheat starch varieties have been reported [209]. If the most crystalline wheat starch is given the value 100, the least crystalline starch of the six investigated would have the value 86. Moreover, the small granules were found to have a higher degree of crystallinity than the large granules. A study using cross polarization/magic angle spinning ^{13}C CP/MAS NMR showed that two types of starch fractions—crystalline and amorphous—completely build up the ^{13}C CP/MAS NMR spectrum [210]. It is possible to combine crystalline starch and amorphous starch in different proportions to match the spectrum from the native starch.

An important feature of the crystalline domains of the starch granule is their small size. The small size of crystallites causes line broadening in the X-ray diffraction pattern, an effect that can be used to estimate the size of the crystallites. Crystallite sizes in the approximate range 100–150 Å [188] as well as smaller starch crystallites (40–50 Å) have been reported [211]. For wheat starch the crystallite size was found to be about 100 Å, with no difference in crystallite size between A- and B-granules [212]. X-ray scattering measurements also indicate some long-range periodicity; a scattering maximum at 103 Å is observed in wheat starch [211]. Spacings just above 100 Å in wheat starch have also been observed by small-angle neutron scattering [213]. However, this periodicity was not observed in dry starch. It was concluded that the periodicity is in the radial direction and that although the gelatinization causes a loss in birefringence and in X-ray crystallinity it does not lead to a major disturbance in the dimensions of the amylopectin molecule. A scattering maximum at about 260–296 Å has also been observed in native starch but disappears after heating [214]. Radii of gyration (R_g) were found to decrease upon heating, from 249 Å at 25°C to 231 Å after 2 h at 88°C. Cooling to 25°C and storage at this temperature caused a further decrease in R_g, to 190 Å after 3 days at 25°C [214].

It has been suggested that the average chain length of the amylopectin molecule determines the crystal polymorph in the native starch granule [184]. The short chain length in wheat (CL = 25) results in the A-polymorph, whereas the much longer chain length in potato (CL = 34) results in the B-polymorph. The C-starches have CL values in between (CL = 28). During retrogradation of wheat starch the B-pattern will form, as already discussed. Heat and moisture treatment might shift a B-pattern into an A-pattern. The reverse transition has not been reported [215]. Furthermore, the presence of alcohols and acids during crystallization might shift the crystalline polymorphic form from B to A [216]. The presence of free fatty acids and lysolecithin has the same effect—the structure is shifted from B to A. For oligosaccharides, the minimum chain length required for crystallization is 10. If longer chains are present, even maltohexose can cocrystallize. Moreover, short chains favor the A-pattern, as do high concentration and high temperature [216, 217].

The most detailed analysis of the structures of the A-, B-, and C-polymorphs has been obtained from crystalline samples of amylose (see, e.g., [218]). Although there are some discrepancies in the X-ray diffraction patterns obtained from starch and amylose, the main conclusions are applicable also for starch.

The crystalline regions are due to amylopectin, and in order to understand the structure of the crystalline domains it is necessary to take another look at the cluster model of amylopectin (Fig. 44). In the crystalline domains, double helices are arranged in an ordered structure. It is evident from the model shown in Fig. 44 and the chain length distribution in Fig. 43 that the length of such a double helix can be only a few turns. The double helix structure is illustrated in Fig. 45. Each turn of the helix contains six glucose residues, and as the average chain length is about 25 this could be enough for four turns. But the shortest chains contain much fewer glucose residues, and there must also be space for the branching point. The height of the double helix therefore corresponds to only two

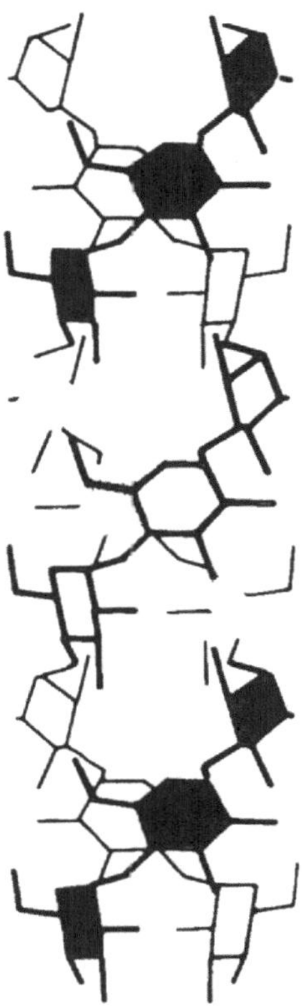

Fig. 45 The double helix structure of amylopectin. (Reprinted with permission from [219].)

to three turns of each chain. This double helix-structure is considered to be the same irrespective of the type of X-ray diffraction pattern. The helices are assumed to be right-handed and parallel-stranded. The different diffraction patterns are due to the packing of these helices, and the packing and the unit cells of the A- and B-polymorphs are illustrated in Fig. 46. The packing of helices into the unit cells has been interpreted as being antiparallel (e.g., [218]). An antiparallel packing, however, seems not to fit in with the biosynthesis, and more recent analysis has not been able to unambiguously determine the packing one way or the other [220].

The unit cell in the A-polymorph is basically hexagonal but is usually described as centered orthorhombic, with the dimensions a = 11.90 Å, b = 17.70 Å, c (fiber axis) = 10.52 Å [221]. The unit cell contains 12 glucose residues and eight molecules of water. The unit cell in the B-polymorph is hexagonal, with the dimensions a = b = 18.50 Å, c (fiber axis) = 10.40 Å [222]. This unit cell also contains 12 glucose residues but 36 water molecules. One very important difference between the A- and B-polymorphs is thus the water content [188, 211, 218].

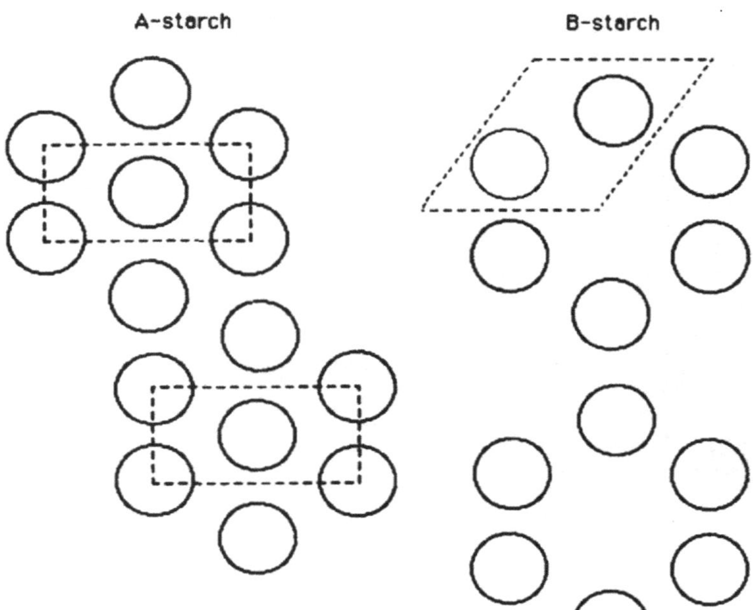

Fig. 46 The unit cells of the A- and B-polymorphs of starch. (Adapted from [218].)

Knowledge of these subcell dimensions defining the packing behavior of the helices is far from enough to construct the whole starch granule. The basic problem is the relation between the chemical linkage of the cluster and the crystallite. Furthermore, we do not know the location of the lipids, we do not know what the amorphous regions look like, etc. Recently we suggested the structure of the starch granule illustrated in Fig. 47. To construct this model we combined data from X-ray diffraction analysis with general knowledge of packing principles in interpenetrating structures [186, 223].

The starch granule has a layered structure, with alternating layers of different refractive index, density, crystallinity, and resistance to enzymatic attack [188]. In the light microscope these layers are visible in potato starch without any further treatment, whereas wheat starch requires enzymatic or acid treatment to visualize such rings. They have been connected with a periodicity of the biosynthesis and are referred to as *growth rings*. The crystalline regions contain the double helices, and their length is about 50 Å. The $\alpha(1\rightarrow6)$ linkage has been suggested to be located in the amorphous region, which has a height of about 20 Å. However, one feature of the packing is that the branching point [the $\alpha(1\rightarrow6)$ linkage] fits in very nicely between the double helices [186]. The brass model in Fig. 48 illustrates the double helices with possible $\alpha(1\rightarrow6)$ linkages. It is thus not necessary to assume that all the branching points are located in the amorphous regions. Instead, the disruption in crystallinity could be due to the termination of chains. The short chains (CL ≈ 17) terminate at the crystallite surface, whereas this cannot be the case for the longer chains. Such a long chain might then be present in several alternating crystalline and amorphous regions. The diameter of the crystalline domain is about 100 Å.

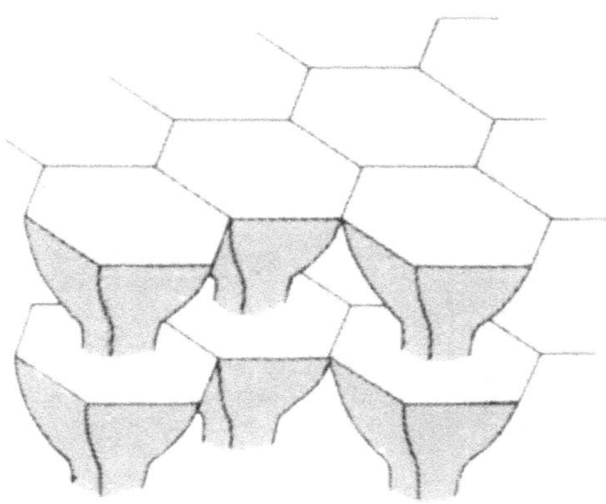

Fig. 47 Proposed cluster arrangement into the concentric crystalline layers of the starch granule. (Reprinted with permission from [186].)

The layers described above surround the central region (hilum) in concentric shells. The starch molecules, on the other hand, are oriented radially, as judged from the birefringence properties [188]. The starch molecules are thus oriented perpendicular to the growth rings and to the surface of the starch granule. The ordered structure is built up from crystalline layers forming a mainly spherically concentric arrangement. Amorphous material might fill in where necessary to avoid empty space. The implication of this is that the starch crystallites have to be curved. Due to the curvature of the starch structure it could be described as a quasicrystal with a two-dimensional dilation of the lattice [224]. The distortions in the packing, as illustrated in Fig. 49, would be at the boundaries of the clusters.

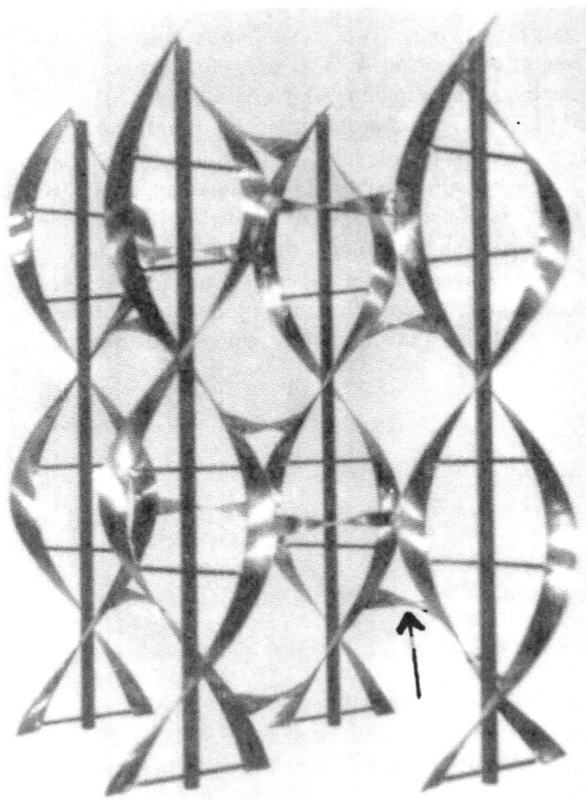

Fig. 48 A brass model showing a possible location of the $\alpha(1\rightarrow6)$ linkage in double helices. (Reprinted with permission from [186].)

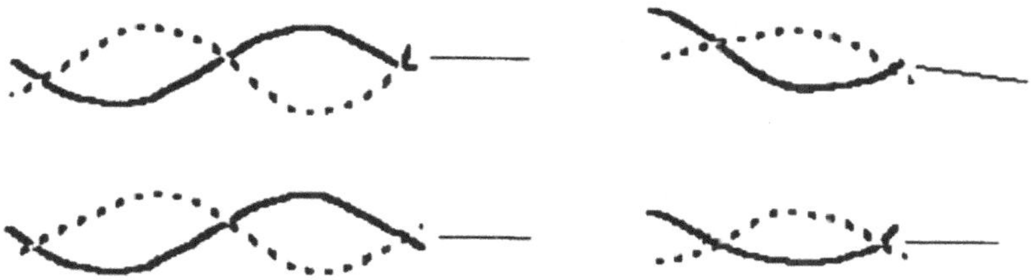

Fig. 49 The curved layers in an amylopectin cluster. (Left) The normal close-packing arrangement of double helices; (right) the proposed translation to account for a change in the direction of the helical axes. (Adapted from [224].)

The lipids must be located in the amorphous region because the crystallographic dimensions indicate that epitaxial growth is not possible [186]. The presence of the lipids in native wheat starch as an amylose–lipid complex has been questioned because of the lack of the V pattern of native starch. The V pattern is observed first after the gelatinization of starch. However, the lack of the V pattern does not prove that the amylose–lipid complex does not exist. It only proves that the amylose–lipid complex does not exist in a crystalline structure that we can analyze with the X-ray powder diffraction tehnique. The crystalline domains could be too small, or the proportion of crystallinity might be too low to be detected.

C. The Gelatinization Phenomenon

When starch is heated in water, certain characteristic changes occur, changes that usually manifest themselves in a change in rheological behavior, such as an increase in viscosity (Fig. 50). The extent of these changes is related not only to the type of starch but also to the water content. The starch concentration has to exceed a certain value for any viscosity increase to be recorded (Table 13). The viscosity (η) depends not only on the concentration bust also on the wheat variety. Wong and Lelievre [225] compared different wheat varieties, and at a shear rate of 20 s^{-1} and a concentration of about 0.5 g/100 mL, η was in the range 0.31–0.86 $N \cdot s/m^2$ for five different starches. They related the observed differences to the number fraction of large granules and to the volume fraction of swollen starch particles.

Rheological changes in starch contribute to the setting of the crumb during baking and might cause an end to the oven spring [226]. They are therefore essential for the crumb structure. *Gelatinization* is used as a collective term to describe several changes that occur in the starch at different temperature intervals. These changes include loss of birefringence, loss of X-ray diffraction pattern, absorption of water and swelling, change of shape and size of starch granules, leaching of amylose (amylopectin) from the granules, and formation of a gel or a paste [227]. The primary event could be described as the melting of the crystalline structure and consequently the disappearance of the X-ray diffraction pattern. All the other events could be regarded as a result of this primary event.

There are several ways to measure the gelatinization temperature range. However, different methods are used to measure different aspects of the gelatinization process. As a result, quite different temperatures are given in the literature for the gelatinization temperature interval of any one starch (Table 14). It is worth noting that the gelatinization temperature interval of a whole starch population is 10–15°C at a water content exceeding 90%. The corresponding interval for a single starch granule is 1–2°C [232]. During recent years differential scanning calorimetry (DSC) has evolved as the preferred method [227]. The DSC thermogram displays the heat flow as a function of temperature during heating (or cooling) of starch. An example of a DSC thermogram is shown in Fig. 51. The events recorded in such a DSC thermogram are endothermic, the first transition being due to the gelatinization of starch. The second transition, at temperatures above 100°C in Fig. 51, is due to the amylose–lipid complex. The gelatinization is irreversible, whereas the transition of the amylose–lipid complex is reversible, and during cooling there will thus be one exothermic transition due to the amylose–lipid complex. The thermogram illustrated in Fig. 51 shows

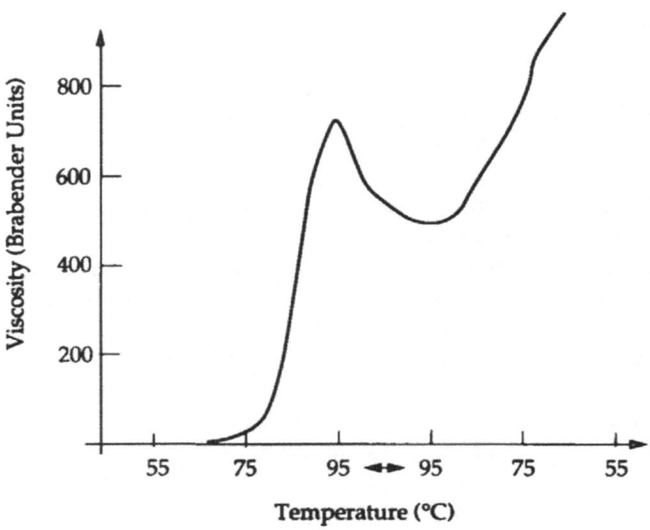

Fig. 50 The increase in viscosity during the gelatinization of wheat starch [8% (w/w)] in a Brabender amylo/viscograph. The heating rate was 1.5 °C/min, followed by a holding period at 95°C and cooling at a rate of 1.5°C/min.

Table 13 Viscosity of Wheat Starch Pastes of Different Concentrations

Starch concentration (g/100 mL)	Apparent viscosity (N·s/m) at $\dot{\gamma}$ = 20 s^{-1}
4.27	0.07
4.71	0.17
5.35	0.31
5.75	0.46
6.15	0.60
6.61	0.73
7.04	1.20
7.60	1.30

Source: Data extracted from [225].

the events at a starch/water ratio approaching the conditions in a dough—starch/water ≈ 1. At a lower starch/water ratio (more water) the double endotherm in Fig. 51 transforms into a single, narrow endotherm. In the following, processes connected with gelatinization will be described in some detail.

1. Birefringence

When wheat starch is heated in water, the birefringence in polarized light is lost. The loss of birefringence is related to the water content, and when the micro-

Table 14 Gelatinization Temperature Range of Wheat Starch

Measurement method[a]	Water content (%)	Temp. (°C)	Ref.
BEPT	Excess	56–64	228
BEPT	Excess	55–65	229
BEPT	Excess	52.0–60.0	170
DSC	57	45.8–81.0	230
DSC	67	54–86	231
DSC	50	55.9–92.2	245
DSC	75	56.5–66.1	164

[a]BEPT = Birefringence end-point temperature; DSC = differential scanning calorimetry.

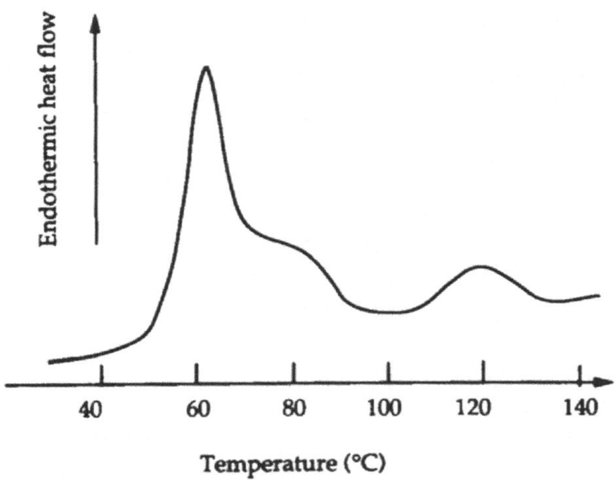

Fig. 51 A DSC thermogram obtained during the gelatinization of wheat starch at a starch/water ratio 1 and a scanning rate of 10°C/min.

scope is used to determine the gelatinization temperature interval, 0.1–0.2% suspensions are used [229]. The loss of birefringence can occur as illustrated in Table 14. In these wheat starch samples the birefringence was completely lost at temperatures of 65°C or less. If the water content is reduced, for example, to 50%, the starch granules are still birefringent at 75°C, and if the water content is as low as 30% the starch granules are birefringent even after a heat treatment at 132°C [233].

The starch/water ratio is reduced in the crust during the baking of bread, owing to the evaporation of water, and it is therefore possible to find birefringent starch granules in the crust. The gelatinization temperature range may also be increased by other ingredients in a recipe, for example, sugar. Birefringent starch granules may also be found in cakes and cookies where the starch/sugar and/or starch/fat ratio is low. The temperature interval during which the birefringence is lost is the same as for the DSC endotherm [234].

2. X-Ray Diffraction Pattern

The native wheat starch shows, as discussed above, an A-type X-ray powder diffraction pattern. When the starch is gelatinized, th V-pattern due to the amylose–lipid complex is observed [235]. The change in crystallinity is gradual and can be followed by the decrease in the intensity of the X-ray pattern as a function of temperature. Such a DPT (diffraction pattern versus temperature) diffractogram has not been published for starch, perhaps owing to the difficulties encountered when long exposure times have to be used. The background scattering is intense, and the diffraction lines are very weak [236]. However, Zobel et al. [211] estimated the intensity of the diffraction lines after collecting the diffractogram at discrete temperatures during the gelatinization interval. They found that both birefringence and the X-ray diffraction pattern are lost during a temperature interval corresponding to the gelatinization temperature range measured by DSC. Zobel et al. [211] further noted that a gradual decrease in intensity of the X-ray pattern occurred across the whole temperature range. This observation rules out such possibilities as transitions between polymorphs and increased crystallinity due to annealing effects. The temperature interval during which the X-ray diffraction pattern is lost is illustrated in Fig. 52. If a wheat starch–water suspension with 29% (w/w) or less water is heated to 95°C, the A-pattern is still

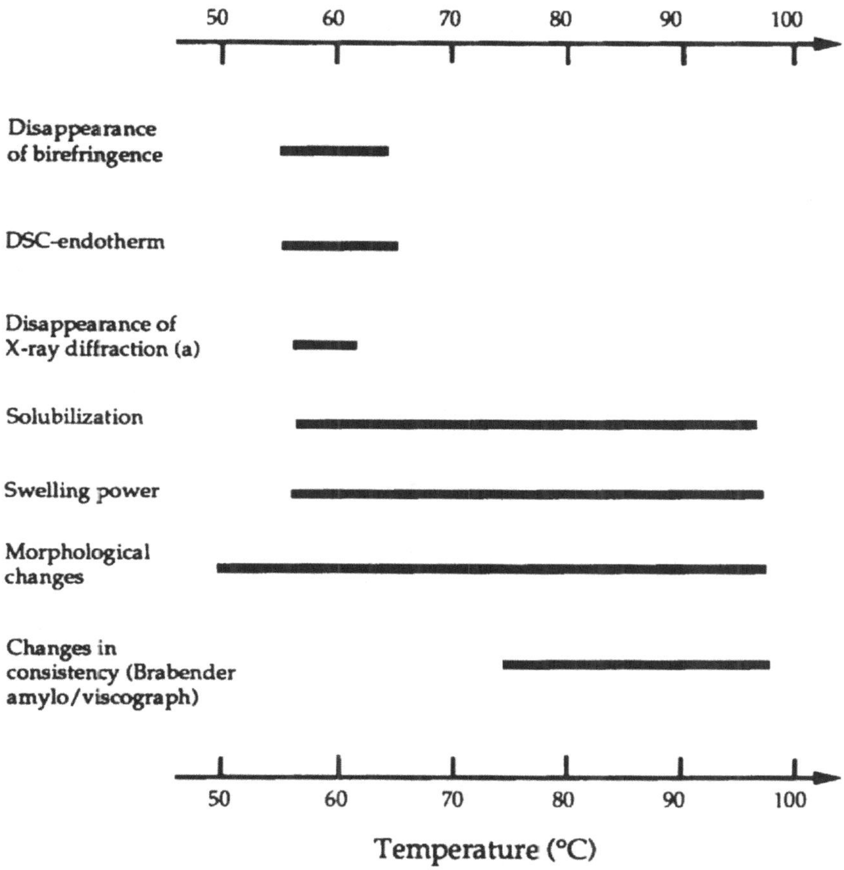

Fig. 52 The temperature interval for the events occurring during the gelatinization of starch. The data on the X-ray diffraction pattern were obtained for potato starch.

recorded [235]. If the water content is above 43% (w/w), only the V-pattern is observed, whereas a mixture of the A- and V-patterns is found in the region of intermediate water contents [29–43% (w/w) water].

3. DSC Thermograms

A DSC thermogram obtained for wheat starch at a starch/water ratio of 1 and a heating rate of 10°C/min is shown in Fig. 51. The first DSC thermograms published illustrated the gelatinization at a high water content, when a single, narrow endotherm is obtained (cf. [231]). It was then found that different starches give rise to endotherms of different sizes (i.e., with different gelatinization enthalpies, ΔH values) and at different temperatures. It was further observed that the gelatinization enthalpy was directly proportional to the level of starch damage—the more damaged the starch (due to milling), the lower the ΔH values. These results more or less confirmed what could be expected from the knowledge obtained with the polarizing microscope. Some years later it was discovered that the appearance of the DSC thermogram was related not only to the type of starch but also to the starch/water ratio [237–239]. An increase in the starch/water ratio caused the single, narrow endotherm to develop a shoulder. The onset temperature (T_o) of the gelatinization was not affected, at least not until very high starch/water ratios, but the conclusion temperature (T_c) was increased. This means that when the starch/water ratio is increased, the temperature must be increased in order to achieve complete gelatinization (see Table 15). This behavior was first related to the Flory equation for the melting of polymers in the presence of a plasticizer [237]. In the case of starch the polymer is, of course, amylopectin, and the plasticizer is water. This approach is useful in showing the

Table 15 Influence of Water Content on Gelatinization Temperature Range of Wheat Starch

Water content (%)	T_o (°C)	T_m (°C)	T_c (°C)
75	57	62	66
60	58	64	85
45	58	64	100
35	54	61	116
30	—[a]	—[a]	—[a]

[a]No endotherm detected below 120°C.
Source: First value taken from [164], and the remaining calculated from [238].

polymeric behavior of starch, but it has its limitations. It has been pointed out that the Flory equation requires equilibrium melting, which is not the case during the DSC scan. In the case of potato starch it has been found that at heating rates below 0.5°C/min the heating rate does not influence the gelatinization temperature and thus that equilibrium melting can be expected only at heating rates below 0.5°C/min [240]. Another aspect is that the volume of the plasticizer is not known; that is, the starch/water composition is easily determined, but knowing this is not the same as knowing the water content the polymer "feels" when it melts. There is a distribution of water to take into account, first between the inside and the outside of the starch granule and then between the different parts of the starch granule [232].

When the first step was taken toward an interpretation of the thermal behavior of starch as a consequence of starch being a polymeric material, the next soon followed. It was the recognition that the melting of starch crystallites is preceded by a glass transition in the amorphous starch [241]. Below the glass transition temperature (T_g) a polymer can be described as brittle, whereas above T_g it is gummy. The melting of the crystallites is a transition of the first order and gives rise to peaks in the DSC thermogram, peaks that are easily detected (see Fig. 51). A glass transition, on the other hand, is a transition of the second order and gives rise to a discontinuity in the heat capacity (C_p). Such a change is much more difficult to detect in the DSC thermogram, especially if the change in C_p is small and/or occurs over a broad temperature interval. Nevertheless, there are several reports on the temperature of the glass transition (T_g) in starch, and its location is somewhere between the freezing of water and the onset of gelatinization [242–244]. T_g is also very sensitive to the water content—not the macroscopic water content, but the local, microscopic water content [243]. In some cases a small discontinuity in the DSC thermogram in close proximity to the gelatinization endotherm has been attributed to T_g [242]. Slade and Levine [244] used an indirect way to show that the initial part of the gelatinization endotherm emerges from the glass transition. They heated starch–water mixtures to different temperatures before and in the gelatinization temperature interval and reheated the sample after an intermediate cooling. In this way shifts in the baseline between the first and second heating could be detected.

The gelatinization temperature interval measured by DSC is given in Table 14 and illustrated in Fig. 52. There are varietal differences among wheat varieties, with winter wheat starches gelatinizing at a lower temperature range than spring wheats [245]. There is also a relation between T_m (the temperature at peak maximum) and loaf specific volume; a correlation coefficient of +0.964 has been found in linear regression analysis [171]. A slight difference in gelatinization parameters between the small and large granules has also been found (see Table 1). Differences in ΔH values between wheat varieties have been found, although a relation between ΔH and quality parameters was not found [245].

4. Swelling

During the gelatinization process the starch granules absorb water and swell. The diameter of the gelatinized wheat starch granule has been determined to be 34 μm, although with considerable variation (±24 μm) [246]. This swelling occurs at temperatures above the gelatinization temperature range as measured by DSC (see Fig. 52). The swelling depends on the starch/water ratio and decreases with increased starch concentration [247]. Swelling is usually determined at a very high excess of water. The starch granule volume as a function of temperature is shown in Fig. 53. The swelling factor (volume of swollen granules divided by the initial volume of the starch) at 70°C was 2.56 at the water/starch weight ratio of 1 : 1, 6.01 at 5 : 1, 6.68 at 10 : 1, and 7.01 at 20 : 1 [249]. The swelling seems to be related to the amylopectin content, with amylose acting as a diluent [249].

5. Solubility

During the heating of starch, molecules from the interior of the granules will leach out. Amylose is the first molecule to leach out, and the degree of leaching is a function of temperature as shown in Fig. 54. The leaching process is a secondary process during gelatinization, as it occurs at a temperature range above the gelatinization range measured by DSC (Fig. 52). The leaching of amylose is not a random process; molecules leached at the lowest temperature have a lower molecular weight than molecules leached at a higher temperature [178]. Amylopectin also leaches out, especially if the mechanical treatment during heating is severe [250]. The amount of polysaccharide leached out at different temperatures is always somewhat higher than the amount of amylose; the amylose content in the leached material is about 80% [249]. When salts (e.g. potassium thiocyanate) are used to swell starch granules, leaching of amylose

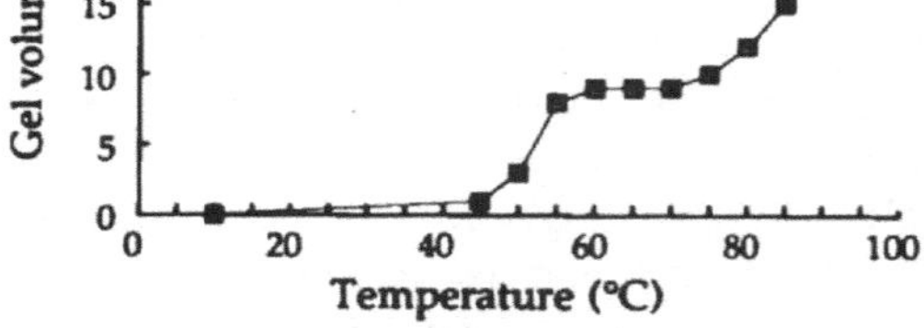

Fig. 53 Wheat starch gel volume as a function of temperature. (Adapted from [248].)

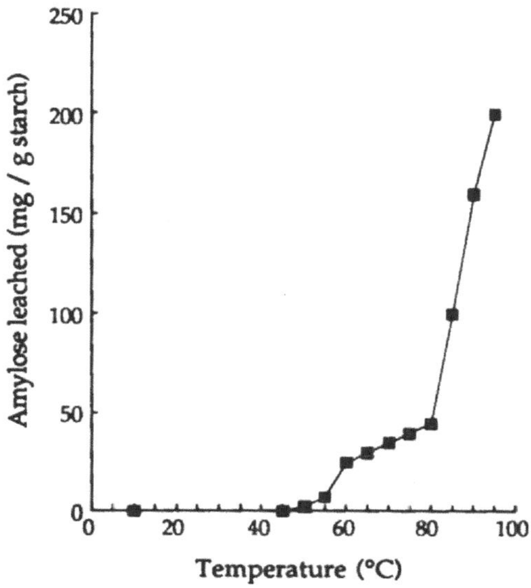

Fig. 54 Amylose leaching during heating of wheat starch. (Adapted from [248].)

seems to be a prerequisite for the swelling and gel formation. If amylose leaching is inhibited, gel formation is also inhibited [251].

At room temperature, or at temperatures below the gelatinization temperature range, the leaching of amylose (or polysaccharide material) is negligible. However, if the starch granule is damaged—for example, by the mechanical treatment during milling—amylose molecules will leach out, or are easily extracted, even at room temperature. This difference forms the basis of the extraction method for the determination of starch damage [252].

6. Shape and Size

The swelling of the starch granules, which is accompanied with the leaching of amylose (and amylopectin), certainly affects the shape and size of starch granules. The granules commonly increase in size due to the absorption of water, and then at some point they collapse [253]. Wheat, rye, and barley starches, however, show a somewhat different pattern. The granules do not expand uniformly when they absorb water. The swelling seems to be more extended in regions close to the equatorial groove, resulting in a typical change in shape of the large granules [254]. The reason for this could be that the amylopectin molecules are asymmetrically distributed [249]. The swelling pattern of wheat A-granules is illustrated in Fig. 55. The swelling of the small B-granules is similar to that of other starches, such as maize starch. They expand uniformly in

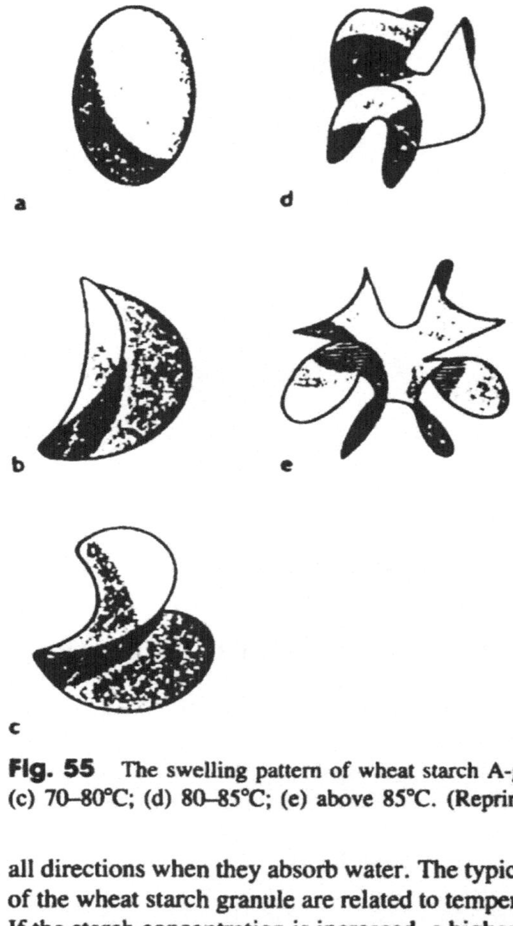

Fig. 55 The swelling pattern of wheat starch A-granules. (a) 50–60°C; (b) 60–70°C; (c) 70–80°C; (d) 80–85°C; (e) above 85°C. (Reprinted with permission from [254].)

all directions when they absorb water. The typical changes occurring in the shape of the wheat starch granule are related to temperature as well as to concentration. If the starch concentration is increased, a higher temperature must be achieved to bring about the same changes as at the lower concentration [248]. When wheat starch is heated during shearing, fragmentation of the outer layers of the starch granule is observed [255].

7. Rheological Behavior

The changes in swelling, solubility, and so on described above affect the rheological behavior of the starch gel. It is evident from Fig. 52 that the primary event and the secondary events that follow it have to occur before we can measure any change in rheological behavior. Several types of starch gel structures are illustrated in Fig. 56. In Chapter 1 gels were divided into macromolecular gels and particle gels, and we can find both of these types of gels represented

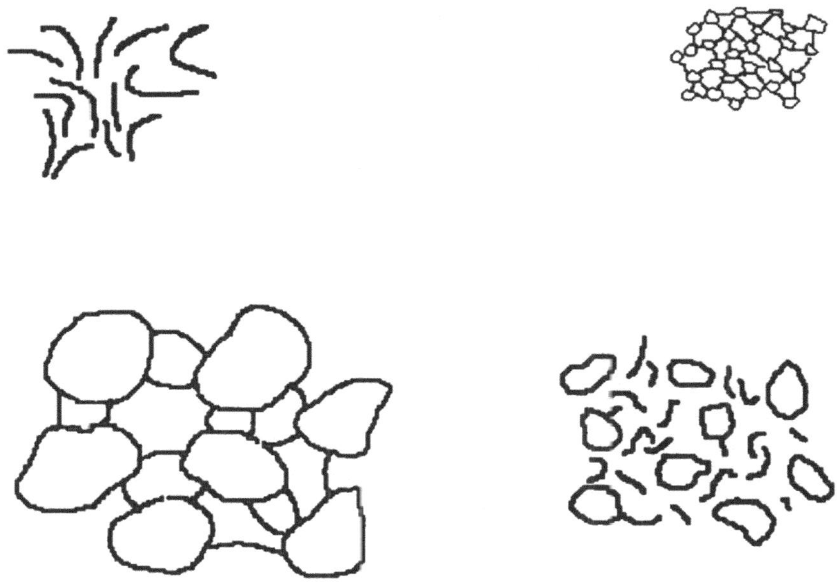

Fig. 56 Starch gel structures. (Adapted from [256].)

in Fig. 56. In a macromolecular gel, complete dispersion of the starch granule and its contents has occurred, so we have a solution of amylose and amylopectin, and the mechanisms governing macromolecular solutions are in effect. When the concentration is above a certain level, gel formation will occur owing to entanglements. For amylose leached from pea starch, the overlap or entanglement concentration, C^*, has been determined to be around 1.5% at 60°C [181]. For waxy maize amylopectin, C^* was found to be 0.9% at 60°C [190]. Below C^* there is a linear relationship between viscosity and concentration, whereas above C^* this dependence changes to $C^{2.0}$ for amylose and $C^{2.7}$ for amylopectin (C being concentration). An alternative mechanism suggested for the aggregation and gelation behavior of amylose is that it depends on the formation of interchain double helices [257].

The other type of gel, the particle gel, is obtained when the starch granules are only slightly swollen but occupy the whole volume. This will result in a weak transient network. We have described these types of starch gels as composite materials built up from a continuous polysaccharide phase with starch granules as fillers [258]. The properties of such a gel depend on the properties of the dispersed phase, on the continuous phase, and on interactions between the phases. It is thus possible to manipulate the properties of the gel by changing the properties of the phases or the interaction between them [259, 260]. One property of wheat starch that is of special interest in this connection is its bimodal size distribution. The rheological properties depend on the volume of the dispersed phase (Φ), but for the same Φ the particle size distribution is also of relevance [261]. At Φ values below 0.2, wheat starch granules that have been gelatinized and washed (to get rid of leached amylose) show Newtonian behavior [262]. The viscosity of the gelatinized starch suspension is about twice that obtained for the starch granules themselves [262].

In the particle gels the swollen starch granules will fill up the volume to the limit for close packing even at very low concentrations; for wheat flour the close-packing concentration is 3.8% [263]. It is then the deformability of the starch granules that is of uttermost importance for the rheological behavior [132]. One obstacle for deformation is the remaining crystallinity in the starch granules. This is observed as a decrease in G when heating continues from T_m to T_c [258]. If partly gelatinized starch granules are added to a continuous amylose phase, reinforcement of the gel, that is, an increase in G, occurs [264]. However, the reinforcement effect depends on the degree of gelatinization and is greatest for the least gelatinized starch granules.

D. Retrogradation—Starch Recrystallization

The gelatinization process causes the melting of the crystallites in the starch granule, and the resulting gel structure is amorphous. However, this gel is not in a state of equilibrium, and changes occur with time. The events occurring in solutions of amylose and amylopectin were described to some extent earlier in this chapter. The changes that occur are collectively described with the term *retrogradation*, and, as in the case with gelatinization, this term is used to describe a range of processes. The macroscopic effect observed is a change in rheological behavior, perhaps together with water syneresis. The rigidity of the starch gel increases during storage as a result of the recrystallization process [265]. Sometimes retrogradation is used to describe the increase in viscosity observed during the cooling sequence in a Brabender amylo/viscograph. As was discussed previously, the retrogradation involves processes such as phase separation, crystallization, and gel formation [227, 266]. The relation between retrogradation (i.e., starch recrystallization) and the staling of bread is discussed in Chapter 7. In the present section some basic concepts of starch recrystallization will be discussed.

1. Amylopectin and Amylose

In the native starch granule the amylopectin is found in the crystalline domains, whereas both amylose and amylopectin are involved in retrogradation. As retrogradation involves a recrystallization process, either the X-ray diffraction technique or DSC can be used to follow it. However, the two methods differ in the way in which they differentiate between the retrogradation of amylose and amylopectin.

By the X-ray diffraction technique the appearance of the B-pattern during the aging of starch gels is followed, and an increase in intensity is observed with time [235, 267]. When the components of starch (amylose and amylopectin) are investigated, it is found that both amylose and amylopectin gels give the B-pattern with time [266, 268]. It could thus be either component, or both, that causes the retrogradation of starch. The crystallization rate, however, differs between amylose and starch [266]. In an amylose gel the crystallinity reaches a maximum value after a couple of days, and thereafter the intensity in the X-ray diffraction pattern does not change. In the starch gel, on the other hand, the intensity of the X-ray diffraction pattern continues to increase even after 14 days. When the amylose gel is heated to 90°C the decrease in X-ray diffraction intensity is quite small, whereas when the starch gel is heated in the same way there is a considerable decrease in the X-ray intensity [266].

Differential scanning calorimetry measurements have shown that recrystallized amylopectin melts in the temperature range 60–80°C [269]. This is the same temperature range in which the staling endotherm is obtained for starch gels and for bread crumb [269, 270]. Amylose gels, on the other hand, do not melt at temperatures below 100°C [269]; a melting endotherm in the temperature range of about 105–150°C has been recorded [271]. Amylose crystals are thus much more stable than amylopectin crystals. One consequence of the high stability is the resistance to enzymatic hydrolysis. The most stable amylose crystals are those formed from short-chain amylose molecules. These crystals are not degraded in the intestinal tract and are termed "resistant starch" [272, 273].

Differential scanning calorimetry and the X-ray diffraction technique thus do not measure the same properties. In the DSC method the retrogradation of amylopectin is followed, whereas the X-ray diffraction technique measures the total crystallinity of both amylopectin and amylose.

The extent of retrogradation (i.e., ΔH_c after infinite time) depends on the amylopectin content of the starch gel and increases in the order amylomaize < wheat < potato < waxy maize [274]. It was found in the same investigation that the rate constant was higher for waxy maize and potato starches than for amylomaize and wheat starches. It was then concluded that amylose (as in the amylomaize starch) and polar lipids (as in wheat starch) decrease the rate of retrogradation [274]. The possibility of cocrystallization has been discussed [274], and in mixtures of amylose and amylopectin it has been observed that the retrogradation of amylopectin does not occur independently of the amylose present [275]. At high levels of amylose the retrogradation of amylopectin occurs to a greater extent than could be expected from the mixing ratio. From the measurement of shear modulus, DSC endotherm, and X-ray intensity on starch

(10–20%) gels and amylose gels it was concluded that the initial crystallization process is the same in starch gels as in an amylose gel, that is, the crystallization of amylose [266]. Whereas the properties of the amylose gel are constant after about 24 h, the starch gel changes continuously over a period of weeks. This latter effect is attributed to the slow crystallization of amylopectin.

2. Factors that Influence the Recrystallization of Amylopectin

In the followng part of this chapter we will be concerned only with the recrystallization of amylopectin. As already mentioned, this is of considerable importance for the staling of bread. Another reason is simply that we know more about the crystallization of amylopectin.

a. Concentration

Starch concentration influences the extent of retrogradation. The most intense B-pattern is obtained for wheat starch gels with a water content of 47–50% (w/w) [235]. For a starch gel with 63% (w/w) water stored 8 days at 24°C, the intensity of the B-pattern is only one-fifth the intensity observed for a starch gel with 50% water treated in the same way. Measurements by DSC have confirmed earlier X-ray results. The maximum ΔH value is obtained in a starch concentration of about 60% (w/w) [276–278]. Also the increase in G' during storage is related to the starch concentration [279].

b. Temperature

The retrogradation process is very sensitive to temperature [269, 279, 280]. This is evident from using DSC and X-ray diffraction as well as rheological measurements. These results are easily understood from classic knowledge about crystallization. The higher the degree of supercooling, the more crystal seeds will be formed, but the growth of crystals is favored at a higher temperature. However, in the case of starch gel the supercooling cannot be brought too far; at some temperature, T_g of the starch as well as the freezing point of water will be passed, and during these conditions no crystallization will occur at all. The best conditions for crystallization of amylopectin therefore seem to be just above 0°C. The retrogradation process might then be avoided or delayed by storage at the correct temperature.

In order to bring about as much starch retrogradation as possible, the correct treatment would be to store the starch gel first at a low temperature (e.g. 4°C) to promote nucleation and then at a higher temperature (e.g., 40°C) to promote crystal growth. Such experiments have been performed, and after 24 h at 4°C starch gels were stored for another 24 h at 4, 12, 25, or 40°C. The highest values of ΔH_c were obtained after storage at 12 and 25°C [281].

The dynamic rheological properties vary with storage temperature [265]. G' increased depending on the storage temperature in the order 30°C < 20°C <

15°C < 10°C. Amylose and starch gels are affected differently by storage temperature [279]. For an amylose gel, G' was found to be more or less independent of the storage temperature at 4–20°C, whereas for a starch gel there was a strong dependence on temperature, with G' increasing at a higher rate at 1°C than at higher temperatures.

c. Source of Starch

The retrogradation differs among starches from different sources. One evident reason, when DSC is used to measure recrystallization (ΔH_c), is the amylopectin content. The higher the amylopectin content, the greater ΔH_c will be [274, 275]. But there are also differences among starches of about the same amylose content. The G value of wheat was found to increase very slowly during storage compared to that of barley [309]. In our laboratory we have used DSC to investigate several cereal starches and found the following order in the extent of retrogradation: oat < rye < wheat [282, 283]. So far there has been no investigation comparing the tendency in retrogradation for different wheat varieties.

d. Influence of Additives

Retrogradation can be delayed by adding suitable substances. The most common additives with this function are the monoacyl lipids, for example, the monoglycerides. Their function will be discussed in Chapter 3. Also the pH value of the starch gel and the amount and type of salt present influences retrogradation [284–286].

3. Changes in the State of Water During Retrogradation

The state of the water in the gel changes during storage. Such changes are observed by both NMR and DSC [269, 287]. An increase in the proportion of "bound" water from 50% to 70% was determined with NMR during the storage of a starch gel [35% (w/w) water] for 150 h [287]. These changes were attributed to changes in the amylopectin. The percentage of unfreezable water (i.e. water that does not freeze when cooled to –50°C), measured by DSC immediately after heating to 100°C, was 40% for a starch–water gel with a total water content of 48% (w/w) [269]. After 7 days of storage at 23°C the level of unfreezable water was 53%.

The mobility of water, determined by NMR, decreases during storage of starch gels. The change depends on the temperature and is highest at 37°C and lower at −5°C and 2°C. It also depends on starch concentration, and the change is greatest at 45% starch [287].

V. NONSTARCH POLYSACCHARIDES

The nonstarch polysaccharides are found in the cell walls and in the matrix that holds cells together. They can be divided into three groups: cellulose, β-glucans, and pentosans. The cellulose content of white flour is only 0.60 g/100 g, whereas the content of nonstarch, noncellulosic polysaccharides is 2.52 g/100 g [200]. Of these latter polysaccharides the β-glucans amount to very low levels in wheat, but the amount is considerable in barley and oats. This section therefore concentrates on the pentosans. They are called pentosans because they are polymers of pentoses—monosaccharides with five carbon atoms in the molecule. Alternatively, they could be described as hemicellulose. The most prevalent pentoses are arabinose and xylose (Table 16). The total pentosan content in the wheat endosperm is in the range of 1.27–2.33% [289, 291, 292], but the content in the wheat kernel is much higher, approaching 7% [292]. Usually, the pentosans are divided according to their solubility in water-soluble (WS) and water-insoluble (WI) pentosans, and of the pentosans in wheat more than 60% are WI pentosans [288, 289, 291]. Their roles and importance in baking have been much discussed, but little is known for sure. The uncertainty about their function is related to the problems in fractionation and analysis of the pentosans. The sugar composition of individual pentosan molecules is known in only a few cases, and the structure is even less known. Pentosan fractions have been prepared, and their impact in the baking and storage of bread has been evaluated. However, these results are difficult to interpret owing to the complex composition of the fractions. Often the pentosan fractions contain protein, and it is not always known whether the protein is covalently linked to the carbohydrate or if the protein should be regarded as a contaminant [293, 294]. If the latter is true, totally wrong conclusions might be drawn when the contaminant content amounts to 25–30% or even higher.

The main interest in pentosans in connection with baking has to do with their role in rye flour, and their importance in wheat flour is often thought to be minor [38]. The pentosans may influence baking in at least two ways. Because of their water-holding capacity they may influence the water distribution in the dough, and they may influence the foam stability owing to their rheological behavior.

Table 16 Monosaccharide Composition of Pentosans

Pentosan fraction	Ara (%)	Xyl (%)	Gal (%)	Glu (%)	Fru (%)
Total pentosan fraction [288]	35–40	29–40	20–24	3–10	trace
Water-insoluble fraction [289]	31–38	49–59		6–20	
Water-soluble fractions [288]					
Arabinoxylan	31–41	58–62		2–10	
Arabinogalactan	32–41	0–2	45–59	0–21	
Water-soluble fractions [290]	11–43	51–72	2–4	2	

The pentosans are known to increase the viscosity of flour extract through oxidative gelation, a property that will be described later in this chapter. An increase in the viscosity of the continuous phase would be expected to increase the gas-holding capacity owing to a decrease in the tendency for the gas cells to coalesce. In a recent study pentosans were found to decrease the surface tension of water [295].

A. Chemical Composition and Structure

1. Water-Soluble Pentosans

The water-soluble (WS) pentosans, pentosans that are extracted from flour with cold water, amount to less than 1.0% of the flour. Values of 0.43–1.01% are reported [289, 291]. The monosaccharide composition of the WS pentosans is given in Table 16, and their chemical structure is illustrated in Fig. 57. The D-xylose residues are linked by $\beta(1\rightarrow4)$ glycosidic bonds to form a chain. Single arabinose residues are attached to the xylose chain as branches. At least in rye, double-branched xylose residues are also found [296].

The WS pentosans can be further fractionated on diethylaminoethyl (DEAE) cellulose (cf [297], [298]) by the use of $(NH_4)_2SO_4$ (cf. [290]) or by the use of 80% ethanol [293]. The composition of the fractions obtained is influenced by the type of flour used and the conditions of preparation. The first fraction eluted on DEAE cellulose is a pure arabinoxylan, essentially without any associated protein. This is the largest fraction. Another fraction is an arabinogalactan, and protein is reported to be covalently bound to this polysaccharide [293]. The

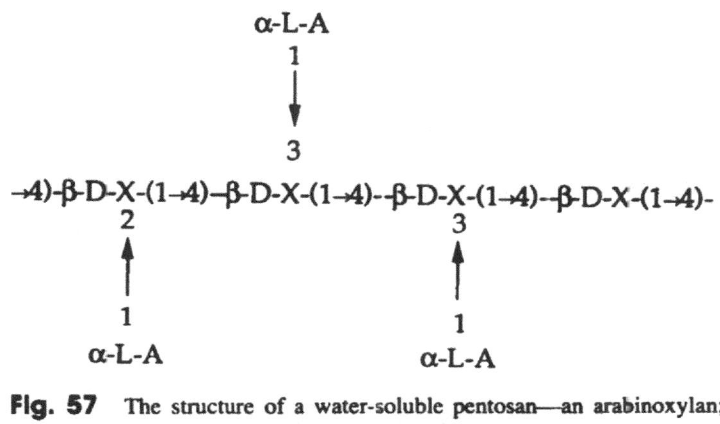

Fig. 57 The structure of a water-soluble pentosan—an arabinoxylan; α-L-A means α-L-arabinofuranosyl and β-D-X means β-D-xylopyranosyl.

amino acid composition of this protein is different from that of wheat storage proteins, the amounts of alanine and hydroxyproline being high and the levels of glutamic acid and proline low [288, 293].

The protein content of the remaining fractions ranges from 1.4 to 46.7%, with the highest protein content being found in the last fraction [200, 298]. The use of $(NH_4)_2SO_4$ to fractionate WS pentosans gave fractions that differed in protein content from 6% to 96% [290]. Recently, the protein content of arabinoxylans from eight different wheats were reported to be 1.3–8.7%, and that of the corresponding arabinogalactans, 6.5–14.3% [288]. These values are considerably lower than the values reported earlier and can be regarded as the upper limit for covalently bound proteins.

The WS pentosans also contain ferulic acid. The ferulic acid is bound only to the largest soluble arabinoxylans [31]. The content of ferulic acid in pentosans extracted from flour and dough is 0.31–0.56 mg/g pentosan [299], and in pentosans extracted from eight wheat varieties the total phenolic acid content was 0.46–0.85 mg/g pentosan [288]. The corresponding values for arabinoxylans were 0.86–1.43 mg/g, whereas ferulic acid was not detected in the arabinogalactans.

2. Water-Insoluble Pentosans

The amount of water-insoluble pentosans in wheat is about 1–1.3% [289]. The main sugars in this fraction are L-arabinose, D-xylose, and D-glucose, with the proportion differing among varieties (Table 16) [289]. One difference between WI and WS pentosans is the degree of branching, with the WI pentosans having the greatest degree of branching. The molecular weight of WI pentosans is higher than that of WS pentosans. The degree of polymerization of the WS pentosans is, depending on the method for determination, about 160–190, and for WI pentosans it is above 900 [200]. The protein content is below 10–15% [200, 289]. Water-insoluble pentosans contain ferulic acid in the range of 0.86–1.10 mg/g, depending on the wheat from which they are isolated.

To solubilize the WI pentosans, alkali has to be used. The poor solubility of WI pentosans may be due to their higher molecular weight as well as to the presence of phenolic acids [200, 293].

B. Properties of the Nonstarch Polysaccharides

1. Oxidative Gelation and Rheological Properties

The pentosans may certainly influence the viscoelastic behavior of the dough not only because of their molecular weight, but also because of their gel-forming ability.

When hydrogen peroxide is added to a flour suspension, the viscosity of the solution increases, and a flour extract may gel under similar conditions. The gelling occurs without heating and cooling, but an oxidizing agent is required. The phenomenon is therefore called oxidative gelation, and the substances involved are the pentosans, more precisely an arabinoxylan that contains ferulic acid [290].

The gelation mechanism involves the ferulic acid, as the oxidative gelation causes a loss of ultraviolet absorbance at $\lambda = 320$ nm, the wavelength where ferulic acid absorbs [31]. The aromatic ring of ferulic acid seems to be involved in the oxidation reactions [300], but other mechanisms have been suggested [297, 301]. There is, however, no correlation between the content of ferulic acid and gel strength [295]. Not all oxidizing agents have the ability to induce oxidative gelatin of pentosans. This is evident from the results presented in Table 17, where the change in viscosity (measured as flow time in a capillary flow viscometer) of a water extract of flour is shown [301]. The common oxidizing agents used in breadmaking, potassium bromate and ascorbic acid, did not cause gelation. This result was interpreted as showing that gelation occurs only with oxidants that cause the formation of free radicals [301]. Hydrogen peroxide does so in the presence of an enzyme, hydrogen peroxidase. The enzyme is present in flour but is inactivated by heating (Table 16). Moreover, sulfhydryl groups are also involved in the gelation (Table 16). The gel formation can be inhibited by the action of proteolytic enzymes [297].

Table 17 Influence of Certain Treatments on the Change in Viscosity of Flour Water-Solubles

Treatment	Change in viscosity measured as change in flow time (s)
Add H_2O_2	+39
Add $KBrO_3$	−1
Add KIO_3	+1
Add ammonium persulfate	+34
Add formamidine disulfide	+32
Add dehydroascorbic acid	±0
Add boiling + H_2O_2	±0
Boiling + peroxidase + H_2O_2	+25
Add H_2O_2 + cysteine	−10

Source: Data compiled from [301].

When WS pentosans and arabinoxylan solutions are treated with horseradish peroxidase and H_2O_2, a very rapid increase in G' occurs (Fig. 58) [295, 302]. Depending on the wheat variety, G' after 5 h was in the range of 2.6–27.6 Pa and δ in the range of 9.7–41.7°. The composition of the gel formed from WS pentosans by oxidative gelation has been determined, and it was found that only about 5–10% of the WS pentosans are immobilized by oxidation [297]. The gel is composed of 25% (dry weight) protein, 1% lipids, and 2.2% ash, and the remainder (61.8%) is polysaccharide. The polysaccharide part contains only xylose and arabinose. The arabinogalactan does not take part in the oxidative gelation [302].

Water-insoluble pentosans are also subject to oxidative gelation (Fig. 59). When hydrogen peroxide and peroxidase are added to aqueous suspensions of WI pentosans, an increase in G' and a decrease in δ has been observed in small-amplitude oscillatory testing, results that indicate gel formation [289]. Differences in δ and G' were noted between WI pentosans from different flours; δ varied between 6.32 and 15.68°, whereas G' was in the interval 0.43–11.32

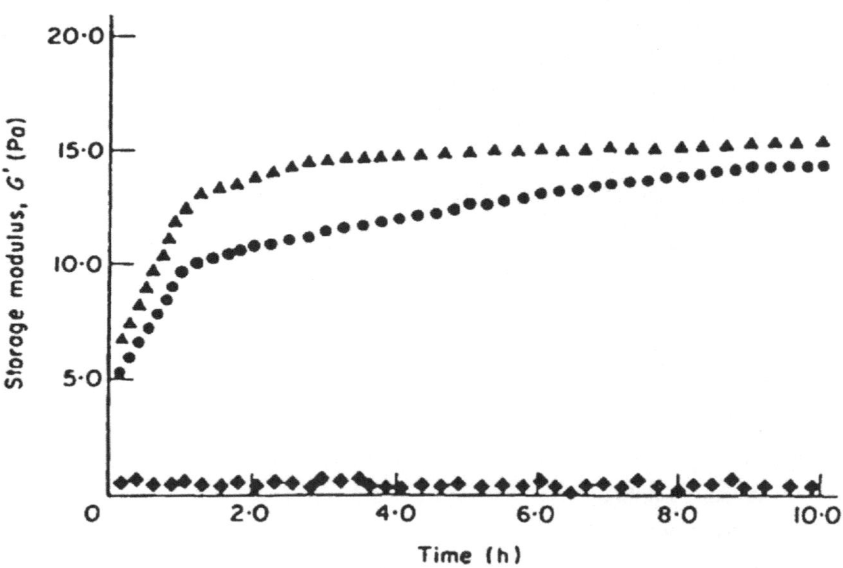

Fig. 58 The oxidative gelation of water-soluble pentosans from wheat. The figure shows storage modulus (G') versus time for 2% (w/v) aqueous solutions of pentosan (●), arabinoxylan (▲), and arabinogalactan (◆) treated with horseradish peroxidase and H_2O_2 at 15°C. The strain was 4%, and the frequency 1.0 Hz. (Reprinted with permission from [302].)

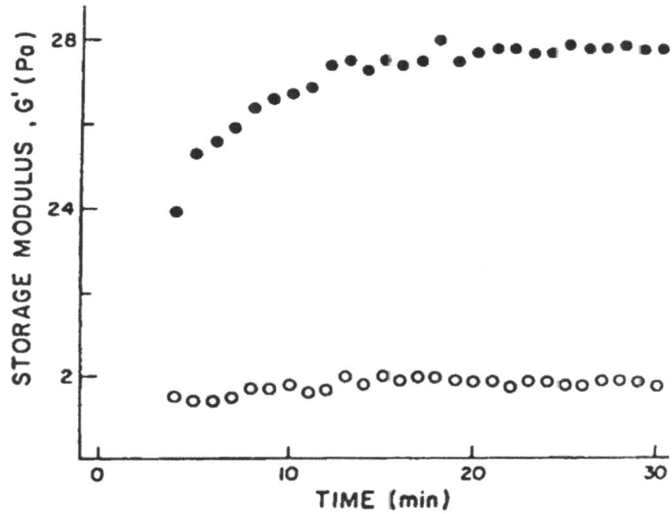

Fig. 59 The oxidative gelation of water-soluble pentosans from wheat. The figure shows storage modulus (G') versus time for 1.5% (w/w) water-insoluble pentosan suspensions in water with (●) or without (○) addition of peroxidase in 0.1 M potassium phosphate buffer (pH 6.0) and 1.2 ppm H_2O. The strain was 4.4%, and the frequency 1.0 Hz. (Reprinted with permission from [289].)

Pa. It was further noted that deesterification of the ferulic acid resulted in δ values above 60°. It was thus concluded that ferulic acid is necessary for oxidative gelation. The amount of ferulic acid was found to be lower in pentosans extracted from an overmixed dough, indicating that pentosans are oxidized during mixing [299].

Pentosans are expected to affect the rheological behavior of dough because of their high molecular weight. As discussed above, WS and WI pentosans differ in molecular weights. There are also differences in the molecular weights of pentosans from different wheat varieties; $[\eta]$ is in the range 2.81–4.23 dL/g for arabinoxylans prepared from different wheat varieties [295]. There is a relation between $[\eta]$ and gel strength; the higher $[\eta]$, the stronger the gel will be [295, 298]. The rheological behavior of aqueous dispersions of pentosans has been studied in small-amplitude oscillatory testing. For aqueous suspensions of WI pentosans [1.5% (w/w)], G' (measured at 1.0 Hz and 4.4% strain) was in the range of 0.16–3.97 Pa, and δ was in the range of 12.7–29.7° [289]. For WS pentosans [2% (w/w)], G' (measured at 1.0 Hz and 4% strain) was in the range of 0.5–25 Pa [295].

2. Water-Holding Capacity

The pentosans have good water-holding capacity, and they greatly influence water distribution in dough as well as in bread. The water uptake of pentosans has been estimated to be 15 g water/g dry basis [303]. This limit, however, is poorly defined and even questionable, as there is no separate phase that swells and coexists in equilibrium with excess water. The values cited above should correspond to 23.4% of the water associated with just 1.5% of the flour in a wheat flour dough [303]. The water-holding capacity of oxidized WS pentosans is 40–60%, and that of oxidized arabinoxylans is 75–90% [295]. Even higher values are observed for small gel particles. When WI pentosans are added to the wheat flour dough, the farinographic water absorption increases [289].

Water absorption is related not only to the WI pentosan preparation, but also to the flour to which it is added. The same WI preparation absorbed 5.6 g water per gram WI when added to one flour and 3.2 g when added to another. The influence of WI pentosans on farinograph absorption is greater than that of WS pentosans [304]. An addition of 1% pentosans caused an increase in water absorption from 63.6% to 68.0% in the case of WS pentosans, and to 73.5% in the case of WI pentosans. An increase in baking absorption from 65% to 69% was observed after the addition of 0.66% WS pentosans [305].

3. Surface Activity

A. Glycoproteins

Glycoproteins are defined as those proteins that contain carbohydrate groups *covalently* linked to the polypeptide chain. They occur frequently at biological interfaces, where they are involved in recognition and adhesion phenomena. When there are many oligosaccharides regularly spaced along the peptide chain, the molecule is sometimes described as a mucin. In these cases the carbohydrate is usually linked to a serine or threonine side chain via an O-glycosidic bond. Such polymers are strongly surface-active, owing to their comblike molecular shape, and they can hardly fold so the peptide chain adopts a regular secondary structure (due to the tendency of all oligosaccharide units to be hydrated).

There are no known association complexes between proteins and carbohydrates based on hydrophobic interaction, such as the association between lipids and proteins (see Chapter 3). This is perhaps not surprising as carbohydrates are not amphiphilic (in exceptional cases they can show a weak amphiphilicity). With this background it seems natural to expect that the protein regions in pentosans are covalently bound.

b. Pentosans

Pentosans exhibit surface activity as shown from surface-tension studies by Izydorczyk et al. [295]. Whether the surface activity is due to the pentosans as such or due to their protein is a basic question related to the mechanism of protein binding to the carbohydrate addressed earlier. It is important to realize in this context that an association complex must be expected to dissociate at the gas/water interface. The interfacial behavior can therefore provide information on the nature of this binding.

The surface tension versus concentration curves of two water-soluble pentosans are shown in Fig. 60. The surface tension of water at 25°C is reduced by about 20 mN/m when the plateau value has been reached. The pentosan concentration in the solution is then about 0.6% (w/v). Arabinogalactan shows a greater surface tension reduction than arabinoxylan, but the concentration behavior is the same (a constant pressure difference below the plateau concentration as well as above).

The surface tension versus concentration curve is typical for amphiphilic molecules showing an association limit but rare among proteins (β-casein is one of the few known).

If we imagine an association complex, it should be expected to behave like two components with regard to reduction of interfacial tension. Presumably, the protein will dominate and, at increasing concentration, successively squeeze out the carbohydrate, and in this case we should expect a surface tension quite different from that observed. To reach the plateau value even higher concentrations would be needed. In fact, the curve found is consistent with one uniform and highly amphiphilic molecule. Thus the interfacial behavior indicates

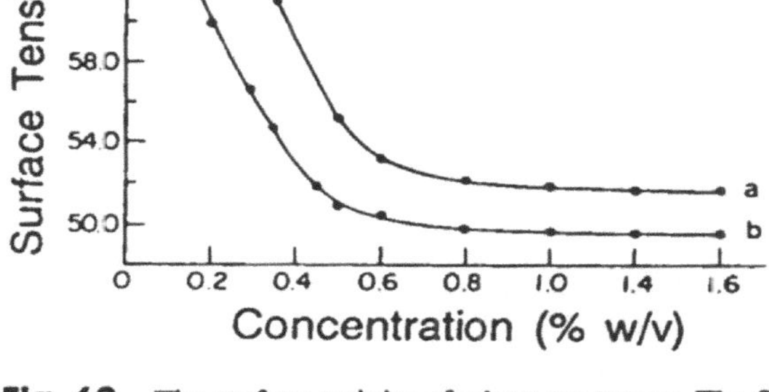

Fig. 60 The surface activity of wheat pentosans. The figure shows the reduction of the surface tension of water at 25°C by (a) arabinoxylan and (b) arabinogalactan at various concentrations. (Reprinted with permission from [295].)

that pentosans are truly linked to proteins. The similarity in behavior between the arabinogalactan and the arabinoxylan is another indication that the proteins are also covalently bound in the arabinoxylans.

C. Significance in Baking

1. Influence on Loaf Volume

Increased as well as decreased and unchanged loaf volumes have been observed after the addition of pentosans to doughs. The discrepancy in results is presumably due to the lack of characterization of the pentosan fractions; fractions that are thought to be comparable might not be so.

Hoseney and coworkers [50] found that if the water-soluble fraction of wheat flour was removed the loaf volume was reduced. The water-soluble pentosan fraction was identified as the fraction responsible. No beneficial effect was found to be due to the addition of WI pentosans [59], whereas others found that the addition of 2% water-insoluble pentosans increased bread volume by 30–45% [200]. In one study, water-soluble as well as water-insoluble pentosans (1%) caused a slight decrease in specific loaf volume, and the mixing time was considerably reduced, from 3.50 min to 2.25–2.50 min [304]. Water-insoluble pentosans exerted the greatest effect. Water-soluble pentosans increased the loaf volume from 62 to 67 mL when added at 0.66% [305]. Loaf volume was found to decrease when the content of WS pentosans was correlated with the loaf volume, but the correlation coefficients were poor (−0.44 or below [291]). The mixing time was also decreased, whereas absorption increased with the WS pentosan content. The gelling capacity of pentosans in the presence of oxidants is lost during dough resting, probably due to enzymatic degradation of the pentosans [306].

2. Influence on the Staling of Bread

It is often stated that pentosans are very important for the keeping qualities of bread. Also, relations between the addition of pentosans and the staling rate have been found [307]. The WI pentosans were more effective than the WS pentosans. It was suggested that the effect was due to a lesser amount of starch components available for crystallization [307]. The pentosans seem not to affect starch retrogradation; DSC measurements show that the recrystallization of starch occurs independently of the pentosan addition [276]. In these experiments WS as well as WI fractions were used. We recently performed similar experiments in our laboratory with a well-characterized pure arabinoxylan, and our results support the previous findings [308]. The addition of pentosans/arabinoxylans to starch does not decrease the retrogradation of starch; in fact, the opposite might be found. The explanation is given in Fig. 61. It is known that the recrystalliza-

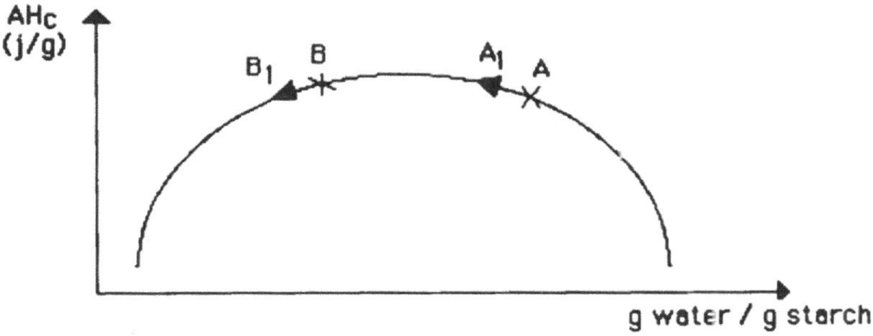

Fig. 61 The effects of pentosans on the retrogradation of starch. Pentosans are added to a starch–water mixture of composition *A*. The new water starch/ratio will be A_1, which results in increased crystallization (ΔH_c increases). If pentosans are added to sample *B* the new water distribution will result in decreased ΔH_c values.

tion of starch depends on the water content [276, 277, 308]. Depending on the initial starch/water ratio, the addition of pentosans might either increase or decrease the retrogradation of starch.

REFERENCES

1. Somerville, C., and Browse, J. (1991). Plant lipids: metabolism mutants and membranes, *Science*, 252:80.
2. Morrison, W. R., Mann, D. L., Soon, W., and Coventry, A. M. (1975). Selective extraction and quantitative analysis of non-starch and starch lipids from wheat flour, *J. Sci. Food Agric.*, 26:507.
3. MacRitchie, F., and Gras, P. W. (1973). The role of flour lipids in baking, *Cereal Chem.*, 50:292.
4. MacRitchie, F. (1981). Flour lipids: theoretical aspects and functional properties, *Cereal Chem.*, 58: 156.
5. MacRitchie, F. (1983). Role of lipids in baking, in *Lipids in Cereal Technology*, P. J. Barnes, Ed., Academic, London, p. 165.
6. McCormack, G., Panozzo, J., and MacRitchie, F. (1991). Contributions to breadmaking of inherent variations in lipid content and composition of wheat cultivars. II. Fractionation and reconstitution studies, *J. Cereal Sci.*, 13:263.
7. Chung, O. K. (1986). Lipid–protein interactions in wheat flour, dough, gluten, and protein fractions, *Cereal Foods World*, 31:242.
8. MacRitchie, F. (1978). Differences in baking quality between wheat flours, *J. Food Technol.*, 13:187.
9. Pomeranz, Y., Rubenthaler, G. L., Daftary, R. D., and Finney, K. F.

(1966). Effects of lipids on bread baked from flours varying widely in bread-making potentialities, *Food Technol.*, 131.
10. McCormack, G., Panozzo, J., Bekes, F., and MacRitchie, F. (1991). Contributions to breadmaking of inherent variations in lipid content and composition of wheat cultivars. I. Results of survey, *J. Cereal Sci.*, 13:255.
11. MacRitchie, F. (1977). Flour lipids and their effects in baking, *J. Sci. Food Agric.*, 28:53.
12. Finney, K. F. (1943). Fractionating and reconstituting techniques as tools in wheat flour research, *Cereal Chem.*, 20:381.
13. Chung, O. K., Pomeranz, Y., and Finney, K. F. (1982). Relation of polar lipid content to mixing requirement and loaf volume potential of hard red winter wheat flour, *Cereal Chem.*, 59:14.
14. Békes, F., Zawistowska, U., Zillman, R. R., and Bushuk, W. (1986). Relationship between lipid content and composition and loaf volume of twenty-six common spring wheats, *Cereal Chem.*, 63:327.
15. Morrison, W. R., Law, C. N., Wylie, L. J., Coventry, A. M., and Seekings, J. (1989). The effect of group 5 chromosomes on the free polar lipids and breadmaking quality of wheat, *J. Cereal Sci.*, 9:41.
16. Morrison, W. R. (1989). Recent progress on the chemistry and functionality of flour lipids, in *Wheat End-Use Properties. Wheat and Flour Characterization for Specific End-Use*, H. Salovaara, Ed., University of Helsinki, p. 131.
17. Larsen, N. G., Humphrey-Taylor, V. J., and Baruch, D. D. (1989). Glycolipid content as a breadmaking quality determinant in flours from New Zealand wheat blends, *J. Cereal Sci.*, 9:149.
18. Carlson, T., Larsson, K., and Miezis, Y. (1978). Phase equilibria and structures in the aqueous system of wheat lipids, *Cereal Chem.*, 55:168.
19. Carlson, T. L.-G., Larsson, K., Miezis, Y., and Poovarodom, S. (1979). Phase equilibria in the aqueous system of wheat gluten lipids and in the aqueous salt system of wheat lipids, *Cereal Chem.*, 56:417.
20. Carlson, T., Larsson, K., and Miezis, Y. (1980). Physical structure and phase properties of aqueous systems of lipids from rye and triticale in realation to wheat lipids, *J. Disp. Sci. Technol.*, 1:197.
21. Carlson, T. L.-G. (1981). Law and order in wheat flour dough. Colloidal aspects of the wheat flour dough and its lipid and protein constituents in aqueous media, thesis, Lund University, Lund.
22. Rajapaksa, D., Eliasson, A.-C., and Larsson, K. (1983). Bread baked from wheat/rice mixed flours using liquid-crystalline lipid phases in order to improve bread volume, *J. Cereal Sci.*, 1:53.
23. Marion, D., Le Roux, C., Akora, S., Tellier, C., and Gallant, D. (1987). Lipid–protein interactions in wheat gluten:a phosphorus nuclear magnetic

resonance spectroscopy and freeze-fracture electron microscopy study, *J. Cereal Sci.*, 5:101.
24. Larsson, K., and Puang-Ngern, S. (1979). The aqueous system of monogalactosyldiglycerides and digalactosyldiglycerides, in *Advances in the Biochemistry and Biophysics of Plant Lipids*, C. Liljenberg and L.-Å. Appelqvist, Eds., Elsevier/North Holland, Amsterdam, p. 27.
25. Gaines, L. G. J. (1966). *Insoluble Monolayers*, Wiley, New York.
26. Wannerberger, L., Eliasson, A.-C., and Miezis, Y. (to be published).
27. Finney, K. F., and Barmore, M. A. (1948). Loaf volume and protein content of hard winter and spring wheats, *Cereal Chem.*, 25:291.
28. Faridi, H., and Finley, J. W. (1989). Improved wheat for baking, *CRC Crit. Rev. Food Sci. Nutrition*, 28:175.
29. Campbell, W. P., Wrigley, C. W., Cressey, P. J., and Slack, C. R. (1987). Statistical correlations between quality attributes and grain-protein composition for 71 hexaploid wheats used as breeding parents, *Cereal Chem.*, 64:293.
30. Wrigley, C. W., and Bietz, J. A. (1988). Protein and amino acids, in *Wheat Chemistry and Technology*, Y. Pomeranz, Ed., American Association of Cereal Chemists, St. Paul, Minn. p. 159.
31. Hoseney, R. C. (1986). *Principles of Cereal Science and Technology*, American Association of Cereal Chemists, St. Paul, Minn.
32. Osborne, T. B. (1907). *The Proteins of the Wheat Kernel*, Carnegie Inst., Washington, D.C., Publ. No. 84.
33. Wieser, H., Seilmeier, W., and Belitz, H.-D. (1980). Vergleichende Untersuchungen über partielle Aminosäuresequenzen von Prolaminen und Glutelinen verschiedener Getreidearten, I. Proteinfraktionierung nach Osborne, *Z. Lebenms. Unters. Forsch.*, 170:17.
34. Ewart, J. A. D. (1968). Fractional extraction of cereal flour proteins, *J. Sci. Food Agric.*, 19:241.
35. Singh, N. K., Donovan, R., and MacRitchie, F. (1990). Use of sonication and size-exclusion high-performance liquid chromatography in the study of wheat flour proteins. II. Relative quantity of glutenin as a measure of breadmaking quality, *Cereal Chem.*, 67:161.
36. Krull, L. H., and Wall, J. S. (1969). Relationship of amino acid composition and wheat protein properties, *Baker's Dig.*, 40(4):30.
37. Branlard, G., and Dardevet, M. (1985). Diversity of grain proteins and bread wheat quality. I. Correlation between gliadin bands and flour quality characteristics, *J. Cereal Sci.*, 3:329.
38. MacRitchie, F. (1984). Baking quality of wheat flours, *Adv. Food Res.*, 29:201.
39. MacRitchie, F. (1985). Studies of the methodology for fractionation and reconstitution of wheat flour, *J. Cereal Sci.*, 3:221.

40. Shewry, P. R., and Miflin, B. J. (1985). Seed storage proteins of economically important cereals, in *Advances in Cereal Science and Technology*, Vol. VII, Y. Pomeranz, Ed., American Association of Cereal Chemists, St. Paul, Minn., p. 1.
41. Kobrehel, K., and Matignon, B. (1980). Solubilization of proteins with soaps in relation to the bread-making properties of wheat flours, *Cereal Chem.*, 57:73.
42. Danno, G. (1981). Extraction of unreduced glutenin from wheat flour with sodium dodecyl sulfate, *Cereal Chem.*, 58:311.
43. Field, J. M., Shewry, P. R., and Miflin, B. J. (1983). Solubilisation and characterisation of wheat gluten proteins:correlations between the amount of aggregated proteins and baking quality, *J. Sci. Food Agric.*, 34:370.
44. Schofield, D. (1986). Flour proteins:structure and functionality in baked products, in *Chemistry and Physics of Baking*, J. M. V. Blanshard, P. J. Frazier, and T. Galliard, Eds., The Royal Society of Chemistry, London, p. 14.
45. McDermott, E. E. (1985). The properties of commercial glutens, *Cereal Foods World*, 32:169.
46. Weegels, P. L., and Hamer, R. J. (1989). Predicting the baking quality of gluten, *Cereal Foods World*, 34:210.
47. Czuchajowska, Z., and Pomeranz, Y. (1990). Quest for a universal test of commercial gluten quality for breadmaking, *Cereal Foods World*, 35:458.
48. Bushuk, W., and Wadhawan, C. (1989). Wheat gluten is good not only for breadmaking, in *Wheat Is Unique*, Y. Pomeranz, Ed., American Association of Cereal Chemists, St. Paul, Minn., p. 263.
49. Kasarda, D. D., Bernardin, J. E., and Nimmo, C. C. (1976). Wheat proteins, in *Advances in Cereal Science and Technology*, Vol. I, Y. Pomeranz, Ed., American Association of Cereal Chemists, St. Paul, Minn., p. 158.
50. Hoseney, R. C., Finney, K. F., Shogren, M. D., and Pomeranz, Y. (1969). Functional (breadmaking) and biochemical properties of wheat flour components. II. Role of water-solubles, *Cereal Chem.*, 46:117.
51. Lindahl, L. (1987). Surface properties of the water soluble fraction of the wheat endosperm, *J. Disp. Sci. Technol*, 8:309.
52. Fullington, J. G. (1974). A protein from wheat flour that binds calcium and stearoyl-2-lactylate ions, *Cereal Chem.*, 51:250.
53. Hoseney, R. C., Pomeranz, Y., and Finney, K. F. (1970). Functional (breadmaking) and biochemical properties of wheat flour components. VII. Petroleum ether-soluble lipoproteins of wheat flour, *Cereal Chem.*, 47:153.
54. Zawistowska, U., Bietz, J. A., and Bushuk, W. (1986). Characterization

of low-molecular-weight protein with high affinity for flour lipids from two wheat classes, *Cereal Chem.*, 63:414.
55. Frazier, P. J., Daniels, N. W. R., and Russell Eggitt, P. W. (1981). Lipid–protein interactions during dough development, *J. Sci. Food Agric.*, 32:877.
56. Schofield, J. D., and Booth, M. R. (1983). Wheat proteins and their technological significance, *Dev. Food Proteins*, 2:1.
57. Lásztity, R. (1984). *The Chemistry of Cereal Proteins*, CRC Press, Boca Raton, Fla.
58. Payne, P. I., and Rhodes, A. P. (1982). Cereal storage proteins:structure and role in agriculture and food technology, *Encycl. Plant Physiol., New Ser.*, 14A:346.
59. Hoseney, R. C., and Finney, K. F. (1971). Functional (breadmaking) and biochemical properties of wheat flour components. XI. A review, *Baker's Dig.*, 41(4):30.
60. Marais, G. F., and D'Appolonia, B. L. (1981). Factors contributing to baking quality differences in hard red spring wheat. II. Bases for different mixing properties, *Cereal Chem.*, 58:448.
61. Marais, G. F., and D'Appolonia, B. L. (1981). Factors contributing to baking quality differences in hard red spring wheat. I. Bases for different loaf volume potentials, *Cereal Chem.*, 58:444.
62. Simmonds, D. H., Barlow, K. K., and Wrigley, C. W. (1973). The biochemical basis of grain hardness in wheat, *Cereal Chem.*, 50:553.
63. Simmonds, D. H. (1974). Chemical basis of hardness and vitreosity in the wheat kernel, *Baker's Dig.*, 48(5):16.
64. Greenwell, P., and Schofield, J. D. (1986). A starch granule protein associated with endosperm softness in wheat, *Cereal Chem.*, 63:379.
65. Hoseney, R. C. (1984). Functional properties of pentosans in baked foods, *Food Technol.*, 38(1):114.
66. Shewry, P. R., Tatham, A. S., Forde, J., Kreis, M., and Miflin, B. J. (1986). The classification and nomenclature of wheat gluten proteins:a reassessment, *J. Cereal Sci.*, 4:97.
67. Woychick, J. H., Boundy, J. A., and Dimler, R. J. (1961). Starch gel electrophoresis of wheat gluten proteins with concentrated urea, *Arch. Biochem. Biophys.*, 94:477.
68. MacRitchie, F., Kasarda, D. D., and Kuzmicky, D. D. (1991). Characterization of wheat protein fractions differing in contributions to breadmaking quality, *Cereal Chem.*, 68:122.
69. Bietz, J. A. (1979). Recent advances in the isolation and characterization of cereal proteins, *Cereal Foods World*, 24:199.
70. Wieser, H., Mödl, A., Seilmeier, W., and Belitz, H.-D. (1987). High-performance liquid chromatography of gliadins from different wheat

varieties:amino acid composition and N-terminal amino acid sequence of components, *Z. Lebensm. Unters. Forsch.*, 185:371.
71. Seilmeier, W., Wieser, H., and Belitz, H.-D. (1987). High-performance liquid chromatography of reduced glutenin:amino acid composition of fractions and components, *Z. Lebensm. Unters. Forsch.*, 185:487.
72. Tatham, A. S., Field, J. M., Smith, S. J., and Shewry, P. R. (1987). The conformation of wheat gluten proteins. II. Aggregated gliadins and low molecular weight subunits of glutenin, *J. Cereal Sci.*, 5:203.
73. Popineau, Y., and Pineau, F. (1988). Changes of conformation and surface hydrophobicity of gliadins, *Lebensm. Wiss. Technol.*, 21:113.
74. Ng, P. K. W., and Bushuk, W. (1989). Concerning the nomenclature of high molecular weight glutenin subunits, *J. Cereal Sci.*, 9:53.
75. Charbonnier, L., Tercé-Laforgue, T., and Mossé, J. (1980). Some physiochemical properties of *Triticum vulgare* β-, γ-, and ω-gliadins, *Ann. Technol. Agric.*, 29:175.
76. Tatham, A. S., and Shewry, P. R. (1985). The conformation of wheat gluten proteins. The secondary structures and thermal stabilities of α-, β-, γ-gliadins, *J. Cereal Sci.*, 3:103.
77. Pomeranz, Y., Finney, K. F., and Hoseney, R. C. (1970). Molecular approach to breadmaking, *Baker's Dig.*, 44(3):22.
78. Chung, O. K., and Pomeranz, Y. (1979). Acid-soluble proteins of wheat flours. II. Binding to hydrophobic gels, *Cereal Chem.*, 56:196.
79. Popineau, Y. (1985). Fractionation of acetic acid-soluble proteins from wheat gluten by hydrophobic interaction chromatography:evidence for different behaviour of gliadin and glutenin proteins, *J. Cereal Sci.*, 3:29.
80. Purcell, J. M., Kasarda, D. D., and Wu, C.-S. C. (1988). Secondary structure of wheat α- and ω-gliadin proteins:Fourier transform infrared spectroscopy, *J. Cereal Sci.*, 7:21.
81. Wrigley, C. W., Robinson, P. J., and Williams, W. T. (1981). Association between electrophoretic patterns of gliadin proteins and quality characteristics of wheat cultivars, *J. Sci. Food Agric.*, 32:433.
82. Bietz, J. A. (1985). High performance liquid chromatography:how proteins look in cereals, *Cereal Chem.*, 62:201.
83. Butaki, R. C., and Dronzek, B. (1979). Comparison of gluten properties of flour wheat varieties, *Cereal Chem.*, 56:159.
84. Huebner, F. R., and Wall, J. S. (1976). Fractionation and quantitative differences of glutenin from wheat varieties varying in baking quality, *Cereal Chem.*, 53:258.
85. Graveland, A., Bosveld, P., Lichtendonk, W. J., Marseille, J. P., Moonen, J. H. E., and Scheepstra, A. (1985). A model for the molecular structure of the glutenins from wheat flour, *J. Cereal Sci.*, 3:1.
86. Graveland, A., and Henderson, M. H. (1987). Structure and functionality

of gluten proteins, in *Proceedings of the 3rd International Workshop on Gluten Proteins*, R. Lasztity and F. Bekés, Eds., World Scientific, Singapore, p. 238.
87. Payne, P. I., Law, C. N., and Mudd, E. E. (1980). Control by homologous group 1 chromosomes of the high-molecular-weight subunits of glutenin, a major protein of wheat endosperm. *Theor. Appl. Genet.*, 58:113.
88. Shewry, P. R., Field, J. M., Faulks, A. J., Parmar, S., Miflin, B. J., Dietler, M. D., Lew, E. J.-L., and Kasarda, D. D. (1984). The purification and N-terminal amino acid sequence analysis of the high molecular weight gluten polypeptides of wheat, *Biochim. Biophys. Acta*, 788: 23.
89. Kasarda, D. D. (1989). Glutenin structure in relation to wheat quality, in *Wheat is Unique*, Y. Pomeranz, Ed., AACC, St. Paul, Minn., p. 277.
90. Tatham, A. S., Miflin, B. J., and Shewry, P. R. (1985). The beta-turn conformation in wheat gluten proteins:relationships to gluten elasticity, *Cereal Chem.*, 62:405.
91. Payne, P. I., Nightingale, M. A., Krattiger, A. F., and Holt, L. M. (1987). The relationship between HMW glutenin subunit composition and the bread-making quality of British-grown wheat varieties, *J. Cereal Sci.*, 40:51.
92. Branlard, G., and Dardevet, M. (1985). Diversity of grain protein and bread wheat quality. II. Correlation between high molecular weight subunits of glutenin and flour quality characteristics, *J. Cereal Sci.*, 3:345.
93. Ewart, J. A. D. (1979). Glutenin structure, *J. Sci. Food Agric.*, 30:482.
94. Ewart, J. A. D. (1988). Studies of disulfide bonds in glutenin, *Cereal Chem.*, 65:95.
95. Ewart, J. A. D. (1972). A modified hypothesis for the structure of rheology of glutelins, *J. Sci. Food Agric.*, 23:687.
96. Lawrence, G. J., and Payne, P. I. (1983). Detection by gel electrophoresis of oligomers formed by the association of high-molecular-weight glutenin protein subunits of wheat endosperm, *J. Exp. Bot.*, 34:254.
97. Belitz, H.-D., Kieffer, R., Seilmeier, W., and Wieser, H. (1986). Structure and function of gluten proteins, *Cereal Chem.*, 63:336.
98. MacRitchie, F. (1979). A relation between gluten protein amide content and baking performance of wheat flours, *J. Food Technol.*, 14:595.
99. Wieser, H., Stempfl, C., and Belitz, H.-D. (1985). Peptidmuster von Glutelinfraktionen verschiedener Weizensorten, *Z. Lebensm. Unters, Forsch.*, 181:1.
100. MacRitchie, F. (1980). Studies of gluten protein from wheat flours, *Cereal Foods World*, 25:382.
101. Lagudah, E. S., O'Brien, L. O., and Halloran, G. M. (1988). Influence

of gliadin composition and high molecular weight subunits of glutenin on dough properties in an F_3 population of a bread wheat cross, *J. Cereal Sci.*, 7:33.
102. Cressey, P. J., Campbell, W. P., Wrigley, C. W., and Griffin, W. B. (1987). Statistical correlations between quality attributes and grain-protein composition for 60 advanced lines of crossbred wheat, *Cereal Chem.*, 64:299.
103. Belitz, H.-D., Kim, J.-J., Kieffer, R., Seilmeier, W., Werbeck, W., and Wieser, H. (1987). Separation and characterization of reduced glutelins from different wheat varieties and importance of the gliadin/glutelin ratio for the strength of gluten, in *Proceedings of the 3rd International Workshop on Gluten Proteins*, R. Lásztity and F. Bekes, Eds., World Scientific, Singapore, p. 189.
104. MacRitchie, F. (1972). The fractionation and properties of gluten proteins. *J. Macromol. Sci. Chem.*, A6:823.
105. MacRitchie, F. (1987). Evaluation of contributions from wheat protein fractions to dough mixing and breadmaking, *J. Cereal Sci.*, 6:259.
106. Gupta, R. B., MacRitchie, F., Shepherd, K. W., and Ellison, F. (1990). Relative contribution of LMW and HMW glutenin subunits to dough strength and dough stickiness of bread wheat, in *Gluten Proteins 1990*, W. Bushuk and R. Tkachuk, Eds., American Association of Cereal Chemists, St. Paul, Minn., p. 71.
107. Payne, P. I., Corfield, K. G., Holt, L. M., and Blackman, J. A. (1981). Correlations between the inheritance of certain high-molecular weight subunits of glutenin and break-making quality in progenies of six crosses of bread wheat, *J. Sci. Food Agric.*, 32:51.
108. Lukow, O. M., Paynes, P. I., and Tkachuk, R. (1989). The HMW glutenin subunit composition of Canadian wheat cultivars and their association with bread-making quality, *J. Sci. Food. Agric.*, 46:451.
109. Khan, K., Tamminga, G., and Lukow, O. (1989). The effect of wheat flour proteins on mixing and baking—correlations with protein fractions and high molecular weight glutenin subunit composition by gel electrophoresis, *Cereal Chem.*, 66:391.
110. Lawrence, G. J., MacRitchie, F., and Wrigley, C. W. (1988). Dough and baking quality of wheat lines deficient in glutenin subunits controlled by the *Glu-A1*, *Glu-B1* and *Glu-D1* loci, *J. Cereal Sci.*, 7:109.
111. Eliasson, A.-C., and Lundh, G. (1989). Rheological and interfacial behaviour of some wheat protein fractions, *J. Text. Stud.*, 20:431.
112. Eliasson, A.-C., and Silverio, J. (1990). Interfacial behaviour of gluten proteins after heat treatment, in *Gluten Proteins 1990*, W. Bushuk and R. Tkachuk, Eds., American Association of Cereal Chemists, St. Paul, Minn., p. 11.

113. Tschoegl, N. W., and Alexander, A. E. (1960). The surface chemistry of wheat gluten, I. Surface pressure measurements, *J. Colloid Sci.*, 15:155.
114. Lundh, G., Eliasson, A.-C., and Larsson, K. (1988). Cross-linking of wheat storage protein monolayers by compression/expansion cycles at the air/water interface, *J. Cereal Sci.*, 7:1.
115. Tao, H. P., Cornell, D. G., and Kasarda, D. D. (1989). Surface and optical properties of wheat glutenin molecules, *J. Cereal Sci.*, 10:5.
116. Eliasson, A.-C., and Silverio, J. (1992). Interfacial and rheological behaviour of some commercial gluten samples, *Acta Agric. Scand.*, Sect. B, *Soil and Plant Sci.*, 42:1.
117. Tornberg, E. (1978). The interfacial behaviour of three food proteins studied by the drop volume technique, *J. Sci. Food Agric.*, 29:762.
118. Henderson, D., Larsson, K., and Rao, Y. K. (1991). A study of wheat storage protein monolayers by Faraday wave damping, *Langmuir*, 7:2737.
119. Mita, T., Nikai, K., Hiraoka, T., Matsuo, S., and Matsumoto, H. (1977). Physiochemical studies on wheat protein foams, *J. Colloid Interface Sci.*, 59:172.
120. Mita, T., and Matsumoto, H. (1981). Flow properties of aqueous gluten and gluten methyl ester dispersions, *Cereal Chem.*, 58:57.
121. Wall, J. S., and Beckwith, A. C. (1969). Relationship between structure and rheological properties of gluten proteins, *Cereal Sci. Today*, 14:16.
122. Taylor, N. W., and Cluskey, J. E. (1962). Wheat gluten and its glutenin component:viscosity, diffusion and sedimentation studies, *Arch. Biochem. Biophys.*, 97:399.
123. Wu, Y. V., and Dimler, R. J. (1963). Hydrogen ion equilibria of wheat glutenin and gliadin, *Arch. Biochem. Biophys.*, 103:310.
124. Wu, Y. V., and Dimler, R. J. (1964). Conformational studies of wheat gluten, glutenin, and gliadin in urea solutions at various pH's, *Arch. Biochem. Biophys.*, 107:435.
125. Wu, Y. V., Cluskey, J. E., and Sexson, K. R. (1967). Effect of ionic strength on the molecular weight and conformation of wheat gluten proteins in 3 M urea solutions, *Biochim. Biophys. Acta*, 133:83
126. Cole, E. W., Kasarda, D. D., and Lafiandra, D. (1984). The conformational structure of A-gliadin. Intrinsic viscosities under conditions approaching the native state and under denaturing conditions, *Biochim. Biophys. Acta*, 787:244.
127. Field, J. M., Tatham, A. S., and Shewry, P. R. (1987). Determination of prolamin configuration and dimensions by viscometric analysis, in *Proceedings of the 3rd International Workshop on Gluten Proteins*, R. Lásztity and F. Békés, Eds., World Scientific, Singapore, p. 478.

128. Ewart, J. A. D. (1980). Loaf volume and the intrinsic viscosity of glutenin, *J. Sci. Food Agric.*, 31:1323.
129. Lerchenthal, C. H., and Funt, C. B. (1970). The yield function in an unstable viscoelastic material (gluten gel), *Isreal J. Technol.*, 8:317.
130. Funt Bar-David, C. B., and Lerchenthal, C. H. (1975). Rheological and thermodynamic properties of gluten, *Cereal Chem.*, 52:154.
131. Bohlin, L., and Carlson, T. L.-G. (1981). Shear stress relaxation of wheat flour dough and gluten, *Colloids Surfaces*, 2:59.
132. Mita, T., and Bohlin, L. (1983). Shear stress relaxation of chemically modified gluten, *Cereal Chem.*, 60:93.
133. Sperling, L. H. (1986). *Introduction to Physical Polymer Science*, Wiley, New York.
134. Lindahl, L. (1990). Rheological properties in wheat flour systems. Molecular and interfacial basis, thesis, Lund University, Lund.
135. Dreese, P. C., Faubion, J. M., and Hoseney, R. C. (1988). Dynamic rheological properties of flour, gluten and gluten-starch doughs. II. Effect of various processing and ingredient changes, *Cereal Chem.*, 65: 354.
136. Attenburrow, G., Barnes, D. J., Davies, A. P., and Ingman, S. J. (1990). Rheological properties of wheat gluten, *J. Cereal Sci.*, 12:1.
137. Dreese, P.C., and Hoseney, R. C. (1990). The effects of water-extracted solubles from gluten on its baking and rheological properties, *Cereal Chem.*, 67:400.
138. Ferry, J. D. (1970). *Viscoelastic Properties of Polymers*, Wiley, New York.
139. Weipert, D. (1990). The benefits of basic rheometry in studying dough rheology, *Cereal Chem.*, 67:311.
140. Smith, J. R., Smith, T. L., and Tschoegl, N. W. (1970). Rheological properties of wheat flour doughs, *Rheol. Acta*, 9:239.
141. Navickis, L. L., Anderson, R. A., Bagley, E. B., and Jasberg, B. K. (1982). Viscoelastic properties of wheat flour doughs:variation of dynamic moduli with water and protein content, *J. Text. Stud.*, 13:249.
142. Szczesniak, A. A., Loh, J., and Wesley, R. (1983). Effect of moisture transfer on dynamic viscoelastic parameters of wheat flour/water systems, *J. Rheol.*, 27:537.
143. Tschoegl, N. W., and Alexander, A. E. (1960). The surface chemistry of wheat gluten. II. Measurements of surface viscoelasticity, *J. Colloid Sci.*, 15:168.
144. Kokelaar, A., and Prins, A. (1990). Dynamic surface behaviour of gluten in an aqueous system, in *Gluten Proteins 1990*, W. Bushuk and R. Tkachuk, Eds., American Association of Cereal Chemists, St. Paul, Minn., p. 57.

145. Ewart, J. A. D. (1972). Recent research and dough visco-elasticity, *Baker's Dig.*, 46(4):22.
146. MacRitchie, F. (1973). Conversion of a weak flour to a strong one by increasing the proportion of its high molecular weight gluten proteins, *J. Sci. Food Agric.*, 24:1325.
147. Bernardin, J. E. (1975). The rheology of concentrated gliadin solutions, *Cereal Chem.*, 52:136.
148. Bernardin, J. E. (1978). Effect of shear on the nematic mesophase of the wheat storage protein A-gliadin, *J. Text. Stud.*, 9:283.
149. Janssen, A. M., Vereijken, J. M., Beintema, J. J., Witholt, B., and van Vliet, T. (1990). Rheological studies on gluten, in *Gluten Proteins 1990*, W. Bushuk and R. Tkachuk, Eds., American Association of Cereal Chemists, St. Paul, Minn., p. 180.
150. Grosskreutz, J. C. (1960). The physical structure of wheat protein, *Biochim. Biophys. Acta*, 38:400.
151. Hermansson, A.-M., and Larsson, K. (1986). The structure of gluten gels, *Food Microstruct.*, 5:233.
152. Bernardin, J. E., and Kasarda, D. D. (1973). Hydrated protein fibrils from wheat endosperm, *Cereal Chem.*, 50:529.
153. Bernardin, J. E., and Kasarda, D. D. (1973). The microstructure of wheat protein fibrills, *Cereal Chem.*, 50:735.
154. Amend, T., and Belitz, H.-D. (1990). The formation of dough and gluten—a study by scanning electron microscopy, *Z. Lebensm. Unters. Forsch.*, 190:401.
155. Amend, T., and Belitz, H.-D. (1990). Gluten formation studied by the transmission electron microscope, *Z. Lebensm. Unters. Forsch.*, 191:184.
156. Amend, T., and Belitz, H.-D. (1989). Microscopical studies of water/flour systems, *Z. Lebensm. Unters. Forsch.*, 189:103.
157. Miles, M. J., Carr, H. J., McMaster, T. J., Belton, P. S., Morris, V. J., Field, J. M., Shewry, P. R., and Tatham, A. S. (1991). Scanning tunneling microscopy of wheat seed storage proteins reveals an unusual supersecondary structure, *Proc. Natl. Acad. Sci. USA*, 88:68.
158. Hoseney, R. C., Finney, K. F., Pomeranz, Y., and Shogren, M. D. (1971). Functional (breadmaking) and biochemical properties of wheat flour components. VIII. Starch, *Cereal Chem.*, 48:191.
159. Sollars, W. F., and Rubenthaler, G. L. (1971). Performance of wheat and other starches in reconstituted flours, *Cereal Chem.*, 48:397.
160. Kulp, K. (1973). Characteristics of small-granule starch of flour and wheat, *Cereal Chem.*, 50:666.
161. D'Appolonia, B. L., and Gilles, K. A. (1971). Effect of various starches in baking, *Cereal Chem.*, 48:625.

162. Sandstedt, R. M. (1961). The function of starch in the baking of bread, *Baker's Dig.*, 35(3):36.
163. Martin, M. L., Zeleznak, K. J., and Hoseney, R. C. (1991). A mechanism of bread firming. I. Role of starch swelling, *Cereal Chem.*, 68:498.
164. Eliasson, A.-C., and Karlsson, R. (1983). Gelatinization properties of different size classes of wheat starch granules measured with differential scanning calorimetry, *Starch/Stärke*, 35:130.
165. Karlsson, R., Olered, R., and Eliasson, A.-C. (1983). Changes in starch granule size distribution and starch gelatinization properties during development and maturation of wheat, barley and rye, *Starch/Stärke*, 35:335.
166. Soulaka, A. B., and Morrison, W. R. (1985). The amylose and lipid contents, dimensions, and gelatinization characteristics of some wheat starches and their A- and B-granule fraction, *J. Sci. Food Agric.*, 36:709.
167. Evers, A. D. (1971). Scanning electron microscopy of wheat starch. III. Granule development in the endosperm, *Starch/Stärke*, 23:157.
168. Duffus, C. M., and Murdoch, S. M. (1979). Variation in starch granule size distribution and amylose content during wheat endosperm development, *Cereal Chem.*, 5:427.
169. Morrison, W. R., and Scott, D. C. (1986). Measurement of the dimensions of wheat starch granule populations using a Coulter counter with 100-channel analyzer, *J. Cereal Sci.*, 4:13.
170. Kulp, K. (1973). Some properties of starches derived from wheat of varied maturity, *Cereal Chem.*, 50:496.
171. Soulaka, A. B., and Morrison, W. R. (1985). The bread baking quality of six wheat starches differing in composition and physical properties, *J. Sci. Food Agric.*, 36:719.
172. Manners, D. J. (1989). Recent developments in our understanding of amylopectin, *Carbohydr. Polym.*, 11:87.
173. Curá, J. A., and Krisman, C. R. (1990). Cereal grains:a study of their α-1,4–α-1,6 glucopolysaccharide composition, *Starch/Stärke*, 42:171.
174. Greenwood, C. T. (1976). Starch, in *Advances in Cereal Science and Technology*, Vol. I, Y. Pomeranz, Ed., American Association of Cereal Chemists, St. Paul, Minn., p. 119.
175. Manners, D. J. (1985). Some aspects of the structure of starch, *Cereal Foods World*, 30:461.
176. Takeda, Y., Hizukuri, S., Takeda, C., and Suzuki, A. (1987). Structures of branched molecules of amylose of various origins, and molar fractions of branched and unbranched molecules, *Carbohydr. Res.*, 165:139.
177. Lii, C.-Y., and Lineback, D. R. (1977). Characterization and comparison of cereal starches, *Cereal Chem.*, 54:138.

178. Banks, W., and Greenwood, C. T. (1975). *Starch and Its Components*, Edinburgh Univ. Press, Edinburgh, Scotland.
179. Ring, S. G., I'Anson, J., and Morris, V. J. (1985). Static and dynamic light scattering studies of amylose solutions, *Macromolecules*, 18:182.
180. Braga, D., Ferracini, E., Ferrero, A., Ripamonti, A., Brant, D. A., Bulga, G. S., and Cesaro, A. (1985). Amylose conformation in aqueous solution:a small angle X-ray scattering study, *Int. J. Biol. Macromol.*, 7:161.
181. Miles, M. J., Morris, V. J., and Ring, S. G. (1985). Gelation of amylose, *Carbohydr. Res.*, 135:257.
182. Lelievre, J., Lewis, J. A., and Marsden, K. (1986). The size and shape of amylopectin:a study using analytical ultracentrifugation, *Carbohydr. Res.*, 153:195.
183. Callaghan, P. T., and Lelievre, J. (1985). The size and shape of amylopectin:a study using pulsed-field gradient nuclear magnetic resonance, *Biopolymers*, 24:441.
184. Hizukuri, S. (1985). Relationship between the distribution of the chain length of amylopectin and the crystalline structure of starch granules, *Carbohydr. Res.*, 141:295.
185. Kobayashi, S., Schwartz, S. J., and Lineback, D. R. (1986). Comparison of the structures of amylopectin from different wheat varieties, *Cereal Chem.*, 63:71.
186. Eliasson, A.-C., Larsson, K., Andersson, S., Hyde, S. T., Nesper, R., and von Schnering, H.-G. (1987). On the structure of native starch—an analogue to the quartz structure, *Starch/Stärke*, 39:147.
187. Robin, J. P., Mercier, R., Charbonniere, R., and Guilbot, A. (1974). Lintnerized starches. Gel filtration and enzymatic studies of insoluble residues from prolonged acid treatment of potato starch, *Cereal Chem.*, 51:389.
188. French, D. (1984). Organization of starch granules, in *Starch: Chemistry and Technology*, Whistler, R. L., BeMiller, J. N. and Paschall, E. F., Eds., Academic, New York, p. 184.
189. Manners, D. J., and Matheson, N. K. (1981). The fine structure of amylopectin, *Carbohydr. Res.*, 90:99.
190. Ring, S. G., Colonna, P., I'Anson, J., Kalichevsky, M. T., Miles, M. J., Morris, V. J., and Orford, P. D. (1987). The gelation and crystallization of amylopectin, *Carbohydr. Res.*, 162:277.
191. Kalichevsky, M. T., and Ring, S. G. (1987). Incompatibility of amylose and amylopectin in aqueous solution, *Carbohydr. Res.*, 162:323.
192. Morrison, W. R. (1981). Starch lipids:a reappraisal, *Starch/Stärke*, 33:408.

193. Acker, L. (1977). Die Lipide der Stärken—ein Forchungsgebiet zwischen Kohlenhydraten und Lipiden, *Fette-Seifen-Anstrichmittel*, 79:1.
194. Carlson, T. L.-G., Eliasson, A.-C., and Nair, B. M. (1980). Note on the determination of maximum amount of cereal storage protein in foods, *J. Sci. Food Agric.*, 31:223.
195. Eliasson, A.-C., Carlson, T. L.-G., Larsson, K., and Miezis, Y. (1981). Some effects of starch lipids on the thermal and rheological properties of wheat starch, *Starch/Stärke*, 33:130.
196. Lowy, G. D. A., Sargeant, J. G., and Schofield, J. D. (1981). Wheat starch granule protein:the isolation and characterization of a salt-extractable protein from starch granules, *J. Sci.Food Agric.*, 32:371.
197. Seguchi, M., and Yamada, Y. (1989). Study of proteins extracted from the surface of wheat starch granules with sodium dodecyl sulfate, *Cereal Chem.*, 66:193.
198. Schäfer, R. (1975). Ausbeutebestimmung im Laboratorium eine Methode zur quantitativen Mehlbeurteilung in der Weizenstärke-industrie, *Starch/Stärke*, 27:257.
199. Cauvain, S. P., Gough, B. M., and Whitehouse, M. E. (1977). The role of starch in baked goods. Part 2. The influence of purification procedure on the surface properties of the granule, *Starch/Stärke*, 29:91.
200. Lineback, D. R. and Rasper, V. F. (1988). Wheat carbohydrates, in *Wheat Chemistry and Technology*, Y. Pomeranz, Ed., American Association of Cereal Chemists, St. Paul, Minn., p. 277.
201. Dronzek, B. L., Hwang, P., and Bushuk, W. (1972). Scanning electron microscopy of starch from sprouted wheat, *Cereal Chem.*, 49:233.
202. Greenwell, P., Evers, A. D., Gough, B. M., and Russell, P. (1985). Amyloglucosidase-catalysed erosion of native, surface-modified and chlorine-treated wheat starch granules. The influence of surface protein, *J. Cereal Sci.*, 3:279.
203. Marsh, R. A., and Waight, S. G. (1982). The effect of pH on the zeta potential of wheat and potato starch, *Starch/Stärke*, 34:149.
204. Russell, P, Gough, B. M., Greenwell, P., Fowler, A., and Munro, H. S. (1987). A study by ESCA of the surface of native and chlorine-treated wheat starch granules:the effects of various surface treatments, *J. Cereal Sci.*, 5:83.
205. Seguchi, M. (1986). Dye binding to the surface of wheat starch granules, *Cereal Chem.*, 63:518.
206. Seguchi, M., and Yamada, Y. (1988). Hydrophobic character of heat-treated wheat starch, *Cereal Chem.*, 65:375.
207. Seguchi, M. (1984). Comparison of oil-binding ability of different chlorinated starches, *Cereal Chem.*, 61:244.

208. Lineback, D. R. (1984). The starch granule organization and properties, *Baker's Dig.*, 58(2):16.
209. Wong, R. B. K., and Lelievre, J. (1982). Comparison of the crystallinities of wheat starches with different swelling capacities, *Starch/Stärke*, 34:159.
210. Gidley, M. J., and Bociek, S. M. (1985). Molecular organization in starches:a 13C CP/MAS NMR study, *J. Am. Chem., Soc.*, 107:7040.
211. Zobel, H. F., Young, S. N., and Rocca, L. A. (1988). Starch gelatinization:an X-ray diffraction study, *Cereal Chem.*, 65:443.
212. Muhrbeck, P. (1991). On crystallinity in cereal and tuber starches, *Starch/Stärke*, 43:347.
213. Blanshard, J. M. V., Bates, D. R., Muhr, A. H., Worcester, D. L., and Higgins, J. S. (1984). Small-angle neutron scattering studies of starch granule structure, *Carbohydr. Polym.*, 4:427.
214. Yang, M., Grider, J., Gordon, J., and Davis, E. A. (1985). Small angle X-ray scattering by hydrated wheat starch granules, *Food Microstruct.*, 4:107.
215. Donovan, J. W., Lorenz, K., and Kulp, K. (1983). Differential scanning calorimetry of heat-moisture treated wheat and potato starches, *Cereal Chem.*, 60:381.
216. Hizukuri, S., Takeda, Y., Usami, S., and Takase, Y. (1980). Effect of aliphatic hydrocarbon groups on the crystallization of amylodextrin: model experiments for starch crystallization, *Carbohydr. Res.*, 83:193.
217. Gidley, M. J., and Bulpin, P. V. (1987). Crystallization of maltooligosaccharides as models of the crystalline forms of starch:minimum chain-length requirements for the formation of double helices, *Carbohydr. Res.*, 161:291.
218. Sarko, A., and Wu, H.-C. H. (1978). The crystal structures of A-, B- and C-polymorphs of amylose and starch, *Starch/Stärke*, 30:73.
219. French, A. D., and Murphy, V. G. (1977). Computer modeling in the study of starch, *Cereal Foods World*, 22:61.
220. Wild, D. L., and Blanshard, J. M. V. (1986). The relationship of the crystal structure of amylose polymorphs to the structure of the starch granule, *Carbohydr. Polym.*, 6:121.
221. Wu, H.-C. H., and Sarko, A. (1978). The double-helical molecular structure of crystalline A-amylose, *Carbohydr. Res.*, 61:
222. Wu, H.-C. H., and Sarko, A. (1978). The double-helical molecular structure of crystalline B-amylose, *Carbohydr. Res.*, 61:7.
223. Andersson, S., Hyde, S. T., Larsson, K., and Lidin, S. (1988). Minimal surfaces and structures:from inorganic and metal crystals to cell membranes and biopolymers, *Chem. Rev.*, 88:221.

224. Larsson, K. (1991). Structure of the starch granule—a curved crystal, *Acta Chem. Scand.*, 45:840.
225. Wong, R. B. K., and Lelievre, J. (1982). Rheological characteristics of wheat starch pastes measured under steady shear conditions, *J. Appl. Polym. Sci.*, 27:1433.
226. Junge, R. C., and Hoseney, R. C. (1981). A mechanism by which shortening and certain surfactants improve loaf volume in bread, *Cereal Chem.*, 58:408.
227. Atwell, W. A., Hood, L. F., Lineback, D. R., Varriano-Marston, E., and Zobel, H.F. (1988). The terminology and methodology associated with basic starch phenomena, *Cereal Foods World*, 33:306.
228. Hill, R. D., and Dronzek, B. L. (1973). Scanning electron microscopy studies of wheat, potato and corn starch during gelatinization, *Starch/Stärke*, 25:367.
229. Goering, K. J., Fritts, D. H., and Allen, K. G. D. (1974). A comparison of loss of birefringence with the percent gelatinization and viscosity on potato, wheat, rice, corn, cow cockle and several barley starches, *Cereal Chem.*, 51:764.
230. Russell, P. L. (1987). Gelatinization of starches of different amylose/amylopectin content. A study by differential scanning calorimetry, *J. Cereal Sci.*, 6:133.
231. Stevens, D. J., and Elton, G. A. H. (1971). Thermal properties of the starch/water system, Part I. Measurement of heat of gelatinisation by differential scanning calorimetry, *Starch/Stärke*, 23:8.
232. Evans, I. D., and Haisman, D. R. (1982). The effects of solutes on the gelatinization temperature of potato starch, *Starch/Stärke*, 34:224.
233. Eliasson, A.-C. (1983). Physical properties of starch in concentrated systems such as dough and bread, Thesis, Lund University, Lund.
234. Ghiasi, K., Hoseney, R. C., and Varriano-Marston, E. (1982). Gelatinization of wheat starch. III. Comparison by differential scanning calorimetry and light microscopy, *Cereal Chem.*, 59:258.
235. Hellman, N. N., Fairchild, B., and Senti, F. R. (1954). The bread staling problem. Molecular organization of starch upon aging of concentrated starch gels at various moisture levels, *Cereal Chem.*, 31:495.
236. Svensson, E., and Eliasson, A.-C. (to be published).
237. Donovan, J. W. (1979). Phase transitions of the starch–water system, *Biopolymers*, 18:263.
238. Eliasson, A.-C. (1980). Effect of water content on the gelatinization of wheat starch, *Starch/Stärke*, 32:270.
239. Biliaderis, C. G., Maurice, T. J., and Vose, J. R. (1980). Starch gelatinization phenomena studied by differential scanning calorimetry, *J. Food Sci.*, 45:1669.

240. Shiotsubo, T., and Takahashi, K. (1984). Differential thermal analysis of potato starch gelatinization, *Agric. Biol. Chem.*, 48:9.
241. Maurice, T. J., Page, C. M., and Slade, L. (1984). Applications of thermal analysis to starch research, *Cereal Foods World*, 29(8):504.
242. Biliaderis, C. G., Page, C. M., Maurice, T. J., and Juliano, B. O. (1986). Thermal characterization of rice starches:a polymeric approach to phase transitions of granular starch, *J. Agric. Food Chem.*, 34:6.
243. Zeleznak, K. J., and Hoseney, R. C. (1987). The glass transition in starch, *Cereal Chem.*, 64:121.
244. Slade, L., and Levine, H. (1988). Non-equilibrium melting of native granular starch. Part I. Temperature location of the glass transition associated with gelatinization of A-type cereal starches, *Carbohydr. Polym.*, 8:183.
245. Eliasson, A.-C. (1989). Some physico-chemical properties of wheat starch, in *Wheat End-Use Properties. Wheat and Flour Characterization for Specific End-Use*, H. Salovaara, Ed., University of Helsinki, p. 355.
246. Ellis, H. S., Ring, S. G., and Whittam, M. A. (1988). Time-dependent changes in the size and volume of gelatinized starch granules on storage, *Food Hydrocolloids*, 2:321.
247. Hansen, L. M., Hoseney, R. C., and Faubion, J. M. (1991). Oscillatory rheometry of starch–water systems:effect of starch concentration and temperature, *Cereal Chem.*, 68:347.
248. Eliasson, A.-C. (1985). Starch gelatinization in the presence of emulsifiers. A morphological study of wheat starch, *Starch/Stärke*, 37:411.
249. Tester, R. F., and Morrison, W. R. (1990). Swelling and gelatinization of cereal starches. I. Effects of amylopectin, amylose, and lipids, *Cereal Chem.*, 67:551.
250. Doublier, J. L. (1981). Rheological studies on starch-flow behaviour of wheat starch pastes, *Starch/Stärke*, 33:415.
251. Lindqvist, I. (1979). Cold gelatinization of starch, *Starch/Stärke*, 31:195.
252. Evers, A. D., Baker, G. J., and Stevens, D. J. (1984). Production and measurement of starch damage in flour. Part I. Damage due to rollermilling of semolina, *Starch/Stärke*, 36:309.
253. Williams, M. R., and Bowler, P. (1982). Starch gelatinization:a morphological study of triticeae and other starches, *Starch/Stärke*, 34:211.
254. Bowler, P., Williams, M. R., and Angold, R. E. (1980). A hypothesis for the morphological changes which occur on heating lenticular wheat starch in water, *Starch/Stärke*, 32:186.
255. Langton, M., and Hermansson, A.-M. (1989). Microstructural changes in wheat starch dispersions during heating and cooling, *Food Microstruct.*, 8:29.

256. Ott, M., and Hester, E. E. (1965). Gel formation as related to concentration of amylose and degree of starch swelling, *Cereal Chem.*, 42:476.
257. Gidley, M. J., and Bulpin, P. V. (1989). Aggregation of amylose in aqueous systems:the effects of chain length on phase behaviour and aggregation kinetics, *Macromolecules*, 22:341.
258. Eliasson, A.-C., and Bohlin, L. (1982). Rheological properties of concentrated wheat starch gels, *Starch/Stärke*, 34:267.
259. Eliasson, A.-C. (1986). Viscoelastic behaviour during the gelatinization of starch. II. Effects of emulsifiers, *J. Text. Stud.*, 17:357.
260. Eliasson, A.-C. (1986). Viscoelastic behaviour during the gelatinization of starch. I. Comparison of wheat, maize, potato and waxy-barley starches, *J. Text. Stud.*, 17:253.
261. Farris, R. J. (1968). Prediction of the viscosity of multimodal suspensions from unimodal viscosity data, *Trans. Soc. Rheol.*, 12:281.
262. Ellis, H. S., Ring, S. G., and Whittam, M. A. (1989). A comparison of the viscous behaviour of wheat and maize starch pastes, *J. Cereal Sci.*, 10:33.
263. Evans, I. D., and Haisman, D. R. (1979). Rheology of gelatinized starch suspensions, *J. Text. Stud.*, 10:347.
264. Ring, S. G. (1985). Some studies on starch gelation, *Starch/Stärke*, 37:80.
265. Wong, R. B. K., and Lelievre, J. (1982). Effects of storage on dynamic rheological properties of wheat starch pastes, *Starch/Stärke*, 34:231.
266. Miles, M. J., Morris, V. J., Orford, P. D., and Ring, S. G. (1985). The roles of amylose and amylopectin in the gelation and retrogradation of starch, *Carbohydr. Res.*, 135:271.
267. Dragsdorf, R. D., and Varriano-Marson, E. (1980). Bread staling:X-ray diffraction studies on bread supplemented with alfa-amylases from different sources, *Cereal Chem.*, 57:310.
268. Ring, S. G. (1985). Observations on the crystallization of amylopectin from aqueous solution, *Int. J. Biol. Macromol.*, 7:253.
269. Eliasson, A.-C. (1985). Retrogradation of starch as measured by differential scanning calorimetry, in *New Approaches to Research on Cereal Carbohydrates*, R. D. Hill and L. Munck, Eds., Elsevier, Amsterdam, p. 93.
270. Russell, P. L. (1983). A kinetic study of bread staling by differential scanning calorimetry and compressibility measurements. The effect of added monoglycerides, *J. Cereal Sci.*, 1:297.
271. Eberstein, K., Höpcke, R., Konieczny-Janda, G., and Stute, R. (1980). DSC-Untersuchungen an Stärken, *Starch/Stärke*, 12:397.
272. Berry, C. S. (1986). Resistant starch:formation and measurement of starch that survives exhaustive digestion with amylolytic enzymes during the determination of dietary fibre, *J. Cereal Sci.*, 4:301.

273. Siljeström, M., and Asp, N.-G. (1985). Resistant starch formation during baking—effect of baking time and temperature and variations in the recipe, *Z. Lebensm. Unters. Forsch.*, 181:4.
274. Russell, P. L. (1987). The ageing of gels from starches of different amylose/amylopectin content studied by differential scanning calorimetry, *J. Cereal Sci.*, 6:147.
275. Gudmundsson, M., and Eliasson, A.-C. (1990). Retrogradation of amylopectin and the effects of amylose and added surfactants/emulsifiers, *Carbohydr. Polym.*, 13:295.
276. Longton, J., and LeGrys, G. A. (1981). Differential scanning calorimetry studies on the crystallization of ageing wheat starch gels, *Starch/Stärke*, 33:410.
277. Eliasson, A.-C. (1983). Differential scanning calorimetry studies on wheat starch–gluten mixtures. II. Effect of gluten and sodium stearoyl lactylate on starch crystallization during ageing of wheat starch gels, *J. Cereal Sci.*, 1:207.
278. Zeleznak, K. J., and Hoseney, R. C. (1986). The role of water in the retrogradation of wheat starch gels and bread crumb, *Cereal Chem.*, 63:407.
279. Biliaderis, C. G., and Zawistowski, J. (1990). Viscoelastic behavior of aging starch gels:effects of concentration, temperature, and starch hydrolysates on network properties, *Cereal Chem.*, 67:240.
280. Colwell, K. H., Axford, D. W. E., Chamberlain, N., and Elton, G. A. H. (1969). Effect of storage temperature on the ageing of concentrated wheat starch gels, *J. Sci. Food Agric.*, 20:550.
281. Zeleznak, K. J., and Hoseney, R. C. (1987). Characterization of starch from bread aged at different temperature, *Starch/Stärke*, 39:231.
282. Gudmundsson, M., and Eliasson, A.-C. (1989). Some physiochemical properties of oat starches extracted from varieties with different oil content, *Acta Agric. Scand.*, 39:101.
283. Gudmundsson, M., and Eliasson, A.-C. (1991). Thermal and viscous properties of rye starch extracted from different varieties, *Cereal Chem.*, 68:172.
284. Lagendijk, J., and Pennings, H. J. (1970). Relation between complex formation of starch with monoglycerides and the firmness of bread, *Cereal Sci. Today*, 15:354.
285. Ciacco, C. F., and Fernandes, J. L. A. (1979). Effects of various ions on the kinetics of retrogradation of concentrated wheat starch gels, *Starch/Stärke*, 31:51.
286. Russell, P. L., and Oliver, G. (1989). The effect of pH and NaCl content on starch gel ageing. A study by differential scanning calorimetry and rheology, *J. Cereal Sci.*, 10:123.
287. Wynne-Jones, S., and Blanshard, J. M. V. (1986). Hydration studies of

wheat starch, amylopectin, amylose gels and bread by proton magnetic resonance, *Carbohydr. Polym.*, 6:289.

288. Izydorczyk, M., Biliaderis, C. G., and Bushuk, W. (1991). Comparison of the structure and composition of water-soluble pentosans from different wheat varieties, *Cereal Chem.*, 68:139.
289. Michniewicz, J., Biliaderis, C. G., and Bushuk, W. (1990). Water-insoluble pentosans of wheat:composition and some physical properties, *Cereal Chem.*, 67:434.
290. Geissman, T., and Neukom, H. (1973). On the composition of the water soluble wheat flour pentosans and their oxidative gelation, *Lebensm. Wiss. Technol.*, 6:59.
291. Shogren, M. D., Hashimoto, S., and Pomeranz, Y. (1987). Cereal pentosans:their estimation and significance. II. Pentosans and breadmaking characteristics of hard red winter wheat flours, *Cereal Chem.*, 64:35.
292. Henry, R. J. (1987). Pentosan and $(1\rightarrow3),(1\rightarrow4)$-$\beta$-glucan concentrations in endosperm and whole grain of wheat, barley, oats and rye, *J. Cereal Sci.*, 6:253.
293. Amadò, R., and Neukom, H. (1985). Minor constituents of wheat flour: the pentosans, in *New Approaches to Research on Cereal Carbohydrates*, R. D. Hill and L. Munck, Eds., Elsevier, Amsterdam, p. 241.
294. Fincher, G. B., and Stone, B. A. (1986). Cell walls and their components in cereal grain technology, in *Advances in Cereal Science and Technology*, Vol. VIII, Y. Pomeranz, Ed., American Association of Cereal Chemists, St. Paul, Minn, p. 207.
295. Izydorczyk, M., Biliaderis, C. G., and Bushuk, W. (1991). Physical properties of water-soluble pentosans from different wheat varieties, *Cereal Chem.*, 68:145.
296. Bengtsson, S., and Åman, P. (1990). Isolation and chemical characterization of water-soluble arabinoxylans in rye grain, *Carbohydr. Polym.*, 12:267.
297. Neukom, H., and Markwalder, H. U. (1978). Oxidative gelation of wheat flour pentosans:a new way of cross-linking polymers, *Cereal Foods World*, 23:374.
298. Ciacco, C. F., and D'Appolonia, B. L. (1982). Characterization of pentosans from different wheat flour classes and of their gelling capacity, *Cereal Chem.*, 59:96.
299. Yeh, Y. F., Hoseney, R. C., and Lineback, D. R. (1980). Changes in wheat flour pentosans as a result of dough mixing and oxidation, *Cereal Chem.*, 57:144.
300. Moore, A. M., Martinez-Muñoz, I., and Hoseney, R. C. (1990). Factors affecting the oxidative gelation of wheat water-solubles, *Cereal Chem.*, 67:81.

301. Hoseney, R. C., and Faubion, J. M. (1981). A mechanism for the oxidative gelation of wheat flour water-soluble pentosans, *Cereal Chem.*, 58:421.
302. Izydorczyk, M. S., Biliaderis, C. G., and Bushuk, W. (1990). Oxidative gelation studies of water-soluble pentosans from wheat, *J. Cereal Sci.*, 11:153.
303. Bushuk, W. (1966). Distribution of water in dough and bread, *Bakers' Dig.*, 40(5):38.
304. Kim, S. K., and D'Apolonia, B. L. (1977). Bread staling studies. III. Effect of pentosans on dough, bread, and bread staling rate, *Cereal Chem.*, 54:225.
305. Patil, S. K., Finney, K. F., Shogren, M. D., and Tsen, C. C. (1976). Water-soluble pentosans of wheat flour. III. Effect of water-soluble pentosans on loaf volume of reconstituted gluten and starch doughs, *Cereal Chem.*, 53:347.
306. Patil, S. K., Tsen, C. C., and Lineback, F. R. (1975). Water-soluble pentosans of wheat flour. I. Viscosity properties and molecular weight estimated by gel filtration, *Cereal Chem.*, 52:44.
307. Kim, S. K., and D'Appolonia, B. L. (1977). Effect of pentosans on the retrogradation of wheat starch gels, *Cereal Chem.*, 54:150.
308. Gudmundsson, M., Eliasson, A.-C., Bengtsson, S., and Åman, P. (1991). The effects of water soluble arabinoxylan on gelatinization and retrogradation of starch, *Starch/Stärke*, 43:5.
309. Kalichevsky, M. T., Orford, P. D., and Ring, S. G. (1990). The retrogradation and gelation of amylopectin from various botanical sources, *Carbohydr. Res.*, 198:49.

3
Interactions Between Components

I. INTRODUCTION

Interactions between any chemical components can be regarded on various levels. Normally we use the term "interactions" to describe noncovalent association phenomena taking place on the *molecular* level. A typical example is the interaction between amylose and lipids. If phospholipids are added to an oil–water mixture we will have an interaction on the *colloidal* level. A dispersion is stabilized as a result of molecular orientation of phospholipids at the interface demonstrated in Chapter 1. When we consider the interactions between wheat flour components in an aqueous environment it is necessary to consider both levels. Whether the particular interaction under discussion refers to the molecular or colloidal level will be clear from the context.

II. PROTEIN–LIPID INTERACTIONS

For a long time lipids in wheat flour have been classified as free or bound on the basis of their extraction properties. Thus nonpolar solvents such as hexane have been regarded as extracting the free lipids, whereas polar ones, such as water-saturated butanol, were supposed to extract bound lipids. Numerous studies exist in the literature of the last three decades on the proportions of free to bound lipids (defined in this way), how the proportion changes upon hydration and various degrees of dough mixing, and finally the effect of lipids on breadmaking. On the basis of the phase properties of wheat lipid–water systems described in Chapter 2 it is obvious that the presence of more or less water will give different phases in which the polar and nonpolar lipids distribute in different ways. These phases will exhibit different rheological properties; thus varying degrees of mixing will disperse them differently. In fact, all the results we have seen in the literature on free and bound lipids are explained simply in terms of the lipid phase behavior. Consequently, considerations of lipid–protein interactions in bread based on this classification into free and bound lipids are fundamentally incorrect.

A. General Aspects of Interactions Between Proteins and Lipids

Usually lipid and protein molecules coexist in an aqueous environment in a living tissue (or in a test tube) without any interaction on the molecular level. Two important exceptions, however, are the cell membrane and certain particles used for lipid transport, the lipoproteins.

The lipid bilayer constitutes the cell membrane matrix in which proteins are embedded. The lipid bilayer is structurally and thermodynamically very similar to that of the L_α phase discussed in Chapter 2. Membrane proteins perform a wide variety of functions, such as active transport, channel formation, or recognition (receptors). One mechanism of membrane association functions via a single α-helix chain with hydrophobic residues extending through the bilayer. The other type occurs via a globular or cylindrical protein domain within the bilayers, often formed by several turns of a peptide chain through the bilayer. A third type of membrane association of proteins has been recently discovered. In it the whole protein is located outside the polar heads (within the aqueous phase) and covalently linked via a spacer arm to a phosopholipid molecule in the bilayer. It should be pointed out that there are important differences in the membrane lipid bilayer structure compared to the L_α phase. One is an asymmetry in the distribution of the lipid species. Another is a general feature of "balancing" on the borderline toward a phase transition ($L_\alpha \rightarrow$ cubic or $L_\alpha \rightarrow H_{II}$) if we consider the bulk behavior of the membrane lipids.

The composition of wheat lipids reflects their biomembrane origin; the thylacoid membrane is also dominated by the same type of galactolipids. Even if milling and water mixing have destroyed the detailed organization of the native membrane, fragments of bilayers of lipids with associated proteins will certainly still occur in a water–flour mixture. The lipoproteins and the protein complexes soluble in organic solvents discussed in Section III of Chapter 2 should be considered with this background in mind.

It is important to keep the biomembrane structure in mind in discussions of wheat flour functional properties. A reason for this is that the biomembrane and a wheat flour dough in different processing stages both represent spontaneously formed self-assembled structures, and the organization is governed by the general forces of surface chemistry discussed in Chapter 1.

Another important group of lipid–protein association structures in living tissues are the lipoproteins. These are particles of various sizes with transport functions. In fact, there are two kinds of complexes. One consists of vesicles—

lipid bilayers with an embedded network of protein molecules that are responsible for the shape (e.g., clathrin). The transport of proteins within the cell from the site of synthesis (at the ribosomes within the endoplasmatic reticulum) to various locations inside or outside the cell is performed by such vesicles. The other particle type of organized lipid–protein complex is best illustrated by the well-known lipoproteins in blood, which are usually classified according to size and density. The most important lipid carrier in the blood is low density lipoprotein (LDL), which is a particle 300 Å in diameter. An LDL particle is similar to an emulsion droplet, with a lipid core and proteins at the surface. Egg yolk lipoproteins are also similar, ranging in size from emulsion-like particles visible under the microscope to small protein–lipid complexes similar in size to large proteins.

B. Protein–Lipid Interaction in Wheat

Lipid–protein interaction is an exceptional feature if we consider all possible systems in which lipids and proteins are present. The reason is perhaps obvious from a biological point of view. Self-assembly into the biological tissues of lipids and proteins takes place independently except for the few cases with specific interactions discussed above.

Lipids and proteins in grain interact mainly in membranes, in accordance with the general principles presented above. With milling, the membranes of the endosperm are fragmented. Oil that is squeezed out from the germ is emulsified by proteins or membrane materials when water is mixed into the flour. The formation of a protein coat on an oil droplet can be regarded as a protein–lipid interaction on the colloidal level. Thus in the dough we can observe emulsified oil droplets and also aggregates of the polar lipids self-assembled as described in Chapter 2. The liquid crystalline aggregates that are formed may contain a minor amount of membrane protein molecules, as fragments of the cell membranes can directly form liposome-like aggregates. In general, however, lipids in dough are distributed in phases separated from proteins and carbohydrates.

In the final bread the situation is the same. There are no protein–lipid composite structures of significance. The oil droplets, however, are redistributed, forming a surface film in the pore structure.

Ligolin (Chapter 2, Section III) has been described as a lipoprotein present in wheat [1]. As was discussed earlier, it should not be regarded as a lipoprotein in the strict meaning of the term; it is more logical to assume that the triglyceride and protein fractions are similar in hydrophobic/hydrophilic behavior and therefore join in the separation process. The classic separation of a gliadin fraction from gluten powder by ethanol will also give wheat lipids (mainly polar ones) in the same solution, and applying an argument similar to the one used earlier, gliadin could be regarded as a lipoprotein. We have not said that wheat lipoproteins do not exist, only that they should be critically evaluated based on the separation techniques used.

C. Protein–Lipid Interaction at the Air/Water Interface

A few years ago we reported on the spreading of protein monolayers from wheat flour on the water surface [2]. Well-defined monolayers were observed and were shown to consist of both low molecular weight (LMW) and high molecular weight (HMW) fractions of the storage proteins. This also demonstrated the significance of surface film behavior in breadmaking. The spreading of wheat flour on water has also been observed in the light microscope, and samples have been examined by electron microscopy [3–6]. The spreading was reported to be fast; within 8 ms of contact with water a protein network has formed [6]. It has been shown by amino acid analysis that the material that spread is proteinaceous [6], and also that the protein has an amino acid composition typical of gluten proteins. The spreading protein is observed in the microscope as strands extending from the flour particle. These strands have a thickness in the micrometer range. It has been observed that these strands can form only at the air/water interface and that the driving force for the extension of protein strands is the reduction in surface energy. If surface-active agents are added to the water before the flour particles are spread on its surface, no protein strands are observed [5].

Hamer and coworkers [7] reported observations from surface balance studies indicating an interaction between a gliadin fraction and lipids. The basis of their conclusion on interaction is that the mixture shows a decrease in surface tension compared to each of the components. Diglycerides were observed to reduce the surface tension from 53 to 43 mN/m. These were values of initial surface tension after spreading, and there is one complication in the interpretation of such measurements. The amount of spread corresponds to a close packing of each component within the monolayer, and the nonequalibrium situation obtained is too complex to allow direct interpretation. It is desirable to form each monolayer in the gaseous state and then compress it.

For several years we have examined in our laboratory wheat protein fractions and lipids with regard to molecular interaction using the surface balance technique. As mentioned above, the main complication is the nonreversible character of these protein monolayers, which means that even the manner of mixing during spreading may influence the results. In simple systems of mixed monolayers, it is possible to determine whether or not there is a molecular interaction from the collapse behavior. If component A shows collapse at pressure $\pi(A)$ and component B at $\pi(B)$, the mixture will show both collapse pressures if there is no interaction, whereas it will show a different collapse pressure with an intermediate value in the case of molecular interaction. The wheat storage proteins show no clear collapse behavior, which limits the possibility of interpreting monolayer behavior in terms of molecular interaction with lipids. A phase transition, however, can be used as an interaction indicator in a similar way. The HMW fraction of the storage proteins shows a transition at 20 mN/m, and the fact that we can observe the same transition pressure when polar lipids such as phospholipids are present proves that there is no molecular interaction between such wheat proteins and lipids in mixed films at the air/water interface.

If there is enough time to reach equilibrium, the polar lipid will squeeze out the proteins and wheat lipids. A probable model of the air/water interface in the presence of an excess of wheat lipids and wheat storage proteins is therefore a double layer with lipid molecules forming an upper layer toward air and proteins a layer below. This is an equilibrium situation, which under certain conditions (for example, in the final stage of fermentation) can occur in a dough.

III. PROTEIN–STARCH INTERACTIONS

In Section IV of Chapter 2, the presence of proteins on the wheat starch granule surface was described. It was also stated that the protein might be involved in the surface hydrophobicity or hydrophilicity of the starch granule. The protein/starch interface in the dough is large, and it is surprising that this interface and possible protein–starch interactions have been given so little attention. Protein–starch interactions may be responsible for some of the quality aspects of wheat, and these will be described in the present section.

As discussed in the introduction to this chapter, interactions can occur at different levels, and in the case of protein–starch interactions we can consider two possibilities. One possibility is that protein in solution interacts with amylose and/or amylopectin in solution. There are no indications of such an interaction other than the specific one when amylases attack these compounds.

In baking, however, we should take into consideration the other possible situation–interactions between proteins and the solid starch granule. A protein molecule is considerably smaller than a starch granule, and the situation can therefore be described as protein adsorption on a solid surface.

A. Protein–Starch Interactions in Wheat Grain, Dough, and Bread

1. Endosperm Hardness

A milling characteristic of wheat grain is its endosperm hardness. The hardest wheat is durum which is used for pasta production, whereas the softest wheats are used for cakes and biscuits [8]. In a soft endosperm, breakage occurs through the cell walls, and there is not much adhesion between cell walls and starch or between starch and proteins. In a hard endosperm, breakage occurs either along the cell walls or through the starch granules. Milling of hard endosperm can therefore result in a high proportion of damaged starch.

The endosperm hardness is attributed to protein–starch interactions, and the adhesion between starch and protein is described as stronger in the hard wheats [9]. There seems to be no difference in hardness between different wheats when the individual components (starch and protein) are compared. Recently, it was found that endosperm hardness is related to the presence of a certain protein on the starch granule surface [10]. The protein, which has a molecular weight of about 15,000, is present in very low levels on starch from hard wheats, and durum wheats seem to have none at all. The protein has been named *friabilin* [10]. It has been speculated that it acts as an antistick agent to prevent strong adhesion between protein and starch in the soft wheats. The hardnes of the endosperm has also been related to the presence of a continuous protein matrix [11].

2. Dough Mixing and Baking

During dough mixing the starch granules are oriented parallel to the shear direction in the continuous gluten matrix as illustrated in Fig. 1 [12]. The interactions between starch and the gluten proteins result in phase separation, described further in Chapter 6. Through interpenetrating networks of the aqueous

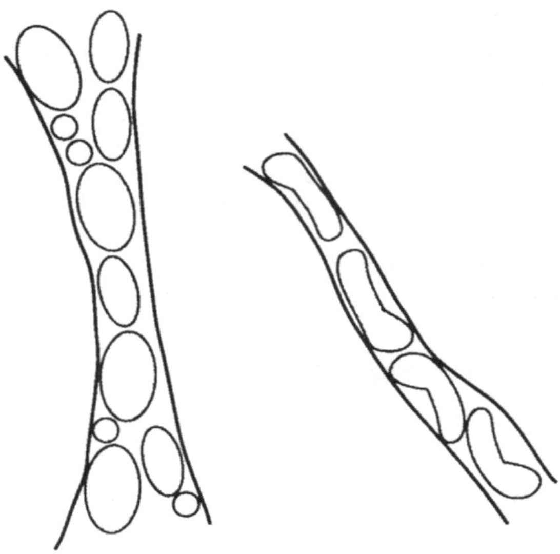

Fig. 1 The orientation of starch granules in the gluten matrix in the dough (left) and the bread (right). (Adapted from [12].)

phases, orientation of the starch–water phase can occur without rupture of protein films. It is, of course, necessary that all the starch granules be uniformly enveloped in the continuous gluten phase; otherwise there will be weak points where the structure can rupture. Uniform distribution of the starch granules in the gluten will be even more critical in the oven during starch gelatinization. It is easy to imagine that the bimodal particle size distribution of the wheat granules is important to achieve close packing in the starch–water phase as described in Fig. 1. Freeze-fracture studies of the ultrastructure of dough and bread have been interpreted to show that there are starch–protein interactions in a fermented dough and that these interactions become stronger during baking. In the bread a tight connection is observed between starch and protein [13]. The starch–protein interactions in the dough are very sensitive to the composition and are affected by the presence of salt, sugar, and shortening.

3. Staling

The deterioration in the quality of bread during storage (staling) is described in Chapter 7. Only the influence of protein–starch interactions on the staling process will be discussed here.

The protein content affects the staling rate, at least when staling is measured as firmness. As protein content increases, the staling rate decreases [14]. Proteins affect the staling process in several ways. They influence the crystallization process of starch directly, they influence the rheological behavior of the continuous phase, and they influence the distribution of water. However, it is difficult to separate one effect from the other. That protein–starch interactions, or at least the presence of proteins, exert an influence on staling is evident from the well-known fact that all gluten-free breads stale very quickly. The influence of protein on staling, measured as firmness, may very well be indirect; a higher protein content could result in a larger loaf volume and thus a softer bread. Recently, protein–amylose interactions were identified as one cause of staling [15].

The retrogradation, or recrystallization, of starch could be affected by the presence of gluten owing to a dilution effect. If the concentration of starch is decreased, there is, of course, a lower concentration of recrystallized starch as well. The recrystallization of starch in the presence of gluten has been studied by differential scanning calorimetry (DSC), and a decrease in the melting enthalpy (ΔH_c) of retrograded starch was observed with increasing levels of gluten [16]. The effect increased with storage time, but there was no difference in the results obtained with various sources of gluten.

This effect of ΔH_c could be explained in the same way as were the effects of pentosans (Chapter 2, Section V)—as related to the distribution of water in the system. The possibility that water migration between starch and gluten is one cause of the staling of bread has been discussed [17]. It is argued that water will migrate from gluten to starch during baking as well as during storage, and a decrease in the water content of gluten could result in the bread feeling "dry." Measurements by DSC of the gelatinization of starch in the presence of gluten can quantify the migration of water from gluten to starch [18].

B. Molecular Interactions Between Amylose or Amylopectin and Protein in Solution

Not much is known about the interactions between proteins and amaylose or amylopectin in solution. When the interactions between amylose or amylopectin and wheat proteins are investigated, the experimental situation is particularly complicated by the poor solubility of the wheat proteins.

Wheat protein–starch interactions have been studied by measuring the absorption of the starch–iodine complex in the presence of protein [19]. The starch investigated was gelatinized wheat starch, and the absorbance was measured at two wavelengths, $\lambda = 650$ nm and $\lambda = 550$ nm, to distinguish between amylose and amylopectin (see Chapter 2, Section IV). When proteins were added into the system, the absorbance decreased at both wavelengths. The decrease was greatest at $\lambda = 650$ nm, which was interpreted as due to a preferential interaction between amylose and protein. pH was found to influence the results, and the absorbance was lowest at pH 6.5. The absorbance increased considerably on the alkaline side, interpreted as showing that the protein–starch interaction occurred to the greatest extent close to neutral pH values. Another explanation might be the interaction between iodide ions and positively charged residues. The pH effects are consistent with such counterion interactions.

If the protein extract was heated before its addition to the starch–iodine solution, there was no change in the absorbance of the starch–iodine complex. Also, this effect of denaturation, which can be expected to "hide" the charged residues, provides arguments against starch–protein interaction. We believe that there is only one way for a molecular interaction to occur between protein and amylose or amylopectin. A side chain along the peptide backbone that is hydrophobic enough, such as leucine, should be expected to form an inclusion complex with an amylose or amylopectin chain, just as lipids or butanol do.

C. Interactions Between Protein and the Starch Granule

1. Adsorption of Protein on the Starch Granule

Proteins are adsorbed from aqueous solutions on all kinds of surfaces. The behavior of proteins at the air/water interface was described in relation to foam stability and gas-holding capacity in Chapter 2. In the present section the protein–starch interaction will be described as protein adsorption at the solid/

water interface. Proteins are amphiphilic molecules, and because of this property they adsorb at interfaces. The surface structure results in a decrease in surface free energy, and this reduction of the interfacial free energy is the driving force behind adsorption [20,21].

We will consider the situation in water (in the seed or at the time of dough mixing). If the surface energy is very high or very low compared to that of water, the protein will be unfolded and irreversibly adsorbed. In the case of hydrophilic surfaces, however, the protein molecule may adsorb in the native state and just orient to adapt to the actual surface forces (cf. Chapter 1). It is thus possible to get reversible adsorption behavior under such conditions. We regard starch as hydrophilic. Whatever the structure of the polyglucan chain at the surface, the glucose unit will render a waterlike character to the overall granule surface.

In any living tissue the surface forces are controlled to avoid any unspecific adsorption at exposed surfaces. This is usually achieved by the membrane structures that normally constitute most interfaces. When an endosperm ripens, however, there is a change from a living tissue to a dry storage of nutrients for the germ. The controlling mechanisms of the interfacial structure during drying will successively be switched off. During this process the protein bodies are forced to fuse, and by drying, a glassy protein continuum is achieved. If we consider this process from a purely colloidal point of view, it is evident that the kinetics of drying will influence the adsorption of proteins on the surface of the starch granules. This can vary on two levels. First, different proteins will compete at the interface, and diffusion properties and amphiphilic character will determine the result. The next level is the degree of reconformation at the interface, where molecular mobility, which is related to water content, is one significant factor. The proteins, which are localized on the starch surface, and their conformation should thus be expected to vary with aqueous conditions during the drying period.

The adsorption of protein on a particulate solid surface can be studied by measuring the protein content before the addition of the solid material and after removal of the particles. Any decrease in protein content is taken as a measure of protein adsorption on the particles [22,23]. An adsorption isotherm can be constructed when the adsorption at different protein concentrations is studied. We have used the same approach to study protein–starch interactions [24]. An adsorption isotherm is constructed from measurements of protein content before and after the addition of the starch granules. Some examples of isotherms are shown in Fig. 2. It was found that the greatest amount of protein adsorbed was obtained with HMW wheat storage proteins; LMW wheat storage proteins were adsorbed to a much lesser extent. Also a non-wheat protein [bovine serum albumin (BSA)] was found to be adsorbed to a very low extent. The adsorption was affected by pH (it decreased when pH was decreased from 5–6 to below 5), by the salt concentration, and by heating of the starch. This latter effect could be

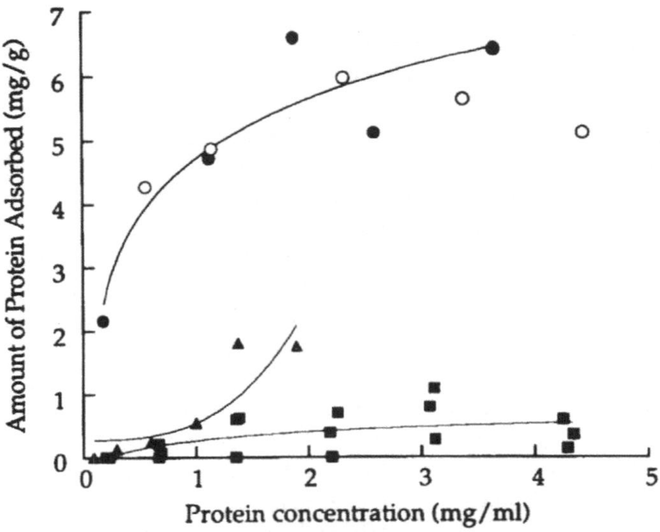

Fig. 2 Adsorption isotherms obtained for the adsorption at wheat starch. (Redrawn from [24].) The following proteins were used: a low molecular fraction (▲), a high molecular fraction (●, ○, two different experiments), bovine serum albumin (■).

due to an increased starch surface area being made available for adsorption. The heating of the protein solutions did not influence the adsorption.

A similar approach was used to evaluate the adsorption of α-amylases onto undamaged starch [25], and it was found that there is a difference between the adsorption of α-amylases onto mature and immature starch granules.

2. Influence of Protein on the Gelatinization of Starch

When gluten is added to wheat starch, the gelatinization temperature of the starch increases (see Fig. 3) [18]. There is also a difference in the DSC parameters obtained for a wheat flour–water mixture and the corresponding starch extracted from the flour; T_o and T_m (onset of gelatinization and temperature at maximum in the gelatinization endotherm, respectively) are lower in the starch than in the flour [26]. The enthalpy, expressed on a dry matter content, is, of course, higher in the starch because of the higher starch content. There is a relation between the gelatinization temperature measured by DSC (T_o and T_m) of a wheat flour and the protein content, so that T_o and T_m increase with the protein content [27]. The influence of proteins other than gluten proteins on the DSC parameters is not easily determined, owing to the possibility of overlapping between denaturation and gelatinization endotherms [28]. Gluten is especially suitable for this type of

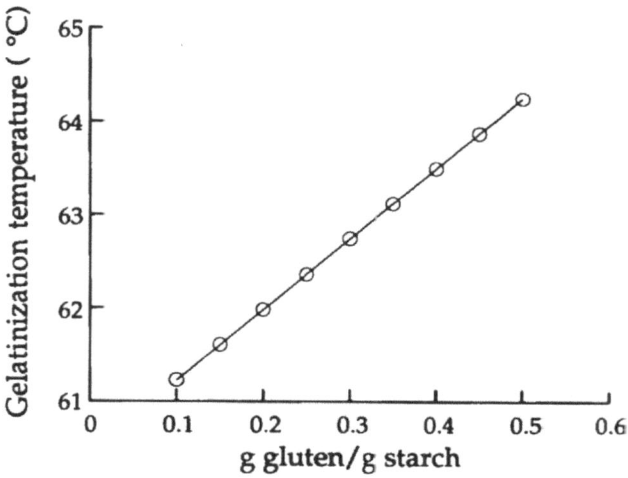

Fig. 3 Influence of gluten on the gelatinization temperature of wheat starch at a water starch ratio of 0.9:1. (Calculated from [18].)

study because it lacks denaturation endotherms (or at least they are very small compared with the starch gelatinization endotherm [29]).

3. Influence of Protein on the Rheological Behavior of Starch

The rheological behavior of starch can be changed by the addition of proteins. When gluten is added to wheat starch, an increase in the storage modulus (G') in small-amplitude oscillatory measurements has been noted [30]. The introduction of starch granules into gluten affects the rheological behavior. Increased modulus values are obtained; that is, the starch granules act as fillers and reinforce the gluten gel [31]. The phase angle (δ) depends strongly on the protein/starch ratio. When this ratio decreases, δ also decreases. Another consequence of the introduction of starch granules into gluten is that the linear viscoelastic behavior of gluten is changed to a nonlinear behavior [32].

IV. STARCH–LIPID INTERACTIONS

Starch and lipids are both present in flour, and the dominant interaction between the components gives the well-known amylose–lipid complex. This type of starch–lipid interaction has received the most attention, although, as will be described later, it is not the only type of starch–lipid interaction. The composition of starch lipids was given in Chapter 2, Section IV. About 80% of the lipids in wheat starch are phospholipids, with lysolecithin the main constituent [33].

High-amylose varieties tend to be high in lipids also, but there is no relation between amylose and lipid content in wheat starch [34].

Although the amount of lipids present in native wheat starch is usually only about 1%, the lipids affect the behavior of starch tremendously. The importance of polar lipids is clearly demonstrated by comparing wheat starch with potato starch, a starch that can be regarded as lipid-free. When polar lipids are added to potato starch, it changes into a starch that is rather similar to wheat starch in behavior [35]. Delipidization of wheat starch increases its viscosity, and the increase occurs at a lower temperature; that is, the delipidized wheat starch becomes more similar to potato starch [36].

In this chapter examples are given of how polar lipids as well as fats and oil affect the behavior of starch. The effects of polar lipids on the behavior of starch are described separately from those of oils and fats, because the mechanisms behind these effects are different. The effects of polar lipids on starch are usually attributed to the formation of a complex between the polar lipid and amylose. First this amylose–lipid complex is described, and then the results demonstrating the existence of an amylopectin–lipid complex are presented. The presence of polar lipids certainly affects the behavior of starch, but the phase behavior of the lipids can also be affected by complexing with starch.

A. Effects of Polar Lipids on the Properties of Starch

1. Gelatinization Behavior

It is possible to study the effect of added monoacyl lipids (MAL) on the gelatinization temperature of starch by DSC. However, there may be one obstacle, and that exists when the polar lipid added has a chain melting temperature in about the same temperature interval as the onset of gelatiniztion [37,38]. Such a case is illustrated in Fig. 4. If the melting temperature of the fatty acid chain is far below the gelatinization temperature, the endotherms are well separated, and there are no problems in determining either T_o and T_m or ΔH. For MAL with unsaturated fatty acid chains, the melting endotherm will not be detected by DSC as the span probably starts at a temperature above this temperature. For long (C_{16}–C_{20}) and saturated fatty acids, the chain melting temperature coincides with T_o of the starch, and the endotherms overlap, making any separate calculations impossible. If the added polar lipid has completely complexed with the starch, there is, of course, no such problem.

The effect of added MAL on T_o and T_m of wheat starch is given in Table 1. It is evident that the effect is rather small. Often a delay in T_o is observed, but at least one surfactant [sodium dodecyl sulfate (SDS)] shifts the gelatinization to a lower temperature. The effects are about the same at a high water content (water/starch ratio 3:1) as at a low water content (water/starch ratio 1:1) [26,38]. The gelatinization enthalpy may decrease owing to the addition of MAL.

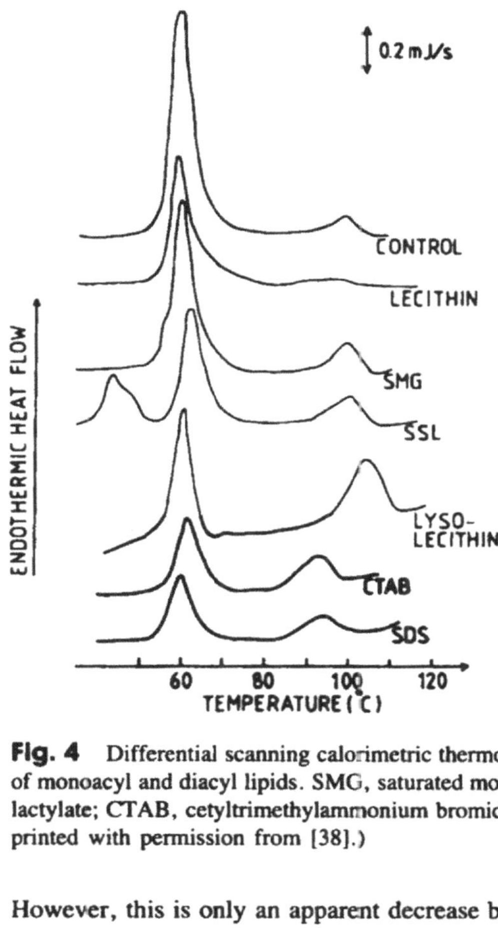

Fig. 4 Differential scanning calorimetric thermograms of wheat starch in the presence of monoacyl and diacyl lipids. SMG, saturated monoglycerides; SSL, sodium stearoyl-2-lactylate; CTAB, cetyltrimethylammonium bromide; SDS, sodium dodecyl sulfate. (Reprinted with permission from [38].)

However, this is only an apparent decrease because of the exothermic complex formation between amylose and the added MAL.

Monoacyl lipids present during gelatinization of starch affect the swelling power, the solubility, and the size and shape of the starch granules [38,40]. Most MAL cause a decrease or delay in swelling and solubility [39,41]. The sediment volume of wheat starch heated to 70°C is lower when monoglycerides are present. However, if the monoglycerides are added after the heating they do not affect the sediment volume [42]. Monoglycerides decrease the solubility more than, for example, sodium stearoyl-2-lactylate (SSL) [39,41]. The swelling power is also lower in the presence of monoglycerides than with SSL. The morphological changes that occur during gelatinization of wheat starch are the same in the presence of MAL as without such additives [39,43]. However,

Table 1 Gelatinization Temperature of Starch in the Presence of Monoacyl and Diacyl Lipids[a]

Lipid additive[b,c]	T_o (°C)	T_m (°C)
Control	57.0±0.2	61.3±0.5
SDS	54.7±0.2	60.1±0.5
CTAB	57.6±0.2	61.7±0.2
SSL	58.4±0.4	62.3±0.5
SMG[d]	56.7±0.5	60.5±1.0
Lysolecithin	55.7±0.2	60.6±0.3
Lecithin	56.7±0.4	60.8±0.2

[a] T_o, temperature of onset of gelatinization; T_m, temperature at maximum in endotherm.
[b] 0.05 g lipid/g starch and 3 g water/g starch.
[c] SDS, sodium dodecyl sulfate; CTAB, cetyltrimethylammonium bromide; SSL, sodium stearoyl-2-lactylate; SMG, saturated monoglycerides.
[d] The melting endotherm of the saturated monoglycerides overlaps the gelatinization endotherm.
Source: Data compiled from [38].

the changes in the size and shape of the granules occur at a somewhat higher temperature in the presence of added lipids [39]. Increased swelling is observed in the presence of SDS [39,44].

After the gelatinization temperature interval has been passed, the V pattern can be recorded by the X-ray diffraction technique [45,46]. This demonstrates that the complex exists as crystalline domains after gelatinization.

2. Rheological Behavior

The rheological behavior of a starch gel is drastically changed by the addition of MAL. The effects of several emulsifiers [distilled monoglycerides (DMG), calcium stearoyl-2-lactylate (CSL), SSL, and diacetyl tartaric acid ester of monoglycerides (DATEM)] have been investigated by the use of the Brabender amylo/viscograf [47], by viscosity measurements [48], and by small-amplitude shear oscillatory measurements [49]. Some results obtained for wheat starch are shown in Fig. 5. Also, surfactants like SDS and cetyltrimethylammonium bromide (CTAB) affect the rheological behavior of starch gels. As described above, defatting of wheat starch causes an increase in viscosity values at lower temperatures [36]. Adding back wheat lipids brings back the behavior of the native starch.

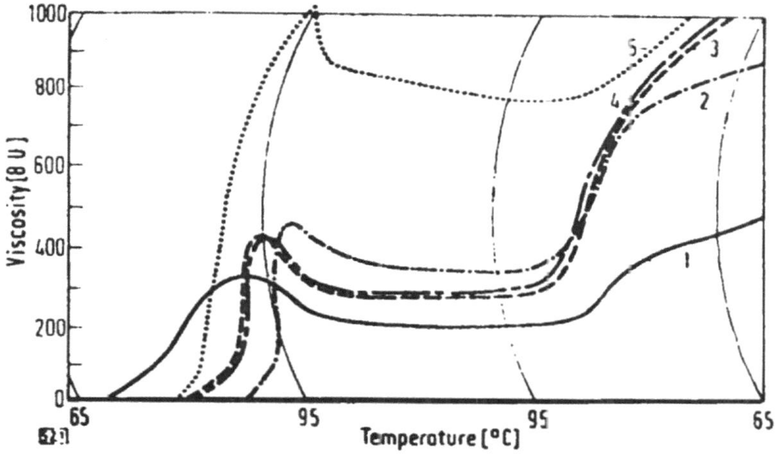

Fig. 5 Viscogram obtained for wheat starch in the presence of monoacyl lipids. (1) control; (2) 0.5% distilled monoglycerides; (3) 0.5% calcium stearoly-2-lactylate; (4) 0.5% sodium stearoyl-2-lactylate; (5) diacetyl tartaric acid esters of monoglycerides. (Reprinted with permission from [47].)

The effects obtained might be understood from the model of a starch gel as a composite material, as given in Chapter 2, Section IV. If, for example, the swelling of the starch granules is delayed by a MAL, then of course the pasting temperature will also be delayed. Owing to an increase in the stiffness of the starch granules, the peak viscosity is increased. But on the other hand, because of a decrease in the amount of material leached from the granules in the presence of MAL, the viscosity decreases. The influence of an added MAL is therefore not easily predicted.

3. Retrogradation

The addition of MAL to starch causes the degree of retrogradation to decrease. An example of results obtained by DSC is given in Fig. 6. The effect of an added emulsifier is possible to observe both in starch systems [16] and in bread [50]. All MALs do not show this effect to the same extent, the long, saturated monoglycerides being the most effective. This result has been attributed to these lipids having a better complex-forming ability because they consist of a straight chain. However, this need not be the only explanation. Instead, the result can be related to the phase behavior of the MAL [51]. Unsaturated monoglycerides form cubic phases under the actual conditions. Such a phase is an equilibrium with a very low level of monomers ($\sim 10^{-6}$ mol/L) [52], and furthermore this phase interacts very poorly with water. The cubic phase is impossible to disperse in an

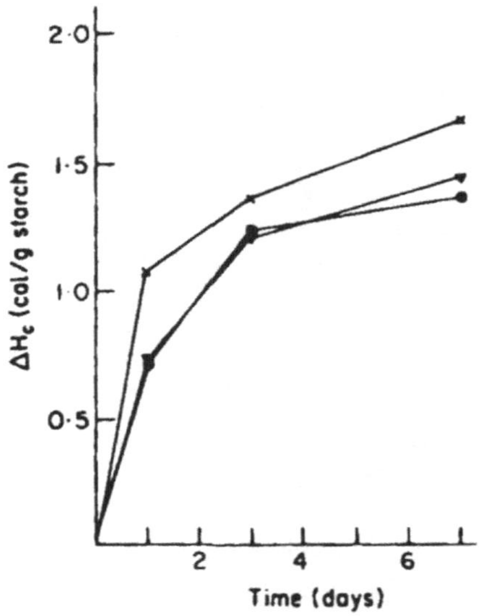

Fig. 6 Retrogradation of wheat starch–gluten gels (0.90 g water/g starch and 0.20 g gluten/g starch) at 21°C in the presence of sodium stearoyl-2-lactylate (SSL). The gels contained no SSL (X); 0.006 g SSL/g starch (●); 0.023 g SSL/g starch (▼). (Reprinted with permission from [16].)

aqueous system. The opposite is true for the lamellar liquid phase, which can be formed by the long saturated monoglycerides under normal conditions.

When MAL is added it is possible to measure an increase in the DSC endotherm of the amylose–lipid complex [50]. This endotherm will be discussed later in this chapter. However, it should further be noted that the retrogradation of starch is diminished by the addition of polar lipids in the case of waxy starches also [53,54].

B. Effects of Oil and Fat on the Properties of Starch

1. Gelatinization Behavior

If starch is heated in oil, no phase transition is observed by DSC [55]. In the presence of both oil and water the gelatinization endotherm is recorded. The gelatinization onset temperatures in the presence of different fats and oils measured by DSC are presented in Table 2. The starch/water/fat ratio differs among the samples, but nevertheless the results show that the added oil or fat does not affect T_o very much. In the low water–high fat system T_o was in the range

Table 2 The Gelatinization Onset Temperature (T_o) of Wheat Starch in the Presence of Fat and Oil

Type of oil	T_o^a (°C)	T_o^b (°C)
Control	56.4±0.7	51.5
Safflower	57.1±0.1	—
Hydrogenated coconut	56.9±1.1	—
Coconut	57.3±0.3	51.3
Cottonseed	57.0±1.7	—
Soybean	56.9±0.1	50.2
Corn	56.7±1.1	—
Butter oil	—	50.3

[a]Starch/water/oil = 1:2:0.66. Data compiled from [56].
[b]Starch/water/oil = 1:1:1 (unpublished results).

50.2–51.5°C, and in the high water–low fat system T_o was in the range 56.7–57.3°C [56]. A small effect on the gelatinization enthalpy was recorded; for example, the ΔH of soybean and corn oils decreased from 7.2 to 6.6 J/g. However, these changes are certainly due to impurities of MAL in the oil. It is hard to get rid of all free fatty acids and partial glycerides during oil refinement, and some FFA will always be present. For a system composed of flour, water, 2% NaCl, and 6% sucrose, a T_o value of 66°C was observed, and when 3% shortening was added T_o increased to 67°C [57].

The morphological appearance of starch granules heated with and without rapeseed oil were quite similar [58]. Other oils, especially corn oil, depress the swelling of starch granules [56]. The differences, although slight, in influence on starch gelatinization seems to be related to the fatty acid composition of the oil; the more saturated fatty acids present, the greater the effect on starch gelatinization [56].

2. Rheological Behavior

Although the gelatinization process, as measured by DSC, is not greatly affected by the presence of fats and oils, the rheological behavior of the starch might nevertheless be different when oil is present in the continuous phase.

For corn starch, a decrease in pasting temperature has been observed (with the Brabender amylograph) when increasing amounts of soybean oil are added [59].

Also the temperature at which peak viscosity was recorded decreased in the presence of soybean oil. Without oil the temperature at maximum viscosity was 92°C, and with 9% oil (calculated as percent by weight of a 6% starch paste) the temperature was 82°C. Further increases in the level of oil did not influence the peak viscosity temperature. No relation to degree of unsaturation or source of fat was found in this investigation [59].

Small-amplitude oscillatory measurements on maize starch gels with added soybean oil [2% (w/w)] showed no significant effect on G' or δ [60], whereas a similar addition to wheat starch caused a slight increase in G' [30].

3. Retrogradation

When soybean oil is added to waxy maize starch, the retrogradation is not significantly affected [53]. When different oils were mixed with wheat flour and water prior to gelatinization, a small decrease in the extent of retrogradation (using DSC) was observed [61]. It is often claimed that fats and oils increase the keeping qualities of bread. This effect must therefore be related to mechanisms other than a direct influence on the retrogradation of starch. The oil will tend to cover the pore system in bread to reduce the surface energy, and this should be expected to affect the rheological properties of the crumb. The influence of oils and fats will be further discussed in Chapter 7.

4. Relation to the Oil/Water Interface

Starch granules exhibit the remarkable property of being wettable by water as well as by oil. When there is a possibility of choice, however, starch will go to the water phase. The fact that starch can be dispersed in fats or oils without aggregation is a most important property in the production of fat-rich baked products. Thus not only can the protein of wheat flour particles adapt to a hydrophobic interface, but flour particles consisting of starch granules or fragments of granules can also. In a water-in-oil emulsion, however, there is a strong tendency of starch granules to fuse with the water droplets.

In a study of the surface of wheat starch granules we found lipids and proteins in amounts of 0.2–0.9 mg/g and 1.5–4.7 mg/g, respectively [62]. Our interpretation was that a biomembrane remains on the surface of the starch granule even after water washing and mechanical mixing of the dough. The lipids were of the polar type corresponding to membrane lipids, and the proteins consisted mainly of gliadin.

From the oil/water wetting property one might expect that starch granules could work as emulsifiers. This is, however, not the case. They contribute somewhat to emulsion stability, which is due only to increased viscosity of the water phase. Efforts in our laboratory to stabilize oil droplets in water, assuming that amylopectin side chains could form an inclusion complex with one acyl chain of the triglyceride molecule, have been unsuccessful. Why is it that the starch granules do not work as particle types of emulsifiers when they are able to expose a hydrophobic surface toward oil as well as a hydrophilic surface toward the water phase? Similar crystals of polar lipids, which can form one hydrophilic and one hydrophobic side, are known to work as very effective emulsifiers. The explanation is probably that the starch granules are too large compared to the oil (or water) droplets, so large that the density difference becomes significant.

The effect of chlorination of starch on the surface character in relation to oil and water has been studied by Seguchi [63]. It was clearly demonstrated that chlorination increases the hydrophobic surface character as shown by oil affinity. The effect was not lost when the granules were treated with proteolytic enzymes or washed with lipophilic agents. It was lost, however, by treatment with amylase or surfactant. This indicates that it is the surface polyglucan structure that has been changed, but the formation of a surfactant complex can reverse this change.

One possible explanation is a complexing reaction between chlorine and surface groups of amylose and/or amylopectin. Chlorine has hydrophobic character similar to that of iodine, even if it is not as lipophilic, and the van der Waals radius is also close enough to that of iodine. A chlorine complex should thus possess the V-amylose conformation. In such a complex a surfactant will tend to replace the chlorine as in the case of iodine complexes. The new surface structure obtained will probably expose the polar heads of the surfactant outwards. Thus a reversal to a more hydrophilic surface structure is obtained. These studies do, in fact, also indicate that the surface structure of starch granules is rich in the amylose–lipid complex. If we start with such a structure, exposure to excess chlorine will result in the replacement of lipids by chlorine, a shift in the reaction equilibrium related to the excess chlorine. Then the chlorine will disappear from the surface complex when the flour is exposed to air, and a new granule surface structure will be obtained, which will be more disordered, also exposing hydrophobic regions outwards.

C. The Amylose–Lipid Complex

The effect of adding monoacyl lipids on the behavior of starch is usually attributed to the formation of a complex between amylose and MAL. Some of the results described above are easily explained in this way (e.g., the decrease in solubility), whereas others are not (e.g., the decrease in the retrogradation tendency). Other results, for example, the rheological behavior, do not directly depend on the presence of an amylose–lipid complex but are due to the consequences of the complex formation, such as a decreased degree of swelling and a smaller amount of soluble material.

1. Structure

The amylose–lipid complex is illustrated in Fig. 7. The amylose molecule forms a single helix, and the ligand occupies the hydrophobic helical cavity. For a polar

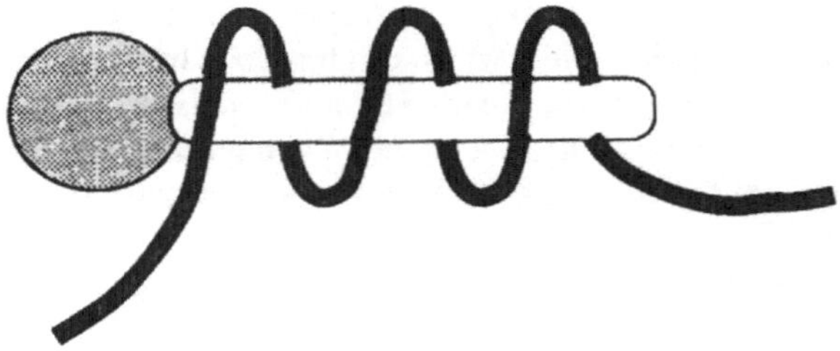

Fig. 7 The amylose–lipid complex. The amylose forms three turns of the helix around the nonpolar part of a monoacyl lipid. The polar head of the lipid is located outside the helical cavity.

lipid such as a monoglyceride, each turn of the helix consists of six anhydroglucose residues, and on the average there are two to three turns for each monoglyceride molecule. On average, about two-thirds of the hydrocarbon chain is involved in the complex [64]. By Raman spectroscopy it has been shown that the hydrocarbon chain in the complex is ordered similarly to the crystalline state [64]. Each amylose molecule may contain several helices with ligands. When the ligand is more bulky than a fatty acid chain, the turn in the helix can consist of seven or eight glucose residues [65,66]. The diameter of the interior of the cavity is 4.8 Å with six glucose 6.3 Å with seven glucose, and 7.8 Å with eight glucose units [65]. The list of ligands that complex with amylose is long and includes fatty acids, monoglycerides, surfactants such as SDS and CTAB, lecithin, alcohols, carboxylic acids, and even cyclic compounds such as cyclohexanol [37,38,44,48,67–73].

Of all the possible complexes, the one with iodine is best known as it has been subjected to detailed X-ray structural analysis. This complex has a deep blue color, which allows it to be used to identify starch. The iodine complex is also essential in several methods for determination of the amylose content in starch (e.g., the Blue Value method [74]). It should, however, be borne in mind that iodine and polar lipids occupy the same sites in the starch granule, and if cereal starches are not delipidized prior to analysis their amylose content will be underestimated [75].

The amylose—lipid complex can be obtained in a crystalline form, and it will then give rise to an X-ray diffraction pattern of the V type [68,76,77]. The complex is not formed only by MAL; lecithin also gives the V pattern [68]. Some typical *d*-spacings are given in Table 3. The unit cell of the complex is orthorombic with the dimensions $a = 13.0$ Å, $b = 23.0$ Å, and $c = 8.05$ Å [67]. The packing of the helices is the same; the ligand is either a fatty acyl or iodine, although some variations in *d*-spacing due to the ligand in the complex have been reported [68]. Moreover, the *d*-spacings observed were also related to the

Table 3 *d* Spacings of the V-Amylose

V, Hydrate (linear alcohol)		V, anhydrous (linear alcohol)	
d Spacing (Å)	Intensity[a]	*d* Spacing (Å)	Intensity[a]
12.0–11.85	m/s	11.3–11.24	m/s
7.56	w	6.61–6.52–6.50	s
6.75–6.83	s	5.62–5.61	w/m
5.80–5.94	w	5.06–5.08	w/m
5.16–5.21	w/m	4.60–4.62	w/m
4.80–4.77	w/m	4.25–4.26	s
4.42–4.48	s	3.77–3.74	w
3.93	w	3.38–3.41	w
3.52–3.53	w/m	2.90	w
3.24–3.26	w		

[a]s, strong; m, medium; w, weak.
Source: Data compiled from [76] and [77].

treatment of the complex—to whether it was extracted with carbon tetrachloride or methanol. The amylose chains in the crystals are folded (like polyethylene) and form a lamellar structure [78].

The circumstances that native starch does not show the V pattern (except in a few special cases [46]) has been taken as evidence that the amylose complex does not exist in the native starch but is formed during the gelatinization of starch. However, the absence of the V pattern does not prove the absence of the amylose–lipid complex, it merely proves that the complex is not arranged in crystalline domains that we can detect by the X-ray deffraction technique.

2. Formation of the Complex

As discussed in the previous subsection, the existence of an amylose–lipid complex in the native starch granule has not been conclusively proved. However, the formation of the amylose–lipid complex during the heating of starch is beyong doubt. The V pattern is recorded after gelatinization, and DSC measurements indicate the presence of the amylose–lipid complex after heating to above the gelatinization temperature.

To achieve the best conditions for the formation of a complex between amylose and MAL, both should be in solution. If a solution of MAL is added to a solution of amylose, the complex forms as a white precipitate, which can be collected by centrifugation for further studies. The presence of uncomplexed lipid can be checked by DSC if the chain melting temperature is suitable (see Fig. 4). It has been noted that the existence of uncomplexed lipid is common when dimethylsulfoxide (DMSO) is used as solvent, perhaps because the MAL and DMSO compete for the same site in the helical cavity [79].

The temperature during complex formation and subsequent crystallization affects the properties of the complex. At low crystallization temperatures (i.e., a high degree of supercooling), a type of complex with a low melting temperature is formed (~96°C in the case of monopalmitin). If the crystallization temperature is increased, a second type of the complex is formed that melts at a somewhat higher temperature (~113–114°C in the case of monopalmitin [80]).

If both components are not in solution, the complex might not form immediately or to the same extent. When amylose is present as starch, a second heating is usually necessary to achieve a more complete complex formation [81].

Complex-forming ability differs among the monoacylglycerides. The best complex-forming ability—measured, for example, as a decrease in iodine binding—is shown by the saturated monoglycerides [68,69,82]; see Table 4. Similar

Table 4 Complex-Forming Ability of Some Lipids and Surfactants

	Complex-forming ability			
	Ref. 82		Ref. 68	
Lipid/surfactant	Mg additive per 100 mg amylose	Complex-forming index	Mg additive per 100 mg amylose	Complex-forming index
Distilled monoglycerides				
Lard, hydrogenated	2.0	18		
	5.0	38		
	10.1	64		
Lard, unhydrogenated	20.0	40		
Monostearins			54	89
Sucrose monostearate	5.0	26	47	99
SDS			43	90
Lecithin	5.0	16	56	86
Sodium stearoyl-2-lactylate	5.0	79		
Sorbitan tristearate	5.0	13		
Sorbitan tetrastearate			50	73

The complex-forming index is calculated from the decrease in iodine binding due to the presence of additive.

results have been obtained in equilibrium dialysis [83]. Monoacyl lipids with a large polar group, or diacyl lipids such as lecithin, were shown to have poor amylose-complexing ability [68, 82]. To form a precipitate, a higher amount is required for a diglyceride than for a monoglyceride [68]. An oil such as soybean oil does not form any precipitate with amylose. The difference in complex-forming ability between saturated and unsaturated monoglycerides can be attributed to the different shapes of the molecules. The straight saturated hydrocarbon chain could be expected to fit more easily into the helical amylose cavity. Another reason for differences in complex-forming ability may be the phase behavior of the lipid.

When the complex-forming abilities of monopalmitin (GMP), monoelaidin (GME), monoolein (GMO), and monolinolein (GML) were compared, GME was found to be the best when the monoglycerides were added as aqueous dispersions at 30°C [51]; see Table 5. At these conditions GMP forms β-crystals, the cis-unsaturated monoglycerides form a cubic phase, and GME is in the border region between the β-crystalline form and the lamellar phase. If the monoglycerides were all added as liposomes (liposomes prepared with the aid of sodium cholate), their complexing ability increased. The effect was not as great for GMP and GME as for GMO and GML because the former have solid chains at 30°C.

Table 5 Influence of Monoglyceride Phase Behavior on Complex Formation with Amylose

Monoglyceride	Temp. (°C)	Phase behavior of added monoglyceride	Amylose complex index	Amylose complex index when monoglyceride is added as sodium cholate liposomes
Monopalmitin	30	β-Crystalline	33	40
	60	Lamellar	99	98
Monoelaidin	30	β-Crystalline/ lamellar	71	81
	60	Lamellar/cubic	82	99
Monoolein	30	Cubic	41	98
	60	Cubic	9	99
Monolinolein	30	Cubic	16	93
	60	Cubic	0	98

Source: Data compiled from [51].

The experiment was repeated at 60°C, and when the monoglycerides were added as aqueous dispersions GMP was the most effective complexing agent. At this temperature GMP forms the lamellar phase. The monoelaidin was somewhat less effective (see Table 5); at this temperature GME is in the border region between the lamellar phase and the cubic phase. The cis-unsaturated monoglycerides were again very poor amylose-complexing agents, and they were present as a cubic phase. When the monoglycerides were added as liposomes at 60°C, they were all very effective.

When monostearin (GMS) was added in the crystalline form, no complex formation was detected [84]. This is certainly related to the low monomer concentration and to the low solubility, which in turn is related to the size of the crystals from which the transport of molecules to the solution takes place.

It is possible to extract lipids from the amylose–lipid complex at ambient temperature with a range of solvents [85]. The best result has been obtained with methanol as a solvent. The yield of lipids was increased when the extraction was performed at 100°C. This could be due to a better extractibility of the lipid when the complex has dissociated at the elevated temperature.

3. Thermal Behavior of the Complex

The use of DSC to study the presence of uncomplexed lipids in the amylose–lipid complex has already been discussed. The amylose–lipid complex itself also shows a thermal transition. The temperature interval of this transition is located above the temperature range of gelatinization (Fig. 8). The exact temperature range depends on the water content as well as on the MAL in the complex. Native wheat starch shows the transition at about 100°C in excess water [38]. The lipid compound in this complex is almost exclusively lysolecithin. The enthalpy of the endotherm is about 1.3 J/g dry starch at a water/starch ratio of 3:1. If MAL is added, this value can be increased, but at a certain amount of addition the enthalpy will level off (Fig. 9). Presumably, the amylose in the wheat starch is then saturated with lipids. The enthalpy of the transition of the amylose–lipid complex (ΔH_{cx}) levels off at a surfactant/starch ratio of just below 0.1 in the case of SDS and CTAB [48].

The ΔH of the gelatinization, also shown in Fig. 9, decreases with increasing addition of lipids. However, if the enthalpy of gelatinization (ΔH) and the enthalpy of the transition due to the amylose–lipid complex (ΔH_{cx}) are summed, the sum is only slightly dependent on the amount of lipid added. It might then be concluded that the complex formation occurs during the gelatinization and that the process is exothermic. The sum of the enthalpies is somewhat lower than ΔH without added lipids (Fig. 9). Evans [48] observed in waxy maize starch a small difference between ΔH in the absence of added lipids and ΔH in the presence of added MAL, the latter being lower. These results could be interpreted as showing that a complex also forms between added MAL and amylopectin.

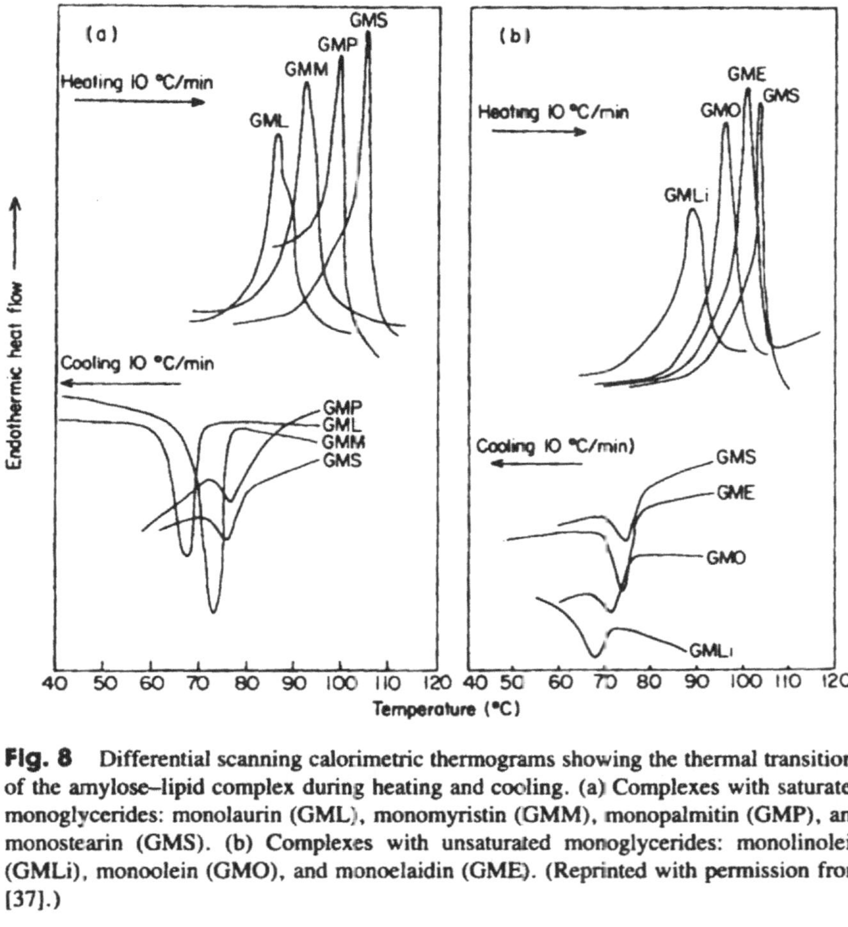

Fig. 8 Differential scanning calorimetric thermograms showing the thermal transitions of the amylose–lipid complex during heating and cooling. (a) Complexes with saturated monoglycerides: monolaurin (GML), monomyristin (GMM), monopalmitin (GMP), and monostearin (GMS). (b) Complexes with unsaturated monoglycerides: monolinolein (GMLi), monoolein (GMO), and monoelaidin (GME). (Reprinted with permission from [37].)

ΔH_{cx} is, of course, proportional to the amount of amylose–lipid complex present and consequently to the amount of amylose present. An attempt has been made to use ΔH_{cx} to determine the amylose content of starch, but without complete success [86]. For some starches (e.g., potato, amylomaize, and waxy maize starches) the amount of amylose calculated from DSC measurements corresponded to the amount determined from iodine binding, whereas for other starches (wheat and maize) the amount of amylose determined by DSC was higher than the amount determined from iodine binding.

The thermal transition of the amylose–lipid complex is reversible, and on cooling an exothermic transition occurs at a temperature about 20°C below the

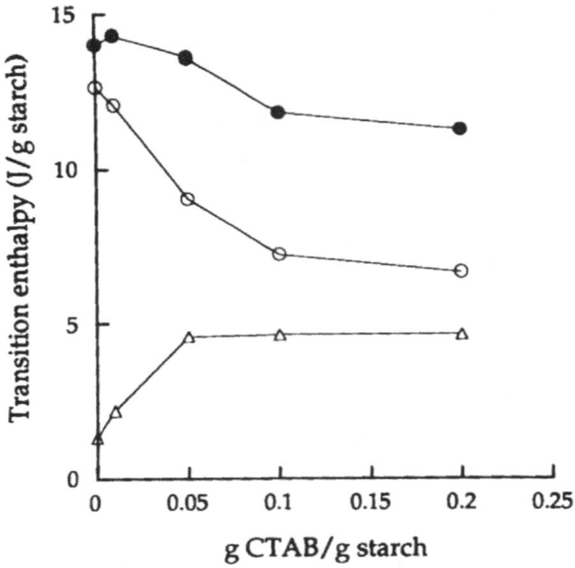

Fig. 9 Transition enthalpy of the gelatinization of wheat starch in the presence of CTAB (○) and of the amylose–lipid complex formed of wheat starch in the presence of CTAB (△). The sum of these two enthalpies is also shown (●). The water/starch ratio is 3:1. (Data calculated from [38].)

endothermic transition. A rescan usually results in an endotherm of a different appearance (Fig. 10). Moreover, the transition enthalpy is usually also changed. If flour is heated, ΔH_{cx} is usually increased during the second scan. This could be due to better possibilities for complex formation during a first heating and subsequent cooling.

The transition temperature (T_{cx}) depends on the structure of the lipid in the complex. The influence of chain length, degree of unsaturation, and charge is shown in Table 6. For complexes between amylose and a series of monoglycerides, T_{cx} increases with increasing chain length, T_{cx} being 85.1°C for monolaurin and 103.5°C for monostearin. The corresponding ΔH_{cx} values did not differ very much, 29.3± 1.3 J/g dry complex for GML compared with 33.5± 3.4 J/g dry complex for GMS. These values are obtained with excess water, the water content obtained in the precipitated complex after centrifugation. The exothermic transition temperatures were lower, 67.7 and 75.1°C, respectively, for the two complexes described above [37]. In the same study the influence of unsaturation was investigated for a series of C_{18} monoglycerides.

As cited above, the complex with the saturated monoglyceride had a T_{cx} value

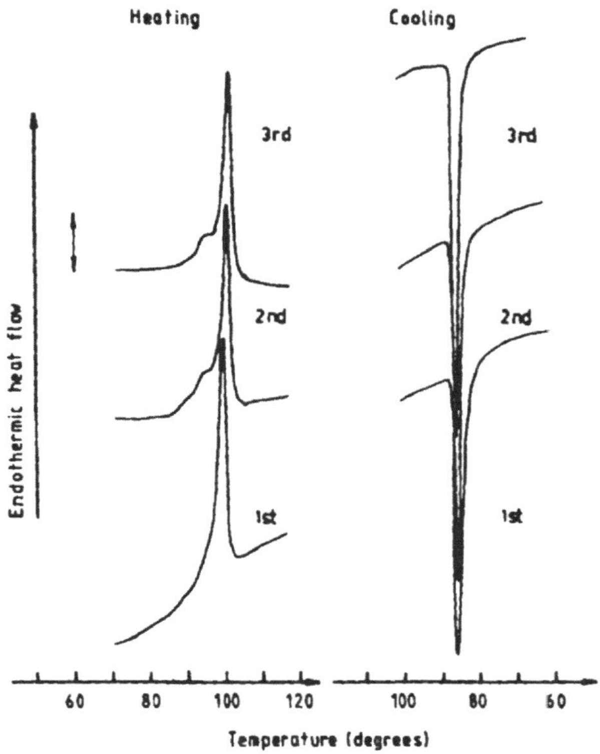

Fig. 10 Differential scanning calorimetric thermograms obtained during repeated heating and cooling of the amylose–CTAB complex in excess water. (Reprinted with permission from [73].)

of 103.5°C. The introduction of *cis* unsaturation (GMO) caused a lowering of T_{cx} to 97.0°C. With a *trans* unsaturation the decrease was smaller; T_{cx} for the complex with GME was 100.8°C. The introduction of another *cis* double bond in the ligand (i.e., GML) resulted in a T_{cx} value of 90.3°C for the complex (Table 6). The trend is the same for amylose–fatty acid complexes, for which T_{cx} increases with increasing chain length and decreases with increasing degree of unsaturation [87]. However, ΔH_{cx} is strongly related to the fatty acid, with ΔH_{cx} = 20 J/g dry matter for palmitin and ΔH_{cx} = 7.5 J/g dry matter for linolein [87]. For complexes with fatty acids, ΔH_{cx} increased with the fatty acid/amylose molar ratio until a saturation ratio was reached. ΔH_{cx} was then 21.0 ± 1.8 J/g amylose for decanoic acid and 33.1±1.2 J/g amylose for arachidic acid [79].

The influence of water content on T_{cx} is shown in Fig. 11. When the water

Table 6 Transition Temperature (T_{CX}) of Some Amylose–Lipid Complexes

Complexing ligand	Ligand/ amylose ratio[a]	T_{CX} (°C)	Ref.
Monolaurin	1:10	85.1±0.4	37
Monomyristin	1:5	91.3±0.8	80
		112.5±1.1	80
Monomyristin	1:10	90.2±0.3	37
Monopalmitin	1:10	98.5±0.3	37
Monopalmitin	1:5	95.0±0.8	80
		112.9±1.0	80
Monostearin	1:10	103.5±0.9	37
Monostearin	1:5	100.1±1.2	80
Monoelaidin	1:10	100.8±0.2	37
Monoolein	1:10	97.0±0.4	37
Monolinolein	1:10	90.3±0.9	37
Lysolecithin	1:5	102.2±0.9	80
Decanoic acid	0.079	87.7±0.6	79
	0.159	108.0±0.3	79
		87.3±0.6	79
Lauric acid	0.040	108.2±0.6	79
		94.1±0.2	79
	0.106	114.0±0.8	79
		96.0±0.7	79
Myristic acid	0.036	114.6±1.2	79
	0.096	93.7±1.0	79
Palmitic acid	0.019	93.0±1.0	79
	0.062	94.1±0.7	79
Stearic acid	0.012	94.4±0.7	79
	0.030	97.3±0.9	79
Arachidic acid	0.020	98.3±0.6	79
	0.039	101.2±0.2	79
		101.0±0.4	79

[a]Weight ratio except for data from [79], where molar ratio is given.

content is decreased, the thermal behavior of the complex is more complicated, and multiple endotherms are observed [79,89]. The thermal transition of the complex, and the resulting enthalpy, probably do involve several thermal events. There is certainly a melting of the crystalline structure, preceded by a glass transition. The thermal transition may also involve the dissociation of the

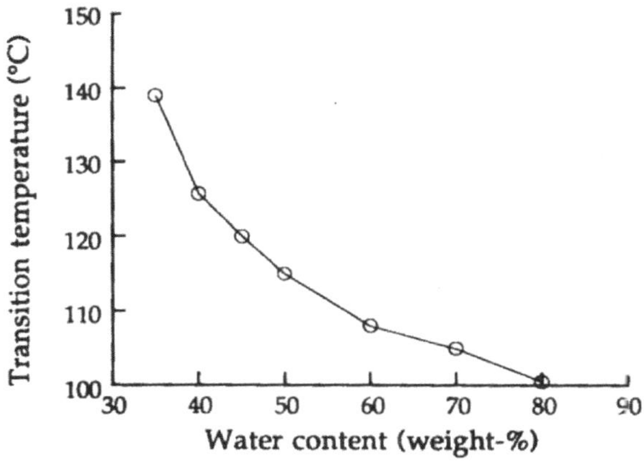

Fig. 11 The influence of water content on the transition temperature of the amylose–lipid complex in wheat starch. (From [88].)

complex. Multiple endotherms are observed in many cases These may be due to the formation of different forms of the complex during the crystallization procedure [80]. The existence of multiple endotherms is also related to the ligand. Lysolecithin does not give two endotherms under the same conditions as GMP [80]. The formation of both a high-temperature and a low-temperature complex is observed only for nonionic ligands [72]. When the water content in the complex is decreased to 50%, even exothermic transitions can be observed [89]. This behavior indicates a partial melting followed by recrystallization and final melting. Such behavior has been observed in complexes with monoglycerides, but not with lysolecithin.

It is evident that the transition temperature of the amylose–lipid complex is so high that it will never be achieved during dough making or baking except for the crust region. The amylose–lipid complex as such might then not directly influence the baking process. However, part of the polar lipids may be withdrawn from the system by the complex formation. During oven spring there might then be a competition for lipids between the amylose and the air/water interface. After the lipids have been complexed, there will be less lipids left to stabilize the air/water interface. This could be one reason for the positive relation between loaf volume and starch gelatinization temperature. Another reason is the delay in the increase in the viscosity of the crumb, which may stop the volume expansion. A redistribution of monoacyl lipids during baking has been observed [90]. Glycolipids were described as associated with the protein in the dough but with the starch in the bread.

4. Enzymatic Degradation of the Complex

The enzymatic degradation of starch gelatinized in the presence of MAL is decreased or at least delayed compared to the degradation when MAL is absent

[42, 91]. Enzymatic hydrolysis is affected also when monoglycerides are added after the gelatinization of the starch [42].

Differential scanning calorimetric measurements show that the thermal stability differs between amylose–lipid complexes depending on the ligands. The stability against enzymatic degradation differs in the same way. The enzymatic hydrolysis by α-amylase plus amyloglucosidase is about the same as that of amylose or of amylose complexed with monolaurin. The stability of the complexed amylose against enzymatic hydrolysis increases in the order monolaurin < monomyristin < monopalmitin and monostearin. For unsaturated monoglycerides in the complex, the stability increases in the order monolinolein < monoolein < monoelaidin [37].

The decrease in enzymatic degradation of complexed amylose can influence the result of starch analysis methods based on enzymatic degradation of starch. In one method, which uses starch degradation with glucoamylase at 60°C, only 39% of complexed amylose was detected [92]. If a thermostable enzyme was included and allowed to work at the temperature of boiling water, 88% of the complexed amylose was accounted for.

D. The Amylopectin–Lipid Complex

For a long time it has been assumed that amylopectin does not form a complex with polar lipids. When emulsifiers are added to an amylopectin solution, no precipitate is formed, as it would with amylose [82]. However, a slight decrease in iodine absorption is observed. On the other hand, when amylopectin and monoglycerides were suspended in water at 60°C for several hours, a precipitate was observed [93]. The least amount of complex was formed with GMS, whereas GMP gave the greatest amount and monomyristin (GMM) a medium amount. The iodine affinity of these complexes was lower than that of amylopectin.

Measurements by DSC have shown that the retrogradation of starch decreases when MAL is added to the starch. This effect has been explained by the formation of the amylose–lipid complex. It should be borne in mind, however, that these DSC measurements show the retrogradation of amylopectin. It is not evident how the amylose–lipid complex can affect the retrogradation of amylopectin. The simplest explanation may be that the amylose–lipid complex does not affect the retrogradation of amylopectin and that it is the polar lipid that directly affects the amylopectin. This explanation is supported by the fact that it is possible to decrease the retrogradation of waxy maize starch or amylopectin by the addition of MAL [53] (Fig. 12). It could be argued that this effect is the result of an unspecific interaction; the MAL may, for example, surround the amylopectin molecules and in this way cause a steric hindrance against crystallization. If such an unspecific interaction is the reason, it could be expected to be

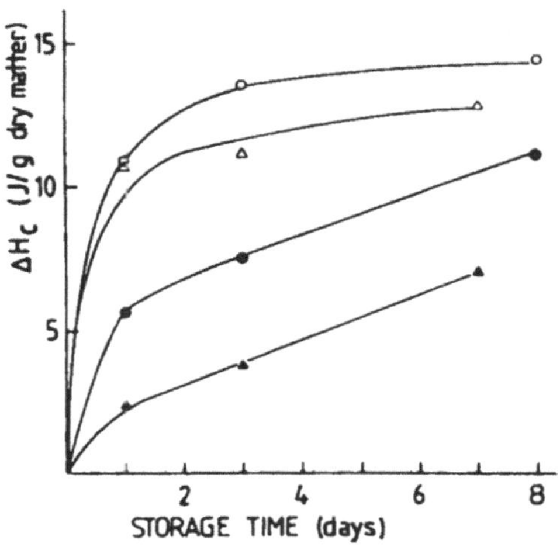

Fig. 12 The influence of a monoacyl lipid (CTAB) on the retrogradation of waxy maize starch and amylopectin. The melting enthalpy of retrograded waxy maize starch without (○) and with (●) CTAB (0.014 g/g starch), and potato amylopectin without (△) and with (▲) CTAB (0.014 g/g starch). The samples were stored at +7°C before heating in the DSC. (Reprinted with permission from [53].)

unrelated to the type of lipid present. However, we have found that the retrogradation of waxy maize starch and that of amylopectin is strongly dependent on the type of lipid present, and that the effectiveness in delaying retrogradation increases in the following order: soybean oil < lecithin < unsaturated monoglycerides < CTAB [53]. This is the same ranking as would be obtained for amylose-complexing ability [82]. Some of the most effective amylose-complexing agents are the series of saturated monoglycerides [82]. However, we could not test them for their influence on the retrogradation of waxy maize starch in a DSC study, because the chain melting temperature of uncomplexed monoglycerides coincides with the melting of recrystallized amylopectin [53].

To gain further insight into the significance of the amylose–lipid complex with respect to the decrease of retrogradation, an investigation was performed in our laboratory in which amylose–CTAB complexes were added to different starches [94]. The conclusion drawn was that the added complex had no effect on the retrogradation unless it was first melted. If the complex is melted and dissociated there is a possibility for direct interaction between amylopectin and the MAL.

Several indirect measurements now indicate that there exists also an amylopectin–lipid complex. As described above, the retrogradation of waxy maize starch and amylopectin is affected by MAL but not by triglycerides [53]. Equilibrium dialysis indicates that stearic acid is bound to waxy maize starch; 0.76 g stearic acid/g amylopectin was bound [83]. The corresponding value for amylose was 5.25 g stearic acid/g amylose. Data from ESR studies showed the formation of a complex in the case of native waxy maize starch and adsorption on the starch granule surface for modified waxy maize starch [95]. The rheological behavior of waxy maize and waxy barley is affected by added MAL [38,48,60]. SSL, DATEM, and DMG appeared to have no effect on the viscosity of waxy maize starch, but SDS and CTAB greatly increased the viscosity, at least in the absence of salt. The increase in viscosity in the absence of salt was ascribed to the formation of a complex between amylopectin and SDS/CTAB. Owing to the charges on them, there will be repulsion between the complexes, causing a strong enhancement of viscosity. When salt is added, screening will eliminate the repulsion effects [48].

When an emulsifier such as SSL is added at the 1% level of waxy maize starch and heated in the calorimeter, two endotherms are observed. The first endotherm, at about 30–45°C, is due to the chain melting of SSL, and the second endotherm, at about 55–80°C, is due to gelatinization [48]. There was no indication of an endotherm at a higher temperature, which could have been interpreted as evidence of an amylopectin–lipid complex. SDS and CTAB, added at the 2% level did not give rise to any endotherms at temperatures above the gelatinization endotherm. However, when ΔH was calculated, it was found that it decreased in the presence of SDS or CTAB in proportion to the amount of surfactant present. ΔH decreased from about 17 J/g starch without any additive to about 12 J/g starch at a surfactant/starch ratio of 0.15:1. Further addition of SDS did not affect ΔH, and SSL had no effect at all on ΔH [48].

Recently, a thermal transition at a temperature interval corresponding to the one for the amylose–lipid complex was observed for waxy maize starches heated in the presence of MAL [54,96]. When waxy maize starch and SSL were heated at 120°C for 15 min at a very low water content (< 10 wt %), a subsequent DSC scan after 24 h at 4°C showed an endotherm at about 70°C. Waxy maize starch treated in the same way produced a featureless thermogram, whereas SSL alone melted at about 50°C [96]. One reason for the difficulties in finding a DSC endotherm that could be related to an amylopectin–lipid complex could be the lack of cooperativity in the melting of such a complex. For a transition to be detected by DSC, it has to involve enough energy, but it should also occur during a short temperature interval (otherwise the only observation will be a shift in heat capacity). If the outer branches of the amylopectin form a crystalline complex with suitable substances, only a few ligand molecules will be involved in the same amylopectin molecule (Fig. 13). Moreover, the degree of crystallinity may be quite low. One amylose molecule, on the other hand, can host a great number of ligand molecules, and several molecules can pack together in cystalline

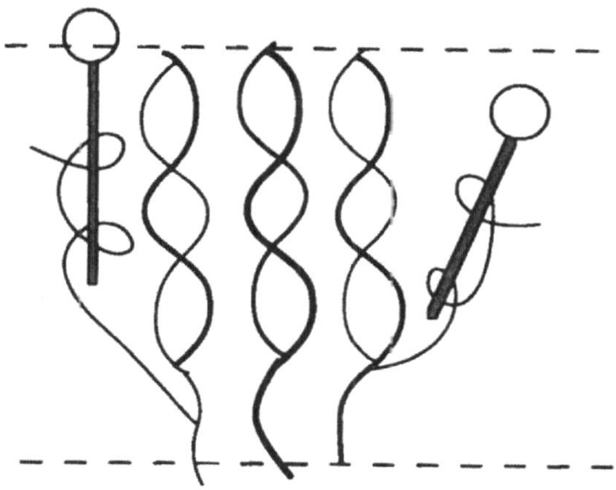

Fig. 13 A proposed model for an amylopectin–lipid complex. The outer branches of the amylopectin molecule might be able to form the complex. If such a complex is formed, the growth of the amylopectin crystal is hindered.

domains [97]. All this will result in a highly cooperative melting process. Recently we observed a V pattern for waxy maize starch in the presence of surfactants [54].

The complex-forming ability of amylose exceeds that of amylopectin, and the amylose–lipid complex will be the complex formed first [54]. A similar conclusion was drawn from experiments using ESR measurements. It was demonstrated that hydrophobic probes are complexed with waxy maize starch, as well as with starches of normal amylose content, although amylose is more effective in its complex formation [98].

Amylose has been determined to bind about six times as much MAL as is bound to amylopectin [69]. From equilibrium dialysis it is calculated that amylose binds about seven times as much stearic acid as is bound by amylopectin [83]. We have used surface tension measurements of SDS solutions of varying concentrations in the absence and presence of amylose and amylopectin to study the binding of MAL to amylose and amylopectin [99]. We found that amylopectin binds two SDS molecules per 1000 glucose residues, whereas amylose binds 35 SDS molecules per 1000 glucose residues; that is, there is a factor of 17.5 between the amount bound by amylose and the amount bound by amylopectin.

E. Influence of Starch on the Properties of Polar Lipids

It has already been illustrated that the formation of an amylose–lipid complex will protect the amylose against enzymatic degradation. However, the lipid composition will also be influenced by formation of the complex, and consequently the physical properties of the lipids will also. Thus the differences between the phase properties of lipids extracted from wheat flour and those from the corresponding gluten, as described in Chapter 2, reflect these lysophospholipids effects on phase properties. The formation of lysophosospholipids and free fatty acids during storage of wheat flour due to lipase and phospholipase activity is one of the most prominent effects of storing flour.

The lysolecithin in wheat starch contains about 50% linoleic acid, and if lysolecithin is isolated from starch it will soon be destroyed by oxidation.

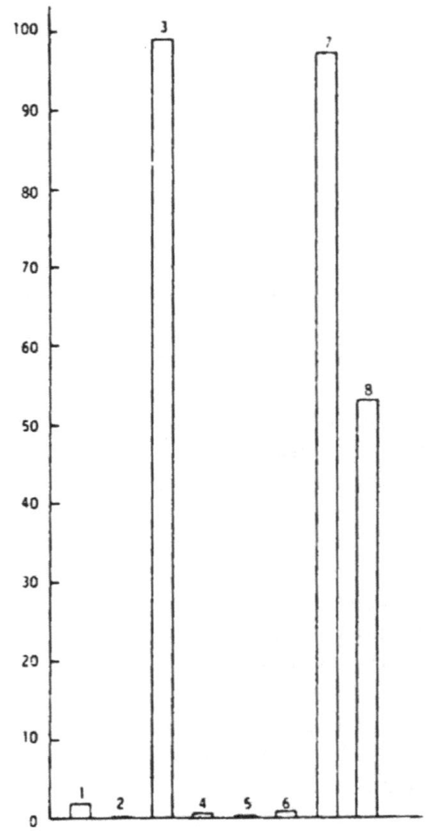

Fig. 14 Hemolytic effect of some emulsifiers used in baking compared to that of lysolecithin (7) [102]. (1) DATE; (2) monostearate; (3) SSL; (4) tetraglycerol monostearate; (5) glycerol monooleate; (6) lecithin; and (8) oleic acid.

However, the lysolecithin in starch is protected against oxidation [100]. This complexing could be applied to protect sensitive lipid types of substances against oxidation.

In medical discussions of dietary lipids, their physical effects on the gastrointestinal surface structure are often neglected. A classic work in this context was reported by Davenport [101], who demonstrated that lysophospholipids in very low concentrations (100 μg/mL) can damage the gastric mucosal barrier. Lysophospholipds are formed in the intestine from all phospholipids in the diet as well as from the phospholipids in the bile. When the food is transported into the intestine, a certain amount may stream back, and this was related to ulceration [101]. It is quite remarkable in this context that the food itself may contain much higher concentrations of lysophospholipids than those reported to cause damage. Oat grains, for example, have a lipid content of almost 10% (w/w), and several percent of these lipids is often lysolecithin (due to the high lipase activity in oats). How can we consume these amounts of lysolecithin when gastric damage can occur at much lower concentrations? There can hardly be any explanation other than the protection due to the presence of starch—the inactivation of the lipid by inclusion complexing into amylose or amylopectin. The cereal starches themselves contain from about 1% (w/w) to several percent by weight lysolecithin. As the starch degradation takes place to only a minor extent before the food enters the intestine, there will be no harmful effects due to free lysophospholipids.

In this connection it should also be mentioned that some of the emulsifiers used in the baking process can have biomembrane-damaging effects as great as that of lysolecithin [102], see Fig. 14. In bread, however, the presence of starch in excess is a safety factor. This can explain why no harmful effects have been observed.

REFERENCES

1. Frazier, P. J., Daniels, N. W. R., and Russell Eggitt, P. W. (1981). Lipid–protein interactions during dough development, *J. Sci. Food Agric.*, 32: 877.
2. Lundh, G., Eliasson, A.-C., and Larsson, K. (1988). Cross-linking of wheat storage protein monolayers by compression/expansion cycles at the air/water interface, *J. Cereal. Sci.*, 7: 1.
3. Bernardin, J. E., and Kasarda, D. D. (1973). Hydrated protein fibrils from wheat endosperm, *Cereal Chem.*, 50: 529.
4. Bernardin, J. E., and Kasarda, D. D. (1973). The microstructure of wheat protein fibrils, *Cereal Chem.*, 50: 735.
5. Amend, T., and Belitz, H.-D. (1989). Microscopical studies of water/flour systems, *Z. Lebensm. Unters. Forsch.*, 189: 103.

6. Amend, T., Belitz, H.-D., and Kurthen, C. (1990). Electron microscopic studies on protein films from wheat and other sources at the air/water interface, *Z. Lebensm. Unters. Forsch.*, 190: 217.
7. Hamer, R. J., van Oort, M., Garritsen, A., and Weegels, P. L. (1990). New methods to understand interactions in wheat flour doughs. Interactions between gliadins and flour lipids, in *Gluten Proteins 1990*, W. Bushuk and R. Tkachuk, Eds., AACC, St. Paul, Minn., p. 296.
8. Simmonds, D. H. (1974). Chemical basis of hardness and vitreosity in the wheat kernel, *Baker's Dig.*, 48(5): 16.
9. Barlow, K. K., Buttrose, M. S., Simmonds, D. H., and Vesk, M. (1973). The nature of the starch–protein interface in wheat endosperm, *Cereal Chem.*, 50: 443.
10. Greenwell, P., and Schofield, J. D. (1986). A starch granule protein associated with endosperm softness in wheat, *Cereal Chem.*, 63: 379.
11. Stenvert, N. L., and Kingswood, K. (1977). The influence of the physical structure of the protein matrix on wheat hardness, *J. Sci. Food. Agric.*, 28: 11.
12. Sandstedt, R. M. (1961). The function of starch in the baking of bread, *Baker's Dig.*, 35(3): 36.
13. Fretzdorff, B., Bechtel, D. B., and Pomeranz, Y. (1982). Freeze-fracture ultrastructure of wheat flour ingredients, dough and bread, *Cereal Chem.*, 59: 113.
14. Kim, S. K., and D'Appolonia, B. L. (1977). Bread staling studies. II. Effect of protein content and storage temperature on the role of starch, *Cereal Chem.*, 54: 216.
15. Martin, M. L., Zeleznak, K. J., and Hoseney, R. C. (1991). A mechanism of bread firming. I. Role of starch swelling. *Cereal Chem.*, 68: 498.
16. Eliasson, A.-C. (1983). Differential scanning calorimetry studies on wheat starch–gluten mixtures. II. Effect of gluten and sodium stearoyl lactylate on starch crystallization during ageing of wheat starch gels, *J. Cereal Sci.*, 1: 207.
17. Willhoft, E. M. A. (1971). Bread staling. I. Experimental study, *J. Sci. Food Agric.*, 22: 176.
18. Eliasson, A.-C. (1983). Differential scanning calorimetry studies on wheat starch–gluten mixtures. I. Effect of gluten on the gelatinization of wheat starch, *J. Cereal Sci.*, 1: 199.
19. Dahle, L. K. (1971). Wheat protein–starch interaction. I. Some starch-binding effects of wheat-flour proteins, *Cereal Chem.*, 48: 706.
20. Arnebrant, T. (1987). Proteins at the metal/water interface. Adsorption in relation to interfacial structure, thesis, Lund University, Lund.
21. Nylander, T. (1987). Proteins at the metal/water interface. Adsorption and solution behaviour, thesis, Lund University, Lund.

22. MacRitchie, F. (1972). The adsorption of proteins at the solid/liquid interface. *J. Colloid Interface Sci.*, 38: 484.
23. Norde, W. (1976). Proteins at interfaces. The adsorption of human plasma albumin and bovine pancreas ribonuclease on polystyrene latices, thesis, Agricultural University, Wageningen.
24. Eliasson, A.-C., and Tjerneld, E. (1990). Adsorption of wheat proteins on wheat starch granules, *Cereal Chem.*, 67: 366.
25. Sargeant, J. G., and Walker, T. S. (1978). Adsorption of wheat alpha-amalyase isoenzymes to wheat starch, *Starch/Stärke*, 30: 160.
26. Eliasson, A.-C. (1989). Some physico-chemical properties of wheat starch, in *Wheat End-Use Properties—Wheat and Flour Characterization for Specific End-Use*, H. Salovaara, Ed., Helsinki, p. 355.
27. Eliasson, A.-C., Silverio, J., and Tjerneld, E. (1991). Surface properties of wheat flour-milling streams and rheological and thermal properties after hydration, *J. Cereal Sci.*, 13: 27.
28. Muhrbeck, P. (1991). Potato starch and its behaviour in proteinaceous systems, thesis, Lund University, Lund.
29. Eliasson, A.-C., and Hegg, P.-O. (1981). Thermal stability of gluten, *Cereal Chem.*, 57: 436.
30. Lindahl, L., and Eliasson, A.-C. (1986). Effects of wheat proteins on the viscoelastic properties of starch gels. *J. Sci. Food Agric.*, 37: 1125.
31. Hibberd, G. E. (1970). Dynamic viscoelastic behaviour of wheat flour doughs. III. The influence of the starch granules, *Rheol. Acta*, 9: 501.
32. Smith, J. R., Smith, T. L., and Tschoegl, N. W. (1970). Rheological properties of wheat flour doughs, *Rheol. Acta*, 9: 239.
33. Morrison, W. R., Mann, D. L., Soon, W., and Coventry, A. M. (1975). Selective extraction and quantitative analysis of non-starch and starch lipids from wheat flour, *J. Sci. Food Agric.*, 26: 507.
34. Soulaka, A. B., and Morrison, W. R. (1985). The amylose and lipid contents, dimensions, and gelatinization characteristics of some wheat starches and their A- and B-granule fraction, *J. Sci. Food Agric.*, 36: 709.
35. Eliasson, A.-C., Larsson, K., and Miezis, Y. (1981). On the possibility of modifying the gelatinization properties of starch by lipid surface coating, *Starch/Stärke*, 33: 231.
36. Melvin, M. A. (1979). The effect of extractable lipid on the viscosity characteristics of corn and wheat starches, *J. Sci. Food Agric.*, 30: 731.
37. Eliasson, A.-C., and Krog, N. (1985). Physical properties of amylose–monoglyceride complexes, *J. Cereal Sci.*, 3: 239.
38. Eliasson, A.-C. (1986). On the effects of surface active agents on the gelatinization of starch—a calorimetric investigation, *Carbohydr. Polym.*, 6: 463.
39. Eliasson, A.-C. (1985). Starch gelatinization in the presence of emulsifiers. A morphological study of wheat starch, *Starch/Stärke*, 37: 411.

40. Tester, R. F., and Morrison, W. R. (1990). Swelling and gelatinization of cereal starches. I. Effects of amylopectin, amylose, and lipids, *Cereal Chem.*, 67: 551.
41. Ghiasi, K., Hoseney, R. C., and Varriano-Marston, E. (1982). Gelatinization of wheat starch. IV. Amylograph viscosity, *Cereal Chem.*, 59: 262.
42. van Lonkhuysen, H., and Blankestijn, J. (1976). Influence of monoglycerides on the gelatinization and enzymatic breakdown of wheat and cassava starch, *Starch/Stärke*, 28: 227.
43. Bowler, P., Williams, M. R., and Angold, R. E. (1980). A hypothesis for the morphological changes which occur on heating lenticular wheat starch in water, *Starch/Stärke*, 32: 186.
44. Collison, R., Campbell, J. D., and Elton, G. A. H. (1960). The effects of surfactants on swelling and gel properties of wheat starch, *Chem. Ind.*, 40: 1230.
45. Hellman, N. N., Fairchild, B., and Senti, F. R. (1954). The bread staling problem. Molecular organization of starch upon aging of concentrated starch gels at various moisture levels, *Cereal Chem.*, 31: 495.
46. Zobel, H. F. (1988). Starch crystal transformations and their industrial importance, *Starch/Stärke*, 40: 1.
47. Krog, N. (1973). Influence of food emulsifiers on pasting temperature and viscosity of various starches, *Starch/Stärke*, 25: 22.
48. Evans, I. D. (1986). An investigation of starch/surfactant interactions using viscometry and differential scanning calorimetry, *Starch/Stärke*, 38: 227.
49. Eliasson, A.-C. (1986). Viscoelastic behaviour during the gelatinization of starch. II. Effects of emulsifiers, *J. Text. Stud.*, 17: 357.
50. Russell, P. L. (1983). A kinetic study of bread staling by differential scanning calorimetry and compressibility measurements. The effect of added monoglycerides, *J. Cereal Sci.*, 1: 297.
51. Riisom, T., Krog, N., and Eriksen, J. (1984). Amylose complexing capacities of cis- and trans-unsaturated monoglycerides in relation to their functionality in bread, *J. Cereal Sci.*, 2: 105.
52. Krog, N. J., Riisom, T. H., and Larsson, K. (1985). Applications in the food industry. I, in *Encylopedia of Emulsion Technology*, P. Becher, Ed., Marcel Dekker, New York, p. 321.
53. Eliasson, A.-C., and Ljunger, G. (1988). Interactions between amylopectin and lipid additives during retrogradation in a model system, *J. Sci. Food Agric.*, 44: 353.
54. Gudmundsson, M., and Eliasson, A.-C. (1990). Retrogradation of amylopectin and the effects of amylose and added surfactants/emulsifiers, *Carbohydr. Polym.*, 13: 295.

55. Cloke, J. D., Gordon, J., and Davis, E. A. (1983). Enthalpy changes in model cake systems containing emulsifiers, *Cereal Chem.*, 60: 143.
56. Davis, E. A., Grider, J., and Gordon, J. (1986). Microstructural evaluation of model starch systems containing different types of oil, *Cereal Chem.*, 63: 427.
57. Ghiasi, K., Hoseney, R. C., and Varriano-Marston, E. (1983). Effect of flour components and dough ingredients on starch gelatinization, *Cereal Chem.*, 60: 58.
58. Leissner, O. (1986). A comparison of the effect of different polymorphic forms of lipids in breadmaking, *Cereal Chem.*, 65: 202.
59. Osman, E. M., and Dix, M. R. (1960). Effects of fats and nonionic surface-active agents on starch pastes, *Cereal Chem.*, 37: 464.
60. Eliasson, A.-C., Finstad, H., and Ljunger, G. (1988). A study of starch–lipid interactions for some native and modified maize starches, *Starch/Stärke*, 40: 95.
61. Silverio, J., and Eliasson, A.-C. (to be published).
62. Eliasson, A.-C., Carlson, T. L.-G., Larsson, K., and Miezis, Y. (1981). Some effects of starch lipids on the thermal and rheological properties of wheat starch, *Starch/Stärke*, 33: 130.
63. Seguchi, M. (1985). Oil-binding ability of gelatinized prime starch granules from chlorinated wheat flour, *Starch/Stärke*, 37: 116.
64. Carlson, T. L.-G., Larsson, K., Dinh-Nguyen, N., and Krog, N. (1979). A study of the amylose–monoglyceride complex by Raman spectroscopy, *Starch/Stärke*, 31: 222.
65. Yamishita, Y.-H., Ryugo, J., and Monobe, K. (1973). An electron microscopic study on crystals of amylose V complexes, *J. Electron Microscopy*, 22: 19.
66. French, A. D., and Murphy, V. G. (1977). Computer modeling in the study of starch, *Cereal Foods World*, 22: 61.
67. Mikus, F. F., Hixon, R. M., and Rundle, R. E. (1946). The complex of fatty acids with amylose, *J. Am. Chem. Soc.*, 68: 1115.
68. Osman, E. M., Leith, S. J., and Fles, M. (1961). Complexes of amylose with surfactants, *Cereal Chem.*, 38: 449.
69. Lagendijk, J., and Pennings, H. J. (1970). Relation between complex formation of starch with monoglycerides and the firmness of bread, *Cereal Sci. Today*, 15: 354.
70. Eberstein, K., Höpcke, R., Koniecyny-Janda, G., an Stute, R. (1980). DSC-Untersuchungen an Stärken, *Starch/Stärke*, 12: 397.
71. Gough, B. M., Greenwell, P., and Russell, P. L. (1985). On the interaction of sodium dodecyl sulphate with starch granules, in *New Approaches to Research on Cereal Carbohydrates*, R. D. Hill and L. Munck, Eds., Elsevier, Amsterdam, p. 99.

72. Kowblansky, M. (1985). Calorimetric investigation of inclusion complexes of amylose with long-chain aliphatic compounds containing different functional groups, *Macromolecules*, 18: 1776.
73. Eliasson, A.-C. (1988). On the thermal transitions of the amylose–cetyltrimethylammonium bromide complex, *Carbohydr. Res.*, 172: 83.
74. Gilbert, G. A., and Spragg, S. P. (1964). Iodometric Determination of Amylose in *Methods in Carbohydrate Chemistry*, Vol. IV, R. L. Whistler, Ed., Academic, New York, p. 168.
75. Morrison, W. R., and Laignelet, B. (1983). An improved colorimetric procedure for determining apparent and total amylose in cereal and other starches, *J. Cereal Sci.*, 1: 9.
76. Zobel, H. F. (1964). X-ray analysis of starch granules, *Methods Carbohydr. Chem.*, 4: 109.
77. Zobel, H. F., French, A. D., and Hinkle, M. E. (1967). X-ray diffraction of oriented amylose fibers. II. Structure of V-amylose, *Biopolymers*, 5: 837.
78. Manley, R. S. J. (1964). Chain folding in amylose crystals, *J. Polym. Sci. A*, 2: 4503.
79. Raphaelides, S., and Karkalas, J. (1988). Thermal dissociation of amylose–fatty acid complexes, *Carbohydr. Res.*, 172: 65.
80. Biliaderis, C. G., and Galloway, G. (1989). Crystallization behaviour of amylose-V complexes: structure–property relationships, *Carbohydr. Res.* 189: 31.
81. Björck, I., Asp, N.-G., Birkhed, D., Eliasson, A.-C., Sjöberg, L.-B., and Lundquist, I. (1984). Effects of processing on starch availability in vitro and in vivo. II. Drum-drying of wheat flour, *J. Cereal Sci.*, 2: 165.
82. Krog, N. (1971). Amylose complexing effect of food grade emulsifiers, *Starch/Stärke*, 23: 206.
83. Hahn, D. E., and Hood, L. F. (1987). Factors influencing corn starch–lipid complexing, *Cereal Chem.*, 64: 81.
84. Conde-Petit, B., and Escher, F. (1991). Rheologische Untersuchungen an Stärke-Emulgator-Systemen, *Getreide Mehl Brot*, 45: 131.
85. Morrison, W. R., and Coventry, A. M. (1989). Solvent extraction of fatty acids from amylose inclusion complexes, *Starch/Stärke*, 41: 24.
86. Kugimiya, M., and Donovan, J. W. (1981). Calorimetric determination of the amylose content of starches based on formation and melting of the amylose–lysolecithin complex, *J. Food Sci.*, 46: 765.
87. Stute, R., and Konieczny-Janda, G. (1983). DSC-Untersuchungen an Stärken. Teil II. Untersuchungen an Stärke-Lipid-Komplexen, *Starch/Stärke*, 35: 340.
88. Eliasson, A.-C. (1980). Effect of water content on the gelatinization of wheat starch, *Starch/Stärke*, 32: 270.
89. Biliaderis, C. G., Page, C. N., Slade, L., and Sirett, R. R. (1985).

Thermal behaviour of amylose–lipid complexes, *Carbohydr. Polym.*, 5: 367.
90. Wehrli, H. P., and Pomeranz, Y. (1970). A note on autoradiography of tritium-labeled galactolipids in dough and bread, *Cereal Chem.*, 47: 221.
91. Larsson, K., and Miezis, Y. (1979). On the possibility of dietary fiber formation by interaction in the intestine between starch and lipids, *Starch/Stärke*, 31: 301.
92. Holm, J., Björck, L., Ostrowska, S., Eliasson, A.-C., Asp, N.-G., Larsson, K., and Lundquist, I. (1983). Digestibility of amylose–lipid complexes in vitro and in vivo, *Starch/Stärke*, 35: 294.
93. Batres, L. R., and White, P. J. (1986). Interaction of amylopectin with monoglycerides in model systems, *J. Amer. Oil Chem. Soc.*, *63:* 1537.
94. Gudmundsson, M. (1992). Effects of an added inclusion–amylose complex on the retrogradation of some starches and amylopectin. *Carbohydr. Polym.*, 17:299.
95. Johnson, J. M., Davis, E. A., and Gordon, J. (1990). Lipid binding of modified corn starches by electron spin resonance, *Cereal Chem.*, 67: 236.
96. Slade, L., and Levine, H. (1987). Recent advances in starch retrogradation, in *Industrial Polysaccharides: The Impact of Biotechnology and Advanced Methodologies* (S. S. Stivala, V. Crescenzi, and I. C. M. Dea, Eds., Gordon and Breach, New York, p. 387.
97. Lehman, G., and Gottschlich, H. (1983). Die Wirkungsweise von Emulgatoren auf Stärke, *Fette-Seifen-Anstrichm.*, 85: 439.
98. Pearce, L. E., Davis, E. A., Gordon, J., and Miller, W. G. (1987). Stearic acid–starch interactions as measured by electron spin resonance, *Cereal Chem.*, 64: 154.
99. Gudmundsson, M., Svensson, E., and Eliasson, A.-C. Binding of Sodium Dodecyl Sulphate to Amylose and Amylopectin Quantified by Surface Tension Measurements (submitted).
100. Acker, L. (1977). Die Lipide der Stärken—Ein Forchungsgebiet zwischen Kohlenhydraten und Lipiden, *Fette-Seifen-Anstrichm.* 79: 1.
101. Davenport, H. W. (1968).Destruction of the gastric mucosa barrier by detergents and urea, *Gastroenterology*, 54: 175.
102. Larsson, K., and Johansson, L.-Å. (1978). Hemolytic effect of some polar lipids used as food additives, *Lebensmittel-Wiss. Technol.*, 11: 206.

4

Components in Other Cereals

I. INTRODUCTION

Wheat is the cereal crop of highest production in the world, followed by rice and maize (Table 1). It is also the main cereal used in breadmaking. Rye is also used for breadmaking, whereas oats, barley, maize, rice, and others are mainly used for other food products, for feed, or for industrial applications. The reason for the dominance of wheat in breadmaking is, of course, its supreme baking performance in comparison with all other cereals. It has even been questioned whether bread without gluten is really bread [4]. Still, bread or breadlike products are baked from composite flours—wheat flour mixed with other cereal flours. It is possible to obtain a gas-holding capacity and porosity of such mixed flours corresponding to that of 100% wheat flour [5].

In this chapter the cereals rye, barley, oats, rice, and maize are compared to wheat in aspects that are relevant to their colloidal behavior in breadmaking. *Triticale*, the hybrid between rye and wheat, behaves like a weak wheat variety from a breadmaking point of view. Therefore we are omitting it from this chapter, which is devoted to cereals *different* from wheat.

The approximate compositions of the selected cereals are given in Table 2, and inspection of the table reveals that these cereals are quite similar in gross composition. The technological differences between them is thus not a question of composition, but rather of the properties of the components. Starch is given special attention in this chapter as it has a dominant influence compared to the other components when these other cereals are introduced in breadmaking.

II. PROPERTIES OF THE COMPONENTS

A. Lipids

Among all cereals, oats and maize are unique with regard to their high lipid content. Oats can contain more then 10% (w/w) of lipids, and maize is remark-

Table 1 World Production of Some Cereals

Cereal	Production (10^6 tonnes)	Year	Ref.
Wheat	540	1989	1
Rice	508	1989	1
Maize	472	1989	1
Barley	169	1989	1
Oats	49	1986	2
Rye	30	1985	3

able in that its lipid content can vary a lot—from 0.4% to 17%. As described in Chapter 2, wheat contains about 2% lipids and the other cereals considered here have a similar amount. There is also a clear difference between maize and oats. Maize lipids are completely dominated by triglycerides, whereas oat lipids are about one-third polar lipids. In fact, oats contains by far the greatest amount of phospho- and galactolipids. We briefly summarize the available data on the non-wheat cereal lipids below, starting with oats, referring to a review by Young [7] and our own data on Swedish cultivars [8].

Young [7] reports 41% triglycerides, 5% free fatty acids (FFA), and 4% partial glycerides and sterols as the nonpolar content, and 12% glycolipids and 10% phospholipids as the polar content of oats. There is, however, a high proportion (28%) of unidentified components. Table 3 shows the results we obtained from two culivars (separation was done on a silica gel column).

We also examined the aqueous systems of lipids from these two varieties.

Table 2 Average Gross Composition of Some Cereals[a]

Cereal	Protein	Fat	Starch	Fiber	Ash
Wheat	12.2	1.9	71.9	1.9	1.7
Rye	11.6	1.7	71.9	1.9	2.0
Barley	10.9	2.3	73.5	4.3	2.4
Oats	11.3	5.8	55.5	10.9	3.2
Maize	10.2	4.6	79.5	2.3	1.3
Rice	8.1	1.2	75.8	0.5	1.4

[a]Data given in percent of dry weight.
Source: Adapted from [6].

Table 3 Lipid Composition[a] of Two Varities of Oats

Lipid	Magne	Chicauhua
Total lipids (% of dry flour)	9.2	9.2
Phospholipids + glycolipids	13.2	14.4
Sterol esters + di- and triglycerides	71.6	63.0
FFA and sterols	10.2	14.8
Polar unidentified	4.1	6.5

[a]Expressed as percentage of total lipid.
Source: Data from [8].

When water is added, spontaneous phase separation takes place. In excess water, a top phase of nonpolar oil, a water layer, and a bottom phase of L2 and liquid crystals are formed. The only liquid crystalline phase observed was the lamellar one. Oat lipids might therefore be of applied interest owing to their ability to supply the lamellar phase to a dough.

Large variations in the lipid content of *barley* is found in the literature, and probable values are 3.3–4.6% according to Morrison [9]. The lipids are rather similar to those in wheat, with about 65–78% nonpolar lipids, 7–13% galactolipids, and 15–26% phospholipids. The starch inclusion portion of the lipids is about 1%, dominated by lysophospholipids. Also the fatty acids pattern is similar to that in wheat, rye, and oats, with about 60% linoleic acid, palmitic acid being second at about 20%.

Rye lipids are most similar to wheat lipids among the cereals considered here [9]. The total content is about 3%, and the endosperm contains about 2% of these lipids. Analytical data from Chung and Tsen [10] suffer from being divided into "free" and "bound." Comparisons of corresponding data of wheat lipids show close relationhips, and the same is true in the case of rye starch inclusion lipids. A phase diagram of rye lipids was presented in Chapter 2.

As mentioned in the beginning of this section, the lipid content of *maize* can vary considerably. Thus when maize is grown as a crop for vegetable oil, the relative proportion of triglycerides is much higher. According to data reported by Weber [11], a maize rich in oil (13.8% "high oil") and one with low lipid content (2.6%; "low oil") show the lipid composition given in Table 4.

Maize starch can vary in amylose/amylopectin ratio, and lipid content has been found to vary accordingly (from 0.2% in waxy maize rich in amylopectin to 1.1% in amylomaize rich in amylose). The fatty acid composition can vary over a wide range. An inverse relation between total oil content and the amount of polyunsaturated FFA has been observed.

Brown rice contains about 3% total lipids, and about 0.5% has been classified

Table 4 Composition of Maize Lipids from Varieties Differing in Oil Content

Type of oil	Triglycerides (%)	FFA (%)	Sterol/sterol esters (%)	Polar lipids (%)
High oil	92.4	0.3	2.3	3.9
Low oil	74.9	0.4	10.4	1.2

Source: Data compiled from [11].

as "bound," the "bound" character reflecting the content of polar lipids. The glycolipids are dominated by sterolglucosides and digalactosyldiglycerides, whereas the phospholipids are dominated by phosphatidylethanolamine.

B. Proteins

1. Composition and Structure

The total protein content of the various cereals is given in Table 5. It is possible to apply the classic Osborne fractionation procedure to proteins from all the cereals, and in this way albumins, globulins, prolamins, and glutelins are ex-

Table 5 Protein Content (%) of Some Cereals and Distribution of Proteins in Solubility Classes According to Osborne

Cereal	Total protein of flour	Albumins	Globulins	Prolamins	Glutelins	Ref.
Wheat	11.0	14.7	7.0	32.6	45.7	12
Rye	6.5	44.4	10.2	20.9	24.5	12
		15.2	18.5	40.2	26.1	6
Barley	8.4	12.1	8.4	25	54.5	12
Oats	13.8	20.2	11.9	14	53.9	12
		11.7	38.4	13.6	27	13
Rice	7.3	10.8	9.7	2.2	77.3	12
Maize	6.5	4.0	2.8	47.9	45.3	12
	8	4	4	47	39	14
High-lysine maize		20.2[a]		14.6–24.4	46.2–53.2	15

[a]The sum of albumins and globulins.

tracted. The proportion of each solubility class obtained for the various cereals (wheat is included for comparison) is given in Table 5. There is a large variation in the distribution of protein in the solubility classes among the non-wheat cereals and also within each species. This latter fact is illustrated with the values cited for rye and oats, respectively. The difference obtained for oats in the case of globulins (11.9 or 38.4%) and glutelins (53.9 or 27%) could be due to the use of different salt solutions for the extraction of globulins. The use of $CaCl_2$ [13] gives a higher yield than NaCl [12]. It is also claimed that the globulins are completely extracted only at salt concentrations above 0.8 M NaCl and at pH below 8.5 [16].

The albumins range from 4.0% in maize to 44.4% in rye, the globulins from 2.8% in maize to 11.9% (or 38.4%) in oats, the prolamins from 2.2% in rice to 47.9% in maize, and the glutelins from 24.5% in rye to 77.3% in rice.

Some typical values for a few amino acids are given in Table 6. When the amino acid compositions of the delipidized flours are compared, it is evident that wheat differs from the others in its very high glutamine/glutamic acid (Glx) content (Glx = 31.7 mol %). High values of Glx are found also in rye and barley (24.1 and 25.3%, respectively). In oats, rice, and maize, the mole percent of Glx is below 20%. When proline is included in the comparison, the close relation between wheat, rye, and barley is even more evident. In these cereals the percentage of Glx and Pro is 40 mol % or higher, whereas in the other cereals this sum is 30 mol % or less. The low lysine content in wheat compared with oats and rice is evident in Table 6.

a. Prolamins and Glutelins

The prolamins, according to the Osborne classification, are extracted by 70% ethanol (60% is preferred for rye) or isopropanol. The prolamins in barley can be prepared by mechanical separation of the protein bodies [6]. If a reducing agent is included, high molecular weight (HMW) prolamins (glutelins) are extracted [17]. The amino acid composition given in Table 4 for the prolamins and glutelins shows that there are great similarities in the proportions of such amino acids as Glx, Pro, Gly, and Cys between wheat, rye, and barley. Rice and maize have the most deviating amino acid compositions, and oat prolamins and glutelins are somewhere in between. Thus, the amino acid composition cannot explain the poorer baking performance of rye and barley compared with wheat. Moreover, there are considerable structural similarities and genetic homology among the storage proteins from wheat, rye, and barley [6, 17]. They have several peptides of similar composition in common [18].

Among the solubility classes of proteins, the prolamins have been most thoroughly investigated, and this is reflected in their nomenclature. They are called *gliadins* in wheat, *secalins* in rye, *hordeins* in barley, *avenins* in oats, and *zeins* in maize [19].

Table 6 Partial Amino Acid Composition of the Protein Solubility Classes of Cereals

Amino acid	Mole percent of amino acid					
	Wheat	Rye	Barley	Oats	Rice	Maize
Flour (delipidized)						
Glx	31.7	24.1	25.3	19.9	15.8	16.3
Pro	12.9	12.4	14.6	6.3	5.3	11.7
Gly	6.2	7.2	6.1	8.4	8.0	7.0
Cys	1.8	1.6	1.5	2.7	1.6	1.8
Lys	1.8	3.2	2.7	3.4	3.4	2.4
Albumins						
Glx	21.4	22.6	14.3	12.7	14.5	12.6
Pro	9.5	12.3	7.7	6.3	4.7	8.7
Gly	7.1	6.7	10.0	12.9	10.0	9.9
Cys	3.3	2.4	3.9	7.0	1.9	1.8
Lys	3.1	3.0	4.4	4.6	5.1	4.0
Globulins						
Glx	15.6	17.4	13.2	16.4	14.9	11.0
Pro	7.1	8.0	7.0	5.4	5.7	5.7
Gly	8.5	8.9	9.7	9.6	10.4	10.6
Cys	3.7	2.1	3.1	2.4	4.2	3.3
Lys	4.1	4.4	4.8	4.5	2.5	4.7
Prolamins						
Glx	37.7	36.0	35.9	34.6	20.0	19.7
Pro	16.9	18.7	23.4	10.4	5.2	10.3
Gly	3.0	4.6	2.3	2.7	5.9	2.6
Cys	2.2	2.2	1.9	3.4	0.8	1.0
Lys	0.8	1.0	0.5	1.0	0.5	trace
Glutelins						
Glx	30.7	20.1	24.7	19.4	15.9	16.3
Pro	12.2	9.6	14.5	5.6	5.2	11.7
Gly	8.1	9.4	6.5	8.1	7.6	7.0
Cys	1.4	0.8	0.5	1.2	1.2	1.8
Lys	2.1	4.1	2.8	3.3	3.4	2.4

Source: Adapted from [12].

Sodium dodedyl sulfate polyacrylamide gel electrophoresis (SDS-PAGE) of prolamins under reducing conditions shows several bands in each cereal protein fraction. There are barley and rye proteins with molecular weights in the range 35,000–75,000. There are also HMW subunits with molecular weights of more than 90,000. The zeins have a much lower molecular weight, values around 20,000 being reported [17]. The molecular weights of the oat prolamins have been determined to be 23,500 and 15,500, together with a minor constituent at 36,000 [13]. The prolamins in rice are also of low molecular weight, 23,000 or below [19]. There seem to be more polypeptides present in wheat than in other cereal prolamins [19]. The SDS-PAGE pattern is unique for each variety and can be used for identification [6].

The C-hordein in barley has no cysteine and contains less than 5 mol % charged amino acid residues. The structure of C-hordein has been determined from measurements of intrinsic viscosity and from circular dichroism studies [20]. The secondary structure contains no α-helix or β-sheet structure but has regularly repeated β-turns. The C-hordein consists of a C-terminal and an N-terminal domain separated by a region with repetitive sequences. The C-hordein has two of these sequences (the pentamer Pro-Gln-Gln-Pro-Tyr and the octamer Pro-Gln-Gln-Pro-Phe-Pro-Gln-Gln) in common with the sulfur-rich prolamins in wheat [21]. The values obtained for the intrinsic viscosity were incompatible with a protein having a compact globular structure. Instead the molecule was proposed to be rod-shaped with the dimensions 301 × 18.5 Å in 0.1 M acetic acid at 15.3°C, and 265 × 19.7 Å at 50.0°C (this conformation might be due to the low pH). It was further assumed that the C-hordein consists of about 440 residues, and that all but eight of these are present as repetitive penta- and octapeptides. A β-spiral calculated at these conditions would result in a molecule of the dimensions 294 × 17 Å, in good agreement with the results experimentally determined from the viscosity measurements. It was therefore suggested that the C-hordein forms an extended β-spiral [20].

When the SDS-PAGE is run without a reducing agent, the patterns differ for all the cereal storage proteins tested (i.e., wheat, rye, barley, and maize) [17]. This means that in all these cereals, protein aggregates are present that are formed via disulfide linkages. The HMW prolamins (according to the definition of Shewry and Miflin [19] constitute less than 5% of the total prolamins in barley and rye and about 7–9% in wheat. The C-hordein in barley, the ω-secalin in rye, and the ω-gliadin in wheat are not affected by the absence or presence of reducing agent and could thus be regarded as being present in a nonaggregated state [17]. This group of proteins lack cysteine, and they account for 10–20% of the total prolamin fraction in wheat, rye, and barley [19]. For these proteins a heavy streaking of the gels was obtained when the SDS-PAGE was performed under nonreducing conditions. In maize, on the other hand, no unspecific staining was obtained, although the SDS-PAGE pattern was affected by reduction.

Gel filtration of the storage proteins from wheat, rye, barley, and maize showed that wheat, rye, and barley contain aggregates that are eluted in the void [17]. These aggregates have a molecular weight of more than 1 million. Maize storage protein does not contain protein aggregates of such a size. It might thus be a common property of prolamins and glutelins from wheat, rye, and barley that they can form large protein aggregates through the formation of intermolecular disulfide bonds. However, all the prolamins and glutelins of the non-wheat cereal proteins have one property in common—they do not form a gluten gel [6, 17]. So even if the prolamins and glutelins of rye and barley have the ability to aggregate into large molecules they still do not have the ability to form a gluten gel.

As discussed previously, the level of glutenins is important for the baking performance of wheat (Chapter 2), and upon inspecting Table 3 we would then expect that the baking performance of rye would be poor and that that of rice would be very good. However, it is well known that this is not the case. The glutelins from the non-wheat cereals evidently lack the properties of the wheat glutenins that are necessary for a good baking performance. In this respect, rye glutelins probably are more similar to wheat glutenins than are the other glutelins.

b. Albumins and Globulins

The albumins and globulins in all cereals except oats are found mainly in the embryo and the aleurone layers [22]. As in wheat, these proteins constitute a complex mixture of many different proteins including the metabolic enzymes. In the oat grain the globulin is found as the storage protein, at least according to the definition used by several workers [16]. The globulins are oligomers, and the aggregates have a molecular weight of over 300,000 [16, 23]. When SDS-PAGE is performed under nonreducing conditions, one band at 56,000–58,000 is found. Under reducing conditions this band disappears, and instead two bands are detected, one at 32,000–36,000 (called the β-band) and one at about 22,000 (the α-band) [16, 23]. These two subunits are linked by a disulfide bond, and six such pairs form the oat globulin, which thus is a hexamer composed of six α and six β subunits. The hexamer is formed through noncovalent linkages [23].

The amino acid composition of cereal globulins (Table 4) is similar to the amino acid composition of the storage proteins of legumes [22].

Rye contains considerable enzymes activity; both amylases and proteases are present [6]. The amylose activity (for example, measured as falling number) is perhaps the most important quality parameter of rye, as rye is extremely sensitive for field sprouting. The falling number is found to correlate with several properties of rye starch [24].

A soluble β-amylase in barley is found to aggregate with another protein, Z-protein. This Z-protein is more or less unaffected during the malting and brewing procedure and is found in beer [25].

2. Physiochemical Properties of Cereal Proteins

a. Rheological Behavior

As it is not possible to prepare gluten from the non-wheat cereals, there are no measurements of the rheological properties of the protein complex from the different cereals. However, the intrinsic viscosity [η] has been determined for hordeins [20, 26]. In 0.1 M acetic acid at 30°C, [η] for C-hordein was determined to be 16.84 mL/g [20]. Barley proteins were dissolved in SDS–urea–acetic acid solvent, and [η] was determined to be 87 mL/g [26]. This value was in the lower range of [η] values determined for wheat gluten (75–121 mL/g).

b. Thermal Behavior

The thermal behavior of oat proteins has been studied by DSC [23]. The prolamins and glutelins are similar to gluten in that they do not give rise to any detectable peaks. The albumins and globulins, on the other hand, were found to denaturate at quite a high temperature (Fig. 1). The denaturation temperature (T_d) of the albumins is about 87°C, and that of the globulins about 110°C [13]. In 0.01 M phosphate buffer at pH 7.4 with 0.1 M NaCl, T_d of the oat globulin is 114.2°C and ΔH = 26.5 J/g [13]. The denaturation process is thought to involve unfolding only; SH/SS exchange is thought to be of minor importance [23]. The thermal stability of oat globulins is demonstrated by their solubility at high temperature. In a solution of 1% of protein in 0.01 M phosphate buffer (pH 7.4) heated to 100°C for 60 min, less than 10% of the protein is precipitated. If the heating temperature is increased to 110°C, 50% of the protein will have precipitated after 15 min, and 70–75% after 60 min [23].

c. Surface Properties

The very quick spreading of protein, in combination with fibril formation, when wheat flour is placed at an air/water interface is also observed for other cereal proteins [27]. Fibrils are observed for rye and triticale, but they form fewer fibrils than wheat. No fibril formation at all has been observed for maize, rice, or barley endosperm sections. Maize flour was found to give rise to a network when spread on water, and this network could not be differentiated from a network formed by gliadin [28].

It can be expected that all the cereal proteins will decrease the surface tension of water, as this is the reason for their spreading at the air/water interface. However, as discussed in Chapter 2, a decrease in surface tension is not enough to stabilize the air/water interface in the dough. The protein film formed must also possess suitable rheological properties.

The surface activity of the rye proteins has been studied in the same way as that of wheat proteins [29]. Rye flour was spread (dry sprinkled) at the air/water interface, and surface-active material was found to spread very quickly. The

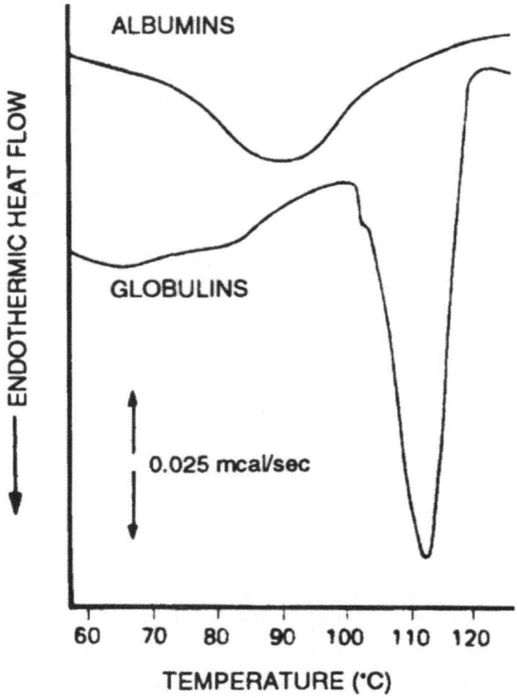

Fig. 1 Differential scanning calorimetric thermograms of albumins and globulins from oats. (Reprinted with permission from [13]. Copyright 1984 American Chemical Society.)

decrease in surface pressure (π) occurred much more rapidly with rye flour than with wheat flour and approached an equilibrium value much faster as well (Fig. 2). Another interesting difference in relation to wheat flour was the response to ascorbic acid. In the case of wheat flour, ascorbic acid was found to depress π compared with when the same amount of flour was spread on distilled water (see Fig. 15 in Chapter 6). When rye flour was spread on ascorbic acid, about the same π value was observed as when it was spread on distilled water. It is interesting in this context to note that differences were found between rye flour mill streams in their response to ascorbic acid, and one stream of very low protein content (3.5%) was greatly affected by ascorbic acid. Another milling stream of high protein content was not at all affected by ascorbic acid [29].

The foaming ability of oat proteins has been investigated, and it was found to be highest for the albumins and lowest for the glutelins. The foaming ability of the albumins is comparable to that of liquid egg white [13]. In this connection it could be of interest to mention the film-forming ability of zein, a

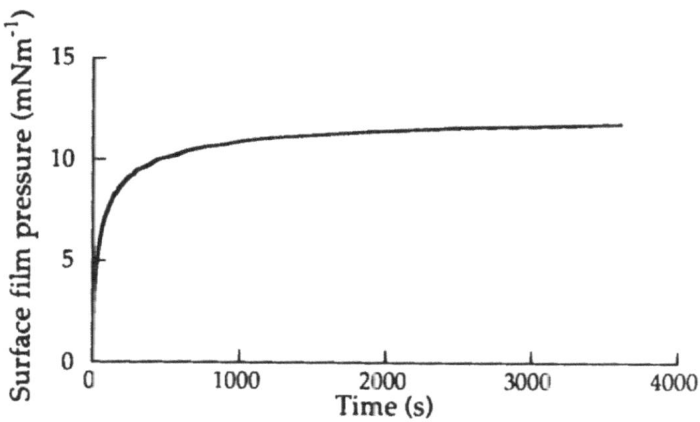

Fig. 2 Pressure–time (π–t) isotherm for the spreading of rye flour on distilled water. (From [29].)

rodlike molecule with a high α-helix content. Its film-forming ability is used in coating applications [30].

C. Starch

Starch comprises the largest proportion of all the cereals (Table 2). Although the compositions of the cereal starches are very similar (~25–30% amylose, ~1% lipids, with the remainder being amylopectin), they behave very differently in baking [31]. Rye and barley starches are almost as good as wheat starch, whereas maize and rice starches are very poor (Table 7). The cereal starches have some

Table 7 Baking Performance of Various Starches

Source of starch	Loaf volume for reconstituted flour (mL)
Wheat	80
Maize	48
Oat	58
Barley	78
Rye	77
Rice	68
Potato	60

Source: Data extracted from [31].

properties in common but differ greatly in others. They all exhibit an X-ray diffraction pattern of the A type, except for the high amylose starches, which give rise to faint B-type patterns [32–34]. Another property in common for cereal starches, which also differentiates them from all other starches, is their lipid content [35]. The cereal starch lipids are an integral part of the starch granule; they are not present as a result of contamination of the granule surface during starch preparation [36]. Most other properties of the cereal starches differ considerably from one species to another, and moreover there may be a large variation in properties among different varieties of the same species. Some of these differences are described in this subsection.

1. Composition and Structure of Cereal Starch Granules

a. Size and Shape

The microscopic appearance differs among cereal starches (Fig. 3), and it is possible to determine the botanical source of a starch by examining it under the microscope [37–39]. Particle sizes and particle size distributions also differ among the cereal starches. Of all starches—cereal and noncereal—only those of wheat, rye, barley, and triticale, the genetic mixture of wheat and rye [41] have a bimodal size distribution (Fig. 4). The particle size of some cereal starches is given in Table 8 together with values of the specific surface area [42]. Some attention has been given to the bimodal size distribution as this is a common property of starches that perform well in baking (See Table 7) [31]. If the particle size distribution contributes to the difference in baking performance between starches, it is not because of such a simple circumstance as a difference in surface area. The large wheat starch granules and maize starch granules have about the same surface area, 0.47 and 0.52 m^2/g, respectively [43, 44]. Moreover, the enlarged surface area due to the presence of the B-granules in a complete wheat flour, compared to the large wheat granule starch fraction, would require more of the gluten matrix for enveloping the starch granules. This in itself could be expected to weaken the gluten matrix. Another implication of the bimodal size distribution is that it makes possible a very effective close packing of the granules.

The proportion of B-granules in barley starch is about 15%, calculated on volume [40, 42], and in rye it is about 35% [40]. The largest cereal starch granules are the A-granules found in rye and barley starches [45]. In rye starch, the A-granules have diameters of 10–35 μm [40], whereas the diameter of the A-granules in barley is in the range of 10–25 μm [46], with an average of 12.5–15.0 μm [42]. Oat and rice starch granules are among the smallest cereal starch granules known (Fig. 3). In the endosperm these granules are present as agglomerates, or clusters. This might explain the polygonal shape of, for example, individual rice starch granules (Fig. 3). The diameter of an in-

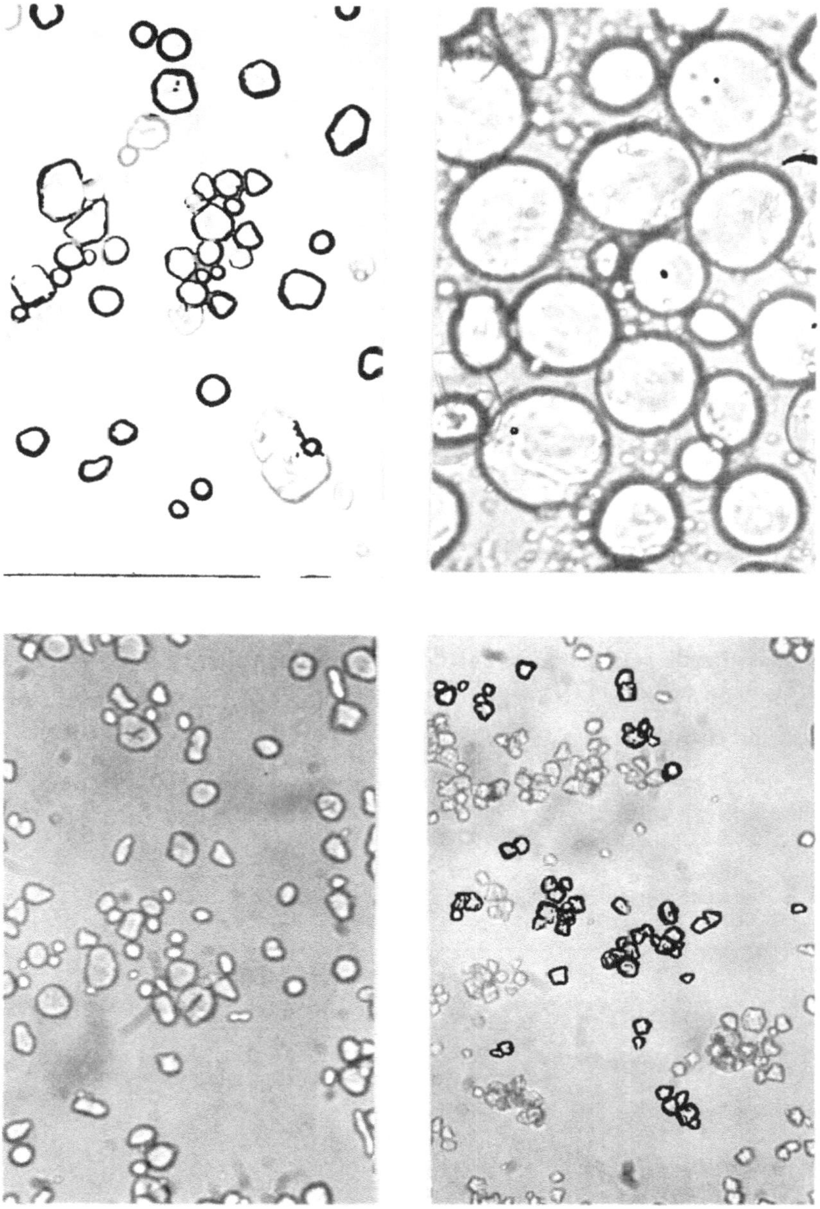

Fig. 3 Some cereal starches in the microscope. Upper row (from the left), waxy maize and rye; lower row (from the left), amylomaize and rice.

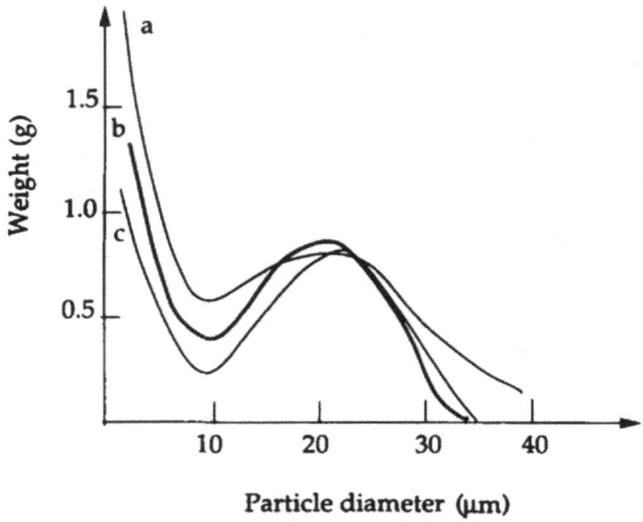

Fig. 4 The particle size distribution of (a) rye starch, (b) wheat starch, and (c) barley starch. (Adapted from [40].)

dividual oat granule is in the range 3–10 μm, whereas the total cluster has a diameter of about 60 μm [47].

b. Chemical Composition

(1) Amylose content. The amylose content and lipid content of some cereal starches are given in Table 9. Waxy varieties and high-amylose varieties are

Table 8 Granule Diameter and Specific Surface Area of Some Cereal Starches

Starch source	Type of granules	Particle diameter (μm)	Specific surface area (m^2/g)
Waxy barley	B	2.1–3.1	1.09–1.80
	A	10.2–13.6	0.285–0.369
Normal barley	B	2.9–3.1	0.99–1.12
	A	12.5–15.0	0.265–0.302
High-amylose barley	B	3.9–4.4	0.75–0.81
	A	10.8–12.2	0.336–0.371

Source: Data compiled from [42].

Table 9 Amylose and Lipid Contents of Some Cereal Starches

Starch source	Amylose (%)	Lipids (mg/100 g starch) FFA[a]	LysoPL[b]
Barley			
waxy	<8.5	28–61	158–460
normal	25–30	31–51	630–984
high-amylose	>38	54–92	864–1171
Oats	25–30	268–462	977–1171
Maize			
waxy	0–3	12–462	5–13
normal	24–32	376–529	183–344
high-amylose	>35	543–667	385–449
Rice			
waxy	<3	15–60	0
normal	10–30	222–367	407–811

[a]Free fatty acids.
[b]Lysophosphatidylcholine.
Source: Data collected from [35], [42], and [48].

found only among the diploid cereal starches, and among these are barley, rice, sorghum, and maize starches [35, 42]. For maize starch there are several genes that affect the composition of the starch—waxy (*wx*), sugary-1 (*su* 1), sugary-2 (*su* 2), dull (*du*), and amylose extender (*ae*). These genes affect the amylose content, but the starches may also contain intermediate material [49].

The waxy varieties are low in amylose (below 8.5% in waxy barley and below 3% in waxy maize and waxy rice starches), but they are also low in lipids (Table 9). The highest lipid values of the waxy starches are obtained for waxy barley starches, which contain quite a considerable amount of lysophospholipids. The high amylose varieties (>35% amylose) are high in lipids (Table 9). Most of the so-called normal starches have a limited range of amylose, about 20–30%. Rice starch is special in this respect, as rice varieties cover the whole range of amylose contents from zero to above 30% [50, 51]. The rice starches could therefore be divided into four groups: high-amylose, intermediate-amylose, low-amylose, and waxy varieties [51]. In each group there are varieties with low, intermediate, and high gelatinization temperatures.

Test baking experiments have shown that waxy varieties do not perform well in baking [31, 52]. This result might be interpreted as showing that retrogradation of amylose is necessary for good crumb structure. The high-amylose varieties seem not to offer any advantages in baking over their normal counterparts. This might be related to restricted swelling of amylose at temperatures below 100°C.

(2) Composition of amylose and amylopectin. The size distribution of chain lengths in amylopectin is found to be bimodal in many starches (waxy rice, waxy maize, rice, maize, and amylomaize), whereas the size distribution is trimodal for wheat starch (53, 54). With better gel filtration equipment, which gives higher resolution of peaks, polymodal distributions have been obtained for amylopectin from several starches, including waxy rice, barley, and wheat [53, 55]. The elution profiles reveal two main fractions (F1 and F2) when polymodal size distributions are obtained. The degree of polyermization of these two main fractions is given in Table 10, together with the average chain length (CL) of the total amylopectin molecule. This latter parameter is in the range 21–28 for all the cereal starches except the high-amylose varieties, which have an average chain length of 44. The shorter chain lengths are typical for starches giving the A-type X-ray diffraction pattern, whereas the chain length of high-amylose starch is typical of a starch giving the B-type pattern [53].

Table 10 Characterization of the Distribution of the Chain Length of Amylopectin

Amylopectin	Average chain length			Ref.
	Whole	F1	F2	
Maize		45	15	55
	28	60	17	53
Waxy	24	51	18	53
High-amylose	44	65	19	53
Barley		50	14	55
A-Granules	21			46
B-Granules	23			46
Waxy barley				
A-Granules	23			46
B-Granules	22			46
Waxy rice	23	48	17	53
Rice	25	56	16	53
	28	59	17	53
Wheat		49	13	55
	25	56	17	53

Inspection of the F1 and F2 fractions shows that these chain lengths are also quite similar, that of F1 being in the range 45–60 for the normal and waxy varieties, whereas that of the F1 fraction of the high-amylose maize starch was somewhat longer (CL = 65). The F2 fraction had chain lengths of 14–18 for the normal and waxy varieties and 19 for the high-amylose maize starch. These intervals are not very different from the intervals obtained for the wheat starches shown in Table 8, 49–56 in the case of the F2 fraction and 13–17 in the case of the F1 fraction. The molecular composition of amylopectin therefore seems not to be very different in, for example, maize than in wheat.

(3) Lipids. In the high-amylose barley starches, the lipid compositions is dominated by the lysophospholipids, whereas the high-amylose maize starches contain a higher proportion of free fatty acids. The results reported in Table 9 seem to indicate that there is a correlation between the amylose content and the lipid content. In fact, if all the starches (waxy, normal, high-amylose) of the same species are included, very good linear regression correlation coefficients are obtained, $r = 0.9347$ for barley starch and 0.9531 for maize starch [35]. However, if only the varieties of normal amylose content are included in the regression anaylsis, no significant correlation is observed. Of the starches with normal amylose content (i.e., about 20–30%), oat starch has an unusually high lipid content (Table 9). For oats it has been found that the lipid content of the starch is related to the lipid content of the grain—the higher the lipid content of the grain, the higher the lipid content of the starch [56].

The lipids present in the starch granules of normal amylose content are lysophospholipids (LysoPL) and FFA. Barley starch is characterized by very low levels of FFA, whereas maize starch (independent of amylose content) contains more FFA than LysoPL. The fatty acid composition does not differ very much between the cereal starches and is dominated by $C_{18:2}$ (35–46% of total fatty acids) and $C_{16:0}$ (37–55% of total fatty acids [35]).

The differences in lipid content between cereal starches are reflected in their DSC parameters, such as the enthalpy of transition due to the amylose–lipid complex, ΔH_{cx} (Table 11). Among the normal starches, the highest values are obtained for the oat starches, and the lowest values are measured for rye and maize starches [24, 56]. As the lipid composition is quite similar for the cereal starches, these differences in ΔH_{cx} may be almost exclusively due to the presence of different amounts of amylose–lipid complex. As wheat starch gives values in between these, the behavior of a starch during baking does not appear to be related to interactions with the polar lipids.

The data in Table 11 show that there are great variations in T_{cx} values. However, as the starch/water ratio also differs among the studies, it is impossible to tell whether the variation is due to the formation of different types of complexes or if it is merely a reflection of the water content.

As expected from the low lipid content in waxy maize starches, no endotherm

Table 11 DSC Parameters of the Amylose–Lipid Complex in Some Cereal Starches

Starch source	ΔH_{CX} (J/g)	T_{CX} (°C)	Experimental conditions	Ref.
Maize				
Waxy	—[a]	—[a]		57,58
Normal	1.2	98.6	Starch/water ratio = 1:3	58
High-amylose	3.6	101.4	Values obtained during a rescan. Starch/water = 43:57	57
	2.7	97.8	Values obtained during a rescan. Starch/water = 1:3	58
Rye	1.6	113	Starch/water = 1:1	40
	0.3–0.6	100.0–106.6	60% water, nine different varieties	24
Barley	2.9	115	Starch/water = 1:1	40
Oats	2.4–3.7	94.1–97.1	75% water, four different varieties	56
	3.52	100.5	= 75% water	59
Wheat	2.1	117	Starch/water = 1:1	40
	2.2	107.7		57

[a]No endotherm detected.

is observed for these starches (Table 11). It might then be expected that ΔH_{cx} would be very high for high-amylose starches. However, these starches are extremely difficult to gelatinize [49], and in DSC measurements there will be an overlap between the gelatinization endotherm and the endotherm due to the amylose–lipid complex [58]. This is the case also in excess water. Any values of ΔH_{cx} for high-amylose varieties are thus only estimates.

(4) Protein. The protein content seems not to differ between wheat starch and other cereal starches. As in the case of wheat starch, the protein content is related to the purification procedure followed. The protein content in oat starch has been reported to be 0.8%, and that of barley starch, 0.5% [60]. For maize starch a value of 0.3% has been reported [61].

(5) Large and small granules. The differences in composition and properties between the large and small granules of wheat starch were described in detail in Chapter 2. A few investigations evaluated differences between large and small granules of barley starch. The amylose content of the large granules was reported to be 22.1%, and that of the small granules, 19.0% [62]. The large granules from a waxy barley starch had an amylose content of 3.6%, and the small granules, 1.8%. In other investigations only very small differences in the amylose content between the large and small barley starch granules were found [42].

c. Crystalline Structure

As discussed above, all the normal cereal starches give rise to an X-ray diffraction pattern of the A type [32–34]. The waxy varieties also give the A-pattern, whereas the high-amylose varieties give the B-pattern [49]. Furthermore, the high-amylose varieties give a lower degree of crystallinity [49]. By hydrolyzing the amorphous starch, the amount of crystalline starch can be determined. For waxy rice with low gelatinization temperature, the degree of crystallinity was determined to be 35%, and in a waxy rice of high gelatinization temperature, 50% [51].

In the normal cereal starches no V-pattern due to the amylose–lipid complex has been observed, not even for oat starch, where the possibility would be the best because of the high lipid content [56]. An interesting observation is that the intensity of the X-ray diffraction pattern seems to be related to the amount of lipid present in the starch [63]. Waxy maize, with a very low level of lipids (Table 9), gives a strong A-pattern, whereas oat starch gives a much weaker pattern.

Steeping barley starches (i.e., keeping the starch at a temperature just below the gelatinization temperature) results in the formation of the V-pattern in normal and high-amylose barley starches [32].

2. Gelatinization

a. Gelatinization Temperature

The gelatinization temperature range has been determined for several cereal starches, either by microscopy or by DSC. Some results are given in Table 12. The corresponding ΔH values are given in Table 13. The results in Table 12 show that there is considerable variation in the gelatinization temperature of different starches. One group of starches gelatinizes at low temperatures—for example, oat starch with T_m in the range 55–61.6°C, barley starch with $T_m \approx$ 60–61°C, and rye with T_m in the range 53.5–61°C. These starches gelatinize at about the same temperature as wheat starch. The other normal starches in Table 10 (maize and rice) gelatinize at higher temperatures; T_m is in the range of 60–71.1°C for maize and 63.6–78.2°C for rice. The waxy varieties gelatinize at about the same temperature as their normal counterparts. Compare, for example, waxy rice (T_m = 64.3–78.8°C) and rice (63.6–78.2°C) in Table 12. The gelatinization temperature range of the high-amylose maize starch is shifted to higher temperatures. However, as discussed above, it is difficult to determine the exact temperature range because of the overlapping endotherm of the amylose–lipid complex. The microscope is not suitable for this starch because of the very weak birefringence [49].

There is great variation in the gelatinization temperature of different varieties

Table 12 Gelatinization Temperature Range of Cereal Starches

Starch	T_o (°C)	T_m (°C)	T_c (°C)	Method	Ref.
Waxy maize	67.2	72.8		DSC	64
	64	68	70	Microscopy	65
Normal maize	69.2	73.2		DSC	64
	49	60	67	Microscopy	49
	65	68	73	Microscopy	31
High-amylose maize	82	86	99	Microscopy	65
	66.0	67.3	110.1	DSC[a]	57
Waxy barley	58	64.5	67.5	Microscopy	32
	43.7	57.6	77.0	DSC	66
Normal barley	55.5	60.5	65	Microscopy	32
	56	60	61	Microscopy	31
	46.7	56.5	73.7	DSC	66
High-amylose barley	58.5	65	69	Microscopy	32
Waxy rice	45.6–62.3	64.3–78.8	81.0–94.8	DSC[b]	51
Rice	69	71	75	Microscopy	31
		63.6–78.2		DSC[c]	67
Rye	50.0	53.5	58	Microscopy	68
		54.8–60.3		DSC[d]	24
	49	61	56	Microscopy	31
Triticale	53.0	56.5	61.5	Microscopy	68
Durum	50.0	55.0	60.5	Microscopy	68
Oats		57.8–61.6		DSC[e]	56
	53	55	59	Microscopy	31
	61.0	66.1		DSC	59
Wheat	54.0	60.5	62.0	Microscopy	68

DSC, differential scanning calorimetry.
[a]The transition of the amylose–lipid complex is included (starch/water 43:57).
[b]12 varieties. [c]17 varieties. [d]Nine varieties. [e]Four varieties.

of the same species (Table 12). For 17 rice starches, T_m was found to be in the range of 63.6–78.2°C [67], and for nine rye varieties, in the range of 54.8–60.3°C [24]. The reason for the variation is certainly due to genetic differences among the varieties, but environmental factors also affect gelatinization behavior [24]. Moreover, there is considerable intrapopulation variability, which has been studied in maize starch [69]. Twenty plants of the same variety were compared, and from each plant, data from five kernels were averaged to give T_o and ΔH. The T_o values were in the range of 59.8–64.7°C, and ΔH was in the range 7.7–11.9 J/g.

Table 13 Gelatinization Enthalpy of Cereal Starches

Starch	ΔH (J/g)	Experimental conditions	Ref.
Waxy maize	17.6	3 mg starch + 10 μL water	64
	16.8	Water/starch ratio 3:1	58
	16.8	52.4% water	65
	14.3	Starch/water ratio 43:57	57
Normal maize	16.0	3 mg starch + 10 μL water	64
	13.2	Starch/water ratio 1:3	58
High-amylose maize	17.6[a]	51.8% water	65
	8.2[a]	Starch/water ratio 1:3	58
	10.2[a]	Starch/water ratio 43:57	57
Rye	10.6–12.0	60% water; nine varieties	24
Oats	9.5–10.6	75% water; four varieties	56
	10.0	≈75% water	59
Waxy barley	12.1	3–4 mg starch + 15 μL water	66
Normal barley	10.5	3–4 mg starch + 15 μL water	66
Waxy rice	12.8–16.0	4 mg starch + 20 mg water; 14 varieties	67
Rice	9.3–16.1	4 mg starch + 20 mg water; 17 varieties	67
Wheat	9.8	Starch/water ratio 43:57	57

[a]The enthalpy due to the transition of the amylose–lipid complex is included.

The gelatinization enthalpy (ΔH) of some starches is given in Table 13. If the high-amylose maize starch is omitted from the comparison, the ΔH values range from 9.3 to 17.6 J/g. The ΔH values are certainly affected by the water content during the measurements, but there are several results in Table 13 that were obtained at such a high water content that the ΔH value is independent of the water content [70, 71]. The waxy maize starches are found toward the upper limit of the interval, whereas rye and oats (together with wheat) are found toward the lower limit. The rice starches might give ΔH values over the whole interval. When the maize starches are compared, it is tempting to suggest that ΔH depends on the amylopectin content. Such a relation is to be expected if the thermal transition is due only to the melting of the cystalline domains of amylopectin. For the rice starches there seems to be no relation between the amylopectin level and ΔH. A waxy rice was no amylose at all gave a ΔH value of 14.7 J/g, which is close to the values obtained for one rice variety of 7.8% amylose (ΔH = 14.2 J/g) and another with 28.1% amylose (ΔH = 14.1 J/g). It is evident that events other than melting contribute to the gelatinization endotherm [72, 73].

The variation in ΔH among oat and rye varieties are not so large as in the case of rice. For oats, ΔH was in the range 9.5–10.6 J/g, and for rye, 10.6–12.0 J/g. The comparatively low values observed for oat starch might be due to the exothermic complex formation that occurs during gelatinization.

b. Swelling and Solubility

The swelling power of cereal starches is illustrated in Fig. 5. The results might be influenced to some extent by different methods used by different investigators but the figures still reveal some typical differences. It is evident that in a series waxy–normal–high-amylose starches, the waxy varieties show the greatest swelling and the high-amylose the least [32, 49]. Of the normal starches compared in Fig. 5, barley starch shows the lowest swelling power whereas maize starch shows the highest. There is a considerable increase in swelling power when the temperature is increased from 90 to 95°C. It could be expected that a low swelling, as in high-amylose starches, should be coupled to a low solubility. This is exactly what is observed for maize starches; at 95°C the solubility of normal maize starch was 18–25%, but for the high-amylose variety it was only 6.4% (Fig. 6). In the case of barley starch the opposite was found. The high-amylose barley starch had a solubility of 10.73% at 90°C, but the normal barley had only 3.74% [32]. If the normal starches are compared at 90°C it is found that wheat, rye, and maize all give high solubilities (>10%; in most cases >15%), whereas barley and oats give very low values (below 10%, and even below 5%).

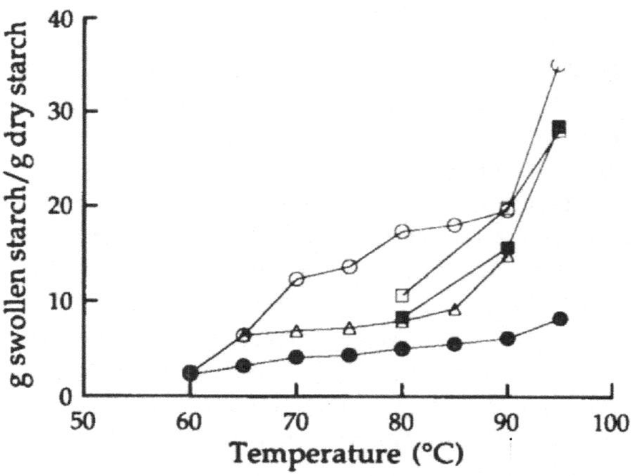

Fig. 5 The swelling power of some cereal starches during heating: wheat (△), maize (○), high-amylose maize (●), oats high in fat (■), oats low in fat (□). (Data taken from [49], [56], and [74].)

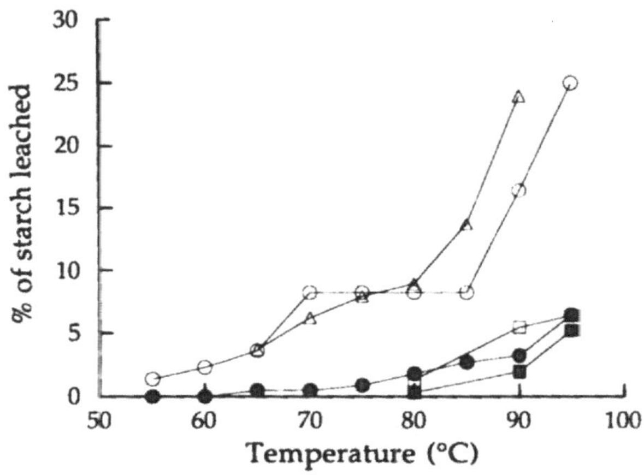

Fig. 6 The solubility of some cereal starches during heating. The legend is the same as in Fig. 5.

The importance of the lipid content to the properties of the starch is clearly seen for oat starch. The swelling power as well as the solubility were restricted in the high-lipid variety compared with the low-lipid variety [56]. If the starch is kept at 96°C, both swelling power and solubility increase for oat starch and will become higher than for wheat or maize starch [61]. The material solubilized from maize at 90°C was found to be 78% amylose, and the corresponding value for wheat was similar, 80% [74]. The composition of the leached material is affected by temperature and by the conditions during heating [75].

c. Rheological Behavior

The rheological behavior of a starch gel is strongly related to the source of the starch, and even during the heating sequence differences are already evident. Pasting temperature (i.e., the temperature at which it is possible to record a viscosity value) measured in the Brabender visco/amylograph increased in the order maize < wheat < rice < barley when suspensions of 12.5% concentration were compared. The maximum viscosity increased in the order wheat≈barley < maize≈rice [76]. The peak viscosity is obtained at different temperatures depending on the starch, and the temperature at peak viscosity increased in the order maize < rice < wheat ≈ barley. The high-amylose varieties give very low viscosities during this kind of measurement [49]. Taking into account the low swelling power and the poor solubility of these starches, such a result is not surprising.

The rheological properties of starch gels reported in the literature are difficult to compare because of different experimental methods. The aim of many of these investigations is to compare the rheological properties so as to be able to substitute one starch for another. Wheat starch has been compared with maize starch [74, 77, 78], oat starch with wheat starch [61], barley starch with oat starch [60], and so on. The rheological behavior of a starch gel depends on the phase volume of the starch granules, their size and shape, their deformability, and the amount and type of leached material [61, 74, 77–85]. The gel formation is easily followed in small-amplitude oscillation measurements as a decrease in phase angle [60, 77]. The gel strength is observed in the G' value. The rheological observations can be followed up with microscopic investigations to determine the degree of fragmentation of starch granules and the location of amylose and amylopectin in the gels [60, 86, 87].

It can easily be seen that the rheological behavior of an oat starch gel differs from, for example, that of a wheat starch gel. But there are also differences among varieties. In the case of oat starch, G' values of 8% gels were compared, and it was found that G' was highest for the starch with the highest lipid content [56]. G' for the strongest gel was about twice that of the weakest gel.

3. Retrogradation Behavior

The retrogradation behavior of different cereal starches has been compared in both rheological measurements and DSC measurements. Some of these results will be summarized in this section.

a. Development of Crystallinity

If DSC data are compared it is found that several starches crystallize to a lower extent and at a lower rate than does wheat starch. This is true for both rye and oat starches [24, 56], whereas maize and barley give higher limiting ΔH values than wheat starch [55]. For starch gel (40%) stored 12 days at 20°C, ΔH_c increased in the order wheat < waxy rice < rice. The crystallization rate was found to increase in the order waxy rice < rice < wheat [88]. For starch gels (30%) stored at 20°C for 7 days, the maize starch gave a higher ΔH_c value than wheat starch [89].

The low retrogradation tendency observed for oat starch might be explained by the high lipid content of this starch. The oat starch of the lowest retrogradation tendency was also the one with the highest lipid content [56]. When the oat starches were delipidized, their retrogradation tendency increased (Fig. 7). However, the retrogradation tendency of the oat starches was still much lower than that of wheat starch. It is interesting to note that the extent of retrogradation of oat starch is lower than that of high-amylose maize, although the amylopectin level is much higher in the oat starch. The lipid content cannot explain the low retrogradation tendency of rye starches, as rye starch is lower in lipids than wheat starch.

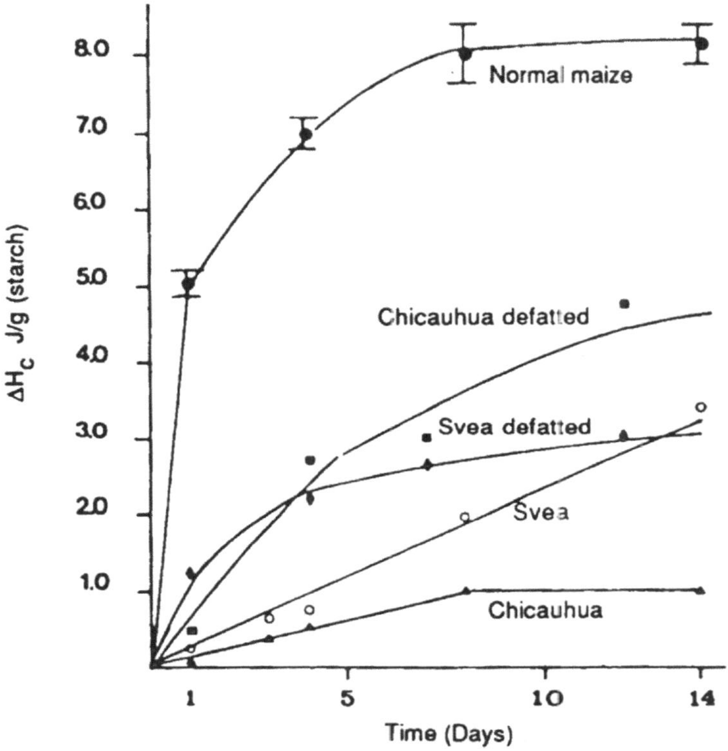

Fig. 7 The retrogradation of two oat starches (Svea with low lipid content and Chicauhua with high lipid content) before and after defatting. The retrogradation of oat starches is compared with maize starches differing in amylose content. (Reprinted with permission from [56].)

When starch gels (57% water) differing in amylose content were stored at 21°C for up to 2 months, the limiting ΔH_c value increased in the order high-amylose maize < wheat < waxy maize [90]. This order correlates with the amylopectin content of the starches. When the limiting ΔH_c value is plotted against the amylopectin content, a straight line is obtained. Extrapolation of this line to $\Delta H_c = 0$ gives an amylopectin content of 24%; that is, with an amylose content of 76% or more, the staling endotherm would not show up. It was suggested that this is due either to amylopectin being distributed in a different way in the starch granule or to the amylopectin crystallites being disrupted owing to interactions with material in other domains of the starch granule [90]. However, the extrapolation of the straight line to $\Delta H_c = 0$ may not be valid. Mixtures of amylopectin and amylose give higher ΔH_c values than expected from the amylopectin content alone at low levels of amylopectin [91].

b. Firmness Measurements

The firmness of a starch gel increases during storage. This is described in Chapters 2 and 7 for wheat starch gels. The extent of retrogradation can be taken from the E modulus or the shear modulus after a long period of storage. The shear modulus was found to increase in the order wheat < barley < maize when 25% starch gels were stored for 30 days at 1°C [55]. The rate of increase was found to increase in the same order. When 40% starch gels were stored at 20°C for 12 days, the E modulus increased in the order waxy rice < wheat < rice, and the initial rate of retrogradation increased in the same order [88]. Wheat and maize were also compared, and it was found that the maize starch gels give higher shear modulus values than the wheat starch gels [89].

It can therefore be concluded that most cereal starches retrograde to a larger extent and at a higher rate than wheat starch. Oat and rye starches, however, are even less prone to retrogradation than wheat starch.

4. Interactions with Other Components

The interaction between cereal starches and lipids has already been described to some extent, at least with respect to the lipids present in the native starch granule. As in the case of wheat starch, the addition of polar lipids might influence the behavior of the starch. The high- and low-amylose varieties have been used to study this type of interaction. Quite a lot of information exists regarding interactions of wheat starch with protein, but very little is known about such interactions involving other cereal starches. In the present section a few examples will be given of interactions with lipids and protein.

a. Interactions Between Cereal Starches and Polar Lipids

When polar lipids are added to a cereal starch of normal amylose content, an increase in the endotherm due to the transition of the amylose–lipid complex is observed. When oat starch was delipidized and then the oat lipids were added back, a considerable increase in ΔH_{cx} was observed [59]. However, these samples had to be run twice to achieve the high ΔH_{cx} values. The first run gave ΔH_{cx} values of 0.472–0.877 J/g, whereas a second run gave 2.68–4.02 J/g. The addition of a surfactant to maize starches differing in amylose content gave the results shown in Table 14. The waxy maize gave no endotherm even in the presence of added lipids under the experimental conditions presented in Table 14. It might then be concluded that no complex formation occurs between the added surfactant and waxy maize starch. The gelatinization enthalpy, however, decreased from 16.8 to 12.7 J/g when the surfactant was added. This result

Table 14 Transition Temperature and Enthalpy Obtained for the Amylose–Lipid Complex in Maize Starches of Different Amylose Content After Addition of a Surfactant[a]

Starch source	Without (−) or with (+) added surfactant	T_{CX} (°C)	ΔH_{CX} (J/g)
Normal maize			
First scan	−	98.6±0.6	1.2±0.3
Second scan	−	96.8±1.4	2.5±0.5
First scan	+	94.4±1.4	4.2±0.3
Second scan	+	98.1±2.1	6.7±0.1
Waxy maize			
First scan	−	No endotherm	
Second scan	−	observed in any	
First scan	+	of the scans	
Second scan	+		
High-amylose maize			
First scan	−	97.1±1.1	3.5±0.9
Second scan	−	97.8±0.7	2.7±1.0
First scan	+	93.6±2.4	4.9±0.6
Second scan	+	97.6±0.6	12.6±1.4

[a]The surfactant added is cetyltrimethylammonium bromide.
Source: Data compiled from [58].

indicates that complex formation does in fact occur [58]. A similar effect occurs when SDS is added to waxy maize starch [92]. In other investigations an endotherm in the temperature range corrsponding to that of the amylose–lipid complex was observed for amylopectin in the presence of added lipids [91, 93]. The results in Table 14 further stress that a second heating is necessary to obtain as much complex as possible.

Another way to study the influence of lipids is to extract the native lipids and compare the properties of the lipid-free starch with the native starch. However, this approach is not as straightforward as might be expected, the reason being the difficulty of extracting the lipids without gelatinizing the starch [36]. The retrogradation of oat starch, as described above, could easily be studied, because in these experiments the starch should already be gelatinized [56]. Maize starch was delipidized by Soxhlet extraction (85% methanol–water for 72 h), which removed 69.2% of the lipids [94]. The low-lipid maize starch gave a higher peak viscosity at a lower temperature in the amylogram than the corresponding native maize starch.

b. Interactions with Proteins

Adsorption of wheat proteins on wheat starch granules [43] was described in Chapter 3, and we have carried out similar experiments for rye, maize, and waxy maize starches. The wheat protein fraction (composed of gluten protein) seems to adsorb to about the same extent on all these cereal starches.

Changes in protein–starch interactions have been suggested to occur during storage of rice grain [95]. The extractability of rice protein decreases during storage owing to changes in the glutelin fraction. The decrease in solubility was coupled to an increase in molecular weight of this protein fraction. It was further found that the adsorption of rice glutelin on starch, amylose, and amylopectin decreased during storage.

The effect of the addition of gluten to starch gels on G' was also investigated [96]. Gluten increased G' for wheat and rye starches, decreased G' for maize starch, and had no influence on barley and triticale starches.

The removal of proteins from rice starch has been found to have a minor, but measurable, effect on starch gelatinization [97].

D. Nonstarch Polysaccharides

Two groups of nonstarch polysaccharides of non-wheat cereal origin are of special interest. These are the pentosans in rye and the β-glucans in oats and barley.

Rye flour is used to a considerable extent in baking, but, as already pointed out, rye proteins do not form a gluten gel. The lack of gluten-forming ability demonstrates the fundamental difference between rye proteins and wheat proteins in aggregation behavior. One possibility could be that the presence of pentosans prevents the protein aggregation. Interactions between pentosans and glutenins have recently been discussed and are suggested to have a negative effect on dough properties [98]. The pentosans are also assigned a positive role in the baking performance of rye. As the proteins of rye lack gluten-forming ability, the pentosans perform some of the roles that gluten performs in wheat.

The β-glucans in barley and oats have come into focus for quite different reasons. The β-glucans of barley are an important quality parameter when barley is selected for malting and brewing. A barley variety suitable in this respect should have as low a β-glucan content as possible. The β-glucans in oats are of interest because of nutritional aspects.

1. Rye Pentosans

Whole-grain rye contains 16.5% dietary fiber distributed as 2.3% β-glucans, 7.6% arabinoxylans, 2.6% cellulose, and 3.0% Klason lignin [99]. The arabinoxylans, or pentosans, thus constitute the main part of the dietary fiber in rye.

Wheat pentosans were described in Chapter 2. One difference between wheat and rye pentosans is the amount found in flour. Whereas the total pentosan content of a wheat flour is below 2%, it is above 3% in rye [100]. The water-soluble (WS) pentosans amount to about 1% or less in a wheat flour [100, 101] but more than 2% in rye [100].

When rye flour is extracted with water (for example, 1 g flour and 20 mL water), a viscous solution is obtained [100]. Rye flour extract has a viscosity three times that of water. The WS pentosans have a high molecular weight, above 170,000 [99].

The WS pentosans were fractionated on DEAE-cellulose in the same way as the WS pentosans from wheat [102]. Five fractions were obtained, and the first one was eluted with water [99]. This fraction contained about 43% of the applied material and can be regarded as a pure arabinoxylan. Its sugar composition was 30.8% arabinose 64.5% xylose, and 3.3% glucose. The ratio of arabinose and xylose residues was 1:2.1. The second fraction obtained could also be regarded as a pure arabinoxylan. It constituted 12% of the applied material and had a sugar composition of 27.3% arabinose, 50.0% xylose, and 10.5% glucose. Of these two fractions, fraction II contained more branched and double-branched xylose residues.

The sugar composition of WS pentosans in wheat and rye is compared in Table 15. The xylose/arabinose ratio is much higher in rye than in wheat. This would imply that there is a lower degree of branching in rye pentosans or that the WS wheat pentosans have longer chains. A lower degree of branching in rye WS pentosans could explain their more viscous solutions compared with wheat [100]. NMR spectroscopy, periodate oxidation, partial acid hydrolysis, and

Table 15 Monosaccharide Composition of DEAE-Cellulose Fractionated Arabinoxylans from Rye and Wheat

Fraction	Yield (%)	Arabinose (%)	Xylose (%)	Glucose (%)	Xylose/ arabinose
Rye[a] I	43	30.8	64.5	3.3	1:0.48
II	10	27.3	50.0	10.5	1:0.55
Durum[b] I	34.5	42.0	58.0	—	1:0.72
II	13.9	42.2	57.8	—	1:0.73
HRS[b,c] I	26.6	40.0	60.0	—	1:0.66
II	18.4	41.0	59.0	—	1:0.69

[a][99]
[b][103]
[c]HRS = hard red spring

enzymatic hydrolysis have been used to reveal details in the structure of water-soluble arabinoxylans in rye [104, 105]. The suggested structures of rye arabinoxylan fractions I and II are illustrated in Fig. 8.

2. β-Glucans in Oats and Barley

It was early realized that β-glucans play an important role in beer production, and most of the research on β-glucans therefore concerns barley. In oats as well as barley the group of nonstarch polysaccharides called β-glucans consists of (1→4) and (1→3) glucosidic links between β-D-glucopyranosyl units. They occur in various places of these plants, even the leaves, but only the β-glucans in the endosperm will be discussed here as being relevant to breadmaking using oat and barley flour. An extensive review of β-glucans in oats has been reported by Wood [106], and some of the data presented here were taken from that review.

Chemically the β-glucans contain both (1→3) and (1→4) linkages in the same molecule, with (1→4) in a ratio of about 3:1 to (1→3) linkages. The structure is illustrated in Fig. 9.

Both acid and alkali extraction have been used to separate β-glucans from oats and barley. The temperature must, of course, be kept below the gelatinization temperature, and the alkali conditions must be mild to avoid starch solubilization. In the case of barley, 1 M sodium hydroxide and 1% (w/w) sodium borohydride have been successfully used as extraction media. Urea solutions

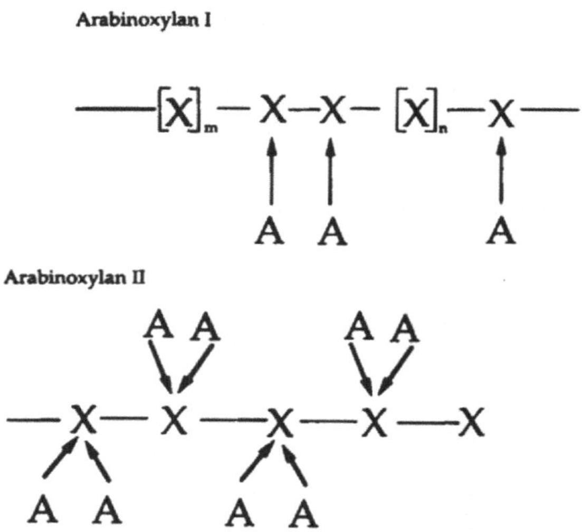

Fig. 8 Structural features of arabinoxylan I and II in rye. (Adapted from [104].) A denotes terminal α-L-arabinofuranosyl residues and X denotes 4-linked β-D-xylopyranosyl residues; $m \geq 1$; $n \geq 1$.

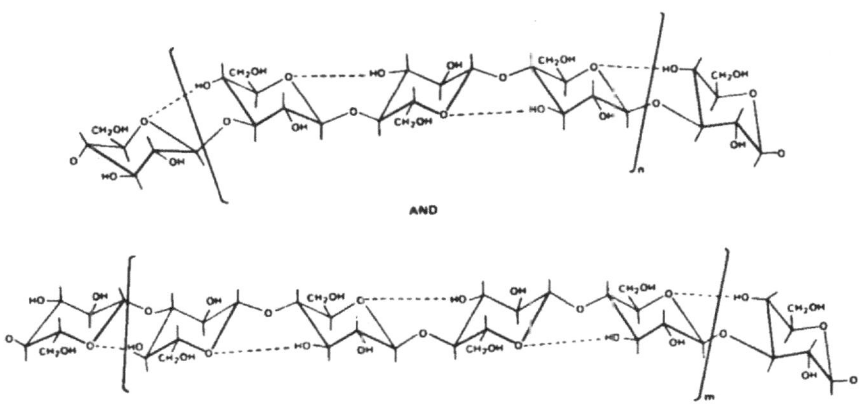

Fig. 9 Main structural units of β-D-glucans of oats. (Adapted from [106].)

(8 M) are also reported to work. In the case of oats, the addition of sodium carbonate to pH 10 has been used. Protein that is also in solution is separated by precipitation, and the β-glucans are then precipitated by treatment with 50% isopropyl alcohol in water [106]. This crude extract also contains pentosans, and by precipitation with 20–30% ammonium sulfate a quite pure β-glucan fraction can be obtained.

β-glucans, both water-soluble and water-insoluble, amount to several percent in the endosperm. There is no simple quantitative analysis. The main contamination is α-glucan, which can be quantified by α-amylase degradation to be subtracted from the total extracted amount.

Variation in β-glucan content was recently reported among six cultivars of *Avena sativa* and eight wild *Avena* species [107]. Among the six cultivars β-glucan content varied between 3.4 and 4.4% (with two different levels of nitrogen fertilizer). The wild species varied much more, from 1.2 to 5.7%.

A recent determination of the molecular weights of β-glucans from oats, barley, and rye has been reported [108]. The oat cultivars showed the highest molecular weights (about 3×10^6).

The main features of interest in β-glucans are their water-binding properties and their physiological effects as dietary fibers.

REFERENCES

1. Anon. (1991). *Jordbruksekon. Medd.*, 53: 37.
2. McMullen, M. S. (1991). Oats, in *Handbook of Cereal Science and Technology*, K. J. Lorenz and K. Kulp, Eds., Marcel Dekker, New York, p. 199.

3. Lorenz, K. J. (1991). Rye, in *Handbook of Cereal Science and Technology*, K. J. Lorenz and K. Kulp, Eds., Marcel Dekker, New York, p. 331.
4. Pomeranz, Y. (1969). Bread without bread. *Baker's Dig.*, 43(2): 22.
5. Rajapaksa, D., Eliasson, A.-C., and Larsson, K. (1983). Bread baked from wheat/rice mixed flours using liquid-crystalline lipid phases in order to improve bread volume, *J. Cereal Sci.*, 1: 53.
6. Lásztity, R. (1984). *The Chemistry of Cereal Proteins*, CRC Press, Boca Raton. Fla.
7. Young, V. L. (1986). Oat lipids and lipid related enzymes, in *Oats—Chemistry and Technology*, F. H. Webster, Ed., AACC, St. Paul, Minn., p. 205.
8. Jayasinghe, G., Larsson, K., Miezis, Y., and Sivik, B. (1991). Oat lipids—interaction with water and characterization of aqueous phases, *J. Disperson Science and Technology*, 12:443.
9. Morrison, W. R. (1978). Cereal lipids, *Adv. Cereal Sci. Technol.*, 2: 221.
10. Chung, O. K., and Tsen, C. C. (1974). Triticale lipids, in *Triticale: First Man-Made Cereal*, C. C. Tsen, Ed., AACC, St. Paul, Minn. p. 191.
11. Weber, E. J. (1969). Lipids from maturing grain of corn. I. Changes in lipid classes and lipid composition, *J. Am. Oil Chem. Soc.*, 46: 485.
12. Wieser, H., Seilmeier, W., and Belitz, H.-D. (1980). Vergleichende Untersuchungen über partielle Aminosäuresequenzen von Prolaminen und Glutelinen verschiedener Getreidearten. I. Proteinfraktionierung nach Osborne, *Z. Lebensm. Unters. Forsch.*, 170: 17.
13. Ma, C.-Y., and Harwalkar, V. R. (1984). Chemical characterization and functionality assessment of oat protein fractions, *J. Agric. Food Chem.*, 32: 144.
14. Johnson, L. A. (1991). Corn: production, processing, and utilization, in *Handbook of Cereal Science and Technology*, K. J. Lorenz and K. Kulp, Eds., Marcel Dekker, New York, p. 55.
15. Paulis, J. W., and Wall, J. S. (1969). Albumins and globulins in extracts of corn grain parts, *Cereal Chem.*, 46: 263.
16. Peterson, D. M. (1978). Subunit structure and composition of oat seed globulin, *Plant Physiol.*, 62: 506.
17. Field, J. M., Shewry, P. R., and Miflin, B. J. (1983). Aggregation states of alcohol-soluble storage proteins of barley, rye, wheat, and maize, *J. Sci. Food Agric.*, 34: 362.
18. Wieser, H., Seilmeier, W., and Belitz, H.-D. (1987). Vergleichende Untersuchungen über partielle Aminosäuresequenzen von Prolaminen und Glutelinen verschiedener Getreidearten, *Z. Lebensm. Unters. Forsch.*, 184: 366.
19. Shewry, P. R., and Mifflin, B. J. (1985). Seed storage proteins of economically important cereals, *Adv. Cereal Sci. Technol.*, 7:1.

20. Field, J. M., Tatham, A. S., Baker, A. M., and Shewry, P. R. (1986). The structure of C hordein, *FEBS Lett.*, 200: 76.
21. Shewry, P. R., Tatham, A. S., Forde, J., Kreis, M., and Miflin, B. J. (1986). The classification and nomenclature of wheat gluten proteins: a reassessment, *J. Cereal Sci.*, 4: 97.
22. Payne, P. I., and Rhodes, A. P. (1982). Cereal storage proteins: structure and role in agriculture and food technology, *Encycl. Plant Physiol., New Ser.*, 14A: 346.
23. Ma, W.-P., and Robyt, J. F. (1987). Preparation and characteriztion of soluble starches having different molecular sizes and composition, by acid hydrolysis in different alcohols, *Carbohydr. Res.*, 166: 283.
24. Gudmundsson, M., and Eliasson, A.-C. (1991). Thermal and viscous properties of rye starch extracted from different varieties, *Cereal Chem.*, 68: 172.
25. Hejgaard, J. (1977). Origin of a dominant beer protein immunochemical identity with a β-amylase-associated protein from barley, *J. Inst. Brew.*, 83: 94.
26. Ewart, J. A. D. (1980). Loaf volume and the intrinsic viscosity of glutenin, *J. Sci. Food Agric.*, 31: 1323.
27. Bernardin, J. E., and Kasarda, D. D. (1973). Hydrated protein fibrils from wheat endosperm, *Cereal Chem.*, 50: 529.
28. Amend, T., Belitz, H.-D., and Kurthen, C. (1990). Electron microscopic studies on protein films from wheat and other sources at the air/water interface, *Z. Lebensm. Unters. Forsch.*, 190: 217.
29. Wannerberger, L., and Eliasson, A.-C. (to be published).
30. Wall, J. S., and Paulis, J. W. (1978). Corn and sorghum grain proteins, *Adv. Cereal Sci. Technol.*, 2: 135.
31. Hoseney, R. C., Finney, K. F., Pomeranz, Y., and Shogren, M. D. (1971). Functional (breadmaking) and biochemical properties of wheat flour components. VIII. Starch, *Cereal Chem.*, 48: 191.
32. Lorenz, K., and Kulp, K. (1984). Steeping of barley starch. Effects on physicochemical properties and functional characteristics, *Starch/Stärke*, 36: 116.
33. Zobel, H. F., Young, S. N., and Rocca, L. A. (1988). Starch gelatinization: an X-ray diffraction study, *Cereal Chem.*, 65: 443.
34. Zobel, H. F. (1988). Starch crystal transformations and their industrial importance, *Starch/Stärke*, 40: 1.
35. Morrison, W. R., Milligan, T. P., and Azudin, M. N. (1984). A relationship between the amylose and lipid contents of starches from diploid cereals, *J. Cereal. Sci.*, 2: 257.
36. Morrison, W. R. (1981). Starch lipids: a reappraisal, *Starch/Stärke*, 33: 408.
37. Wivinis, G. P., and Maywald, E. C. (1967). Photographs of starches, in

Starch: Chemistry and Technology, Vol. 2, *Industrial Aspects*, R. I. Whistler, E. F. Paschall, J. N. BeMiller, and H. J. Roberts, Eds., Academic, New York, p. 649.

38. Williams, M. R., and Bowler, P. (1982). Starch gelatinization: a morphological study of triticeae and other starches, *Starch/Stärke*, 34: 211.
39. Bowler, P., Williams, M. R., and Angold, R. E. (1980). A hypothesis for the morphological changes which occur on heating lenticular wheat starch in water, *Starch/Stärke*, 32: 186.
40. Karlsson, R., Olered, R., and Eliasson, A.-C. (1983). Changes in starch granule size distribution and starch gelatinization properties during development and maturation of wheat, barley and rye, *Starch/Stärke*, 35: 335.
41. Lorenz, K., Yetter, M., and Saunders, R. M. (1978). Scanning electron microscopy of maturing triticale, wheat and rye, *Cereal Chem.*, 55: 66.
42. Morrison, W. R., Scott, D. C., and Karkalas, J. (1986). Variation in the composition and physical properties of barley starches, *Starch/Stärke*, 38: 374.
43. Eliasson, A.-C., and Tjerneld, E. (1990). Adsorption of wheat proteins on wheat starch granules, *Cereal Chem.*, 67: 366.
44. Soulaka, A. B., and Morrison, W. R. (1985). The bread baking quality of six wheat starches differing in composition and physical properties, *J. Sci. Food Agric.*, 36: 719.
45. Berry, C. P., D'Appolonia, B. L., and Gilles, K. A. (1972). The characterization of triticale starch and its comparison with starches of rye, durum, and HRS wheat, *Cereal Chem.*, 48: 415.
46. MacGregor, A. W., and Morgan, J. E. (1984). Structure of amylopectins isolated from large and small starch granules of normal and waxy barley, *Cereal Chem.*, 61: 222.
47. Caldwell, E. F., and Pomeranz, Y. (1973). Industrial uses of cereals—oats, in *Industrial Uses of Cereals*, Y. Pomeranz, Ed., AACC, St. Paul, Minn., p. 393.
48. MacArthur, L. A., and D'Appolonia, B. L. (1979). Comparison of oat and wheat carbohydrates. II. Starch, *Cereal Chem.*, 56: 458.
49. Colonna, P., and Mercier, C. (1985). Gelatinization and melting of maize and pea starches with normal and high-amylose genotypes, *Phytochemistry*, 24: 1667.
50. Biliaderis, C. G., Page, C. M., Maurice, T. J., and Juliano, B. O. (1986). Thermal characterization of rice starches: a polymeric approach to phase transitions of granular starch, *J. Agric. Food Chem.*, 34: 6.
51. Tester, R. F., and Morrison, W. R. (1990). Swelling and gelatinization of cereal starches. II. Waxy rice starches, *Cereal Chem.*, 67: 558.
52. Ghiasi, K., Hoseney, R. C., Zeleznak, K., and Rogers, D. E. (1984).

Effect of waxy barley starch and reheating on firmness of bread crumb, *Cereal Chem.*, 61: 281.
53. Hizukuri, S. (1985). Relationship between the distribution of the chain length of amylopectin and the crystalline structure of starch granules, *Carbohydr. Res.*, 141: 295.
54. Hizukuri, S. (1986). Polymodal distribution of the chain lengths of amylopectins, and its significance, *Carbohydr. Res.*, 147: 342.
55. Kalichevsky, M. T., Orford, P. D., and Ring, S. G. (1990). The retrogradation and gelation of amylopectin from various botanical sources, *Carbohydr. Res.*, 198: 49.
56. Gudmundsson, M., and Eliasson, A.-C. (1989). Some physicochemical properties of oat starches extracted from varieties with different oil content, *Acta Agric. Scand.*, 39: 101.
57. Russell, P. L. (1987). Gelatinization of starches of different amylose/amylopectin content. A study by differential scanning calorimetry, *J. Cereal Sci.*, 6: 133.
58. Eliasson, A.-C., Finstad, H., and Ljunger, G. (1988). A study of starch–lipid interactions for some native and modified maize starches, *Starch/Stärke*, 49: 95.
59. Paton, D. (1987). Differential scanning calorimetry of oat pastes, *Cereal Chem.*, 64: 394.
60. Autio, K. (1990). Rheological and microstructural changes of oat and barley starches during heating and cooling, *Food Struct.*, 9: 297.
61. Doublier, J.-L., Paton, D., and Llamas, G. (1987). A rheological investigation of oat starch pastes, *Cereal Chem.*, 64: 21.
62. MacGregor, A. W. (1979). Isolation of large and small granules of barley starch and a study of factors influencing the adsorption of barley malt α-amylase by these granules, *Cereal Chem.*, 56: 430.
63. Svensson, E., and Eliasson, A.-C. (to be published).
64. Knutson, C. A. (1990). Annealing of maize starches at elevated temperatures, *Cereal Chem.*, 67: 376.
65. Biliaderis, C. G., Maurice, T. J., and Vose, J. R. (1980). Starch gelatinization phenomena studied by differential scanning calorimetry, *J. Food Sci.*, 45: 1669.
66. Tester, R. F., and Morrison, W. R. (1990). Swelling and gelatinization of cereal starches. I. Effects of amylopectin, amylose, and lipids, *Cereal Chem.*, 67: 551.
67. Morrison, W. R., and Nazir Azudin, M. (1987). Variation in the amylose and lipid contents and some physical properties of rice starches, *J. Cereal Sci.*, 5: 45.
68. Lii, C.-Y., and Lineback, D. R. (1977). Characterization and comparison of cereal starches, *Cereal Chem.*, 54: 138.
69. White, P., Abbas, I., Pollak, L., and Johnson, L. (1990). Intra- and

interpopulation variability of thermal properties of maize starch, *Cereal Chem.*, 67: 70.
70. Donovan, J. W. (1979). Phase transitions of the starch–water system, *Biopolymers*, 18: 263.
71. Eliasson, A.-C. (1980). Effect of water content on the gelatinization of wheat starch, *Starch/Stärke*, 32: 270.
72. Slade, L., and Levine, H. (1988). Non-equilibrium melting of native granular starch. Part I. Temperature location of the glass transition associated with gelatinization of A-type cereal starches, *Carbohydr. Polym.*, 8: 183.
73. Muhrbeck, P., Eliasson, A.-C., and Salomonsson, A.-C. (1990). Physical characterization of bromine oxidised potato starch, *Starch/Stärke*, 42: 418.
74. Ellis, H. S., Ring, S. G., and Whittam, M. A. (1989). A comparison of the viscous behavior of wheat and maize starch pastes, *J. Cereal Sci.*, 10: 33.
75. Doublier, J. L. (1981). Rheological studies on starch-flow behaviour of wheat starch pastes, *Starch/Stärke*, 33: 415.
76. Goering, K. J., Fritts, D. H., and Allen, K. G. D. (1974). A comparison of loss of birefringence with the percent gelatinization and viscosity on potato, wheat, rice, corn, cow cockle and several barley starches, *Cereal Chem.*, 51: 764.
77. Eliasson, A.-C. (1986). Viscoelastic behaviour during the gelatinization of starch. II. Effects of emulsifiers, *J. Text. Stud.*, 17: 357.
78. Doublier, J. L. (1987). A rheological comparison of wheat, maize, faba bean and smooth pea starches, *J. Cereal Sci.*, 5: 247.
79. Eliasson, A.-C., and Bohlin, L. (1982). Rheological properties of concentrated wheat starch gels, *Starch/Stärke*, 34: 267.
80. Bagley, E. B., and Christianson, D. D. (1982). Swelling capacity of starch and its relationship to suspension viscosity—effect of cooking time, temperature and concentration, *J. Texture Stud.*, 13: 115.
81. Bagley, E. B., and Christianson, D. D. (1983). Yield stresses in cooked wheat starch dispersions, *Starch/Stärke*, 35: 81.
82. Christianson, D. D., and Bagley, E. B. (1984). Yield stresses in dispersions of swollen, deformable cornstarch granules, *Cereal Chem.*, 61: 500.
83. Christianson, D. D., Casiraghi, E. M., and Bagley, E. B. (1986). Deformation and fracture of wheat, corn, and rice starch gels in lubricated and bonded uniaxial compression, *Carbohydr. Polym.*, 6: 335.
84. Miles, M. J., Morris, V. J., Orford, P. D., and Ring, S. G. (1985). The roles of amylose and amylopectin in the gelation and retrogradation of starch, *Carbohydr. Res.*, 135: 271.

85. Ring. S. G. (1985). Some studies on starch gelation. *Starch/Stärke*, 37: 80.
86. Christianson, D. D., Baker, F. L., Loffredo, A. R., and Bagley, E. B. (1982). Correlation of microscopic structure of corn starch granules with rheological properties of cooked pastes, *Food Microstruct.*, 1: 13.
87. Langton, M., and Hermansson, A.-M. (1989). Microstructural changes in wheat starch dispersions during heating and cooling, *Food Microstruct.*, 8: 29.
88. Roulet, P., MacInnes, W. M., Gumy, D., and Wiirsh, P. (1990). Retrogradation kinetics of eight starches, *Starch/Stärke*, 42: 99.
89. Orford, P. D., Ring, S. G., Carroll, V., Miles, M. J., and Morris, V. J. (1987). The effect of concentration and botanical source on the gelation and retrogradation of starch, *J. Sci. Food Agric.*, 39: 169.
90. Russell, P. L. (1987). The ageing of gels from starches of different amylose/amylopectin content studied by differential scanning calorimetry, *J. Cereal Sci.*, 6: 147.
91. Gudmundsson, M., and Eliasson, A.-C. (1990). Retrogradation of amylopectin and the effects of amylose and added surfactants/emulsifiers, *Carbohydr. Polym.*, 13: 295.
92. Evans, I. D. (1986). An investigation of starch/surfactant interactions using viscometry and differential scanning calorimetry, *Starch/Stärke*, 38: 227.
93. Slade, L., and Levine, H. (1987). Recent advances in starch retrogradation, in *Industrial Polysaccharides—The Impact of Biotechnology and Advanced Methodologies*, S. S. Stivala, V. Crescenzi, and I. C. M. Dea, Eds., Gordon and Breach, New York, p. 387.
94. Melvin, M. A. (1979). The effect of extractable lipid on the viscosity characteristics of corn and wheat starches, *J. Sci. Food Agric.*, 30: 731.
95. Chrastil, J. (1990). Protein–starch interactions in rice grains. Influence of storage on oryzenin and starch, *J. Agric. Food Chem.*, 38: 1804.
96. Lindahl, L., and Eliasson, A.-C. (1986). Effects of wheat proteins on the viscoelastic properties of starch gels, *J. Sci. Food Agric.*, 37: 1125.
97. Marshall, W. E., Normand, F. L., and Goynes, W. R. (1990). Effects of lipid and protein removal on starch gelatinization in whole grain milled rice, *Cereal Chem.*, 67: 458.
98. Hamer, R. J., and Lichtendonk, W. J. (1987). Structure–function studies on gluten proteins. Reassembly of glutenin proteins after mixing, in *Proceedings of the 3rd International Workshop on Gluten Proteins*, R. Lasztity and F. Bekés, Eds., World Scientific, Singapore, p. 227.
99. Bengtsson, S., and Åman, P. (1990). Isolation and chemical characterization of water-soluble arabinoxylans in rye grain, *Carbohydr. Polym.*, 12: 267.

100. Fengler, A. I., and Marquardt, R. R. (1988). Water-soluble pentosans from rye. I. Isolation, partial purification and characterization, *Cereal Chem.*, 65: 291.
101. Michniewicz, J., Biliaderis, C. G., and Bushuk, W. (1990). Water-insoluble pentosans of wheat: composition and some physical properties, *Cereal Chem.*, 67: 434.
102. Neukom, H., Deuel, H., Heri, W. J., and Kündig, W. (1960). Chromatographische Fraktionierung von Polysacchariden an Cellulose-Anionenaustauschern, *Helv. Chim. Acta*, 43: 64.
103. Ciacco, C. F., and D'Appolonia, B. L. (1982). Characterization of pentosans from different wheat flour classes and of their gelling capacity, *Cereal Chem.*, 59: 96.
104. Bengtsson, S. (1990). Structural studies on water-soluble arabinoxylans in rye, thesis, Swedish Univ. Agricultural Sciences, Uppsala.
105. Åman, P., and Bengtsson, S. (1991). Periodate oxidation and degradation studies on the major water-soluble arabinoxylans in rye grain, *Carbohydr. Polym.*, 15: 405.
106. Wood, P. J. (1986). Oat β-glucan: structure, location and properties, in *Oats—Chemistry and Technology*, F. H. Webster, Ed., AACC, St. Paul, Minn., p. 121.
107. Welch, R. W., Leggert, J. M., and Lloyd, J. D. (1991). Variation in the kernel (1→3)(1→4)β-glucan content of oat cultivars and wild *Avena* species and its relationship to other characteristics, *J. Cereal Sci.*, 13: 173.
108. Wood, P. J., Weisz, J., and Mahn, W. (1991). Molecular characterization of cereal β-glucans. II. Size-exclusion chromatography for comparison of molecular weights, *Cereal Chem.*, 68: 530.

5
Flour

I. INTRODUCTION

Before wheat grains are used in the production of foods for human consumption they are ground or milled. The purpose of the milling process is to separate the endosperm from the bran and germ and deliver it as a flour. In this chapter the milling process will be summarized, but first we describe the structure of the grain. Knowledge concerning the structure is necessary to an understanding of both the milling process and the interactions that may take place during baking. Finally, the composition of the flours obtained in the milling process is discussed.

II. GRAIN STRUCTURE

The wheat grain or the wheat kernel is a one-seeded fruit, called caryopsis by the botanist [1–3]. The seed consists of the germ, or embryo, and the endosperm enclosed by a nucellar epidermis and a seed coat (Fig. 1). A fruit coat or pericarp surrounds the seed and adheres closely to the seed coat. The caryopsis is characteristic of the grass family (Gramineae). A wheat kernel is about 5–8 mm in length and 2.5–4.5 mm in width [4]. The wheat kernel has a longitudinal crease that extends almost to the center of the kernel. The existence of the crease makes much of the handling of the grain more difficult. It has been suggested that one way to improve wheat would be to eliminate the longitudinal crease [5]. The wheat kernel is somewhat thicker toward the end where the embryo is located. At the opposite end, hairs of brush are located.

From a technological point of view, the kernel is composed of three parts: germ (embryo), endosperm, and bran. The relation between these "technological" structures and the physiological structures is given in Fig. 2. The

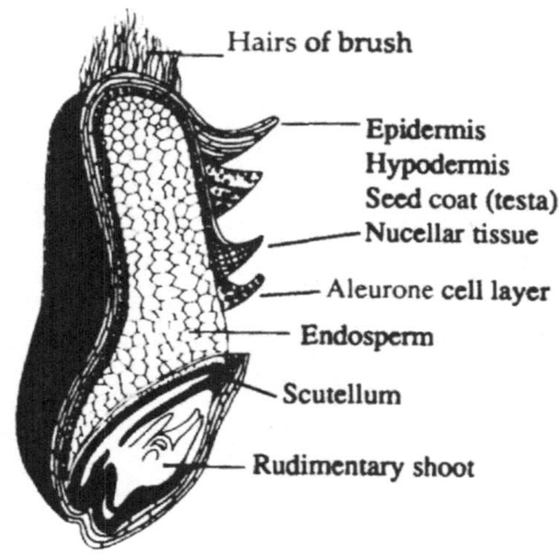

Fig. 1 The wheat kernel in cross section. (Adapted after an original from AB Cerealia.)

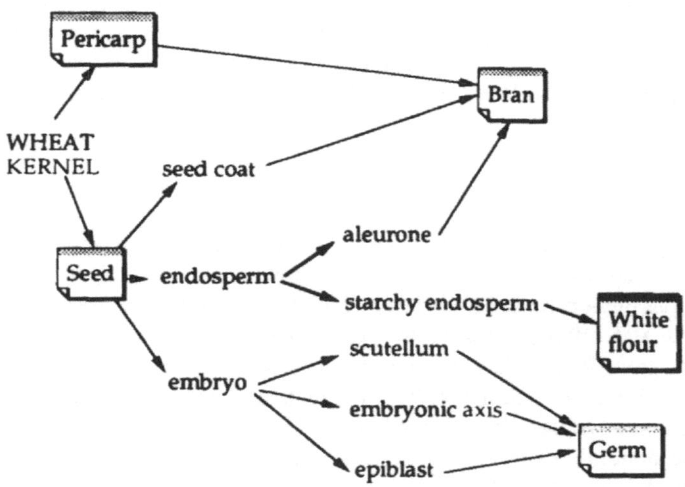

Fig. 2 The relationships between the physiological and technological parts of the wheat kernel. (Adapted after [1].)

germ constitutes 0.99–3.8%, the bran 10.4–20.0%, and the endosperm 74.9–86.5% of the wheat kernel, based on 14% moisture content [1]. These figures show considerable variation, at least in part related to the method of determination. Typical values could be taken as 2–3% for the germ and 78–84% for the endosperm, the bran thus being around 15% calculated on the basis of dry matter [6]. In the milling process we usually want to extract as much of the endosperm as possible as a white flour without losing too much of the endosperm in the bran fraction. The single most important property of the grain that affects this goal is the kernel hardness. This property is described in some detail later in this chapter.

One property of the wheat kernel that is evident immediately on visual inspection is its vitreousness or mealiness. Traditionally, vitreousness is associated with high-protein hard wheats, whereas opaque or mealy kernels are associated with softness and low protein content [7]. The proportion of vitreous kernels has been used as an indication of kernel hardness. There is a weak positive correlation between the percentage of vitreous kernels and kernel hardness, but Svensson [8] has advised that this relation should be used only when no better information is available. High-protein soft wheats are often vitreous [9].

Recently, thin sections of endosperm were used to differentiate between hard and soft varieties [10]. Sections as thin as 1 μm or less (mean 0.8 μm) can be obtained from durum and hard wheat varieties, whereas with soft wheats the sections have to be 4–12 μm thick (mean 9.0 μm) in order to be cohesive. When the sections (or rather the remaining part of the kernel) were viewed in the scanning electron microscope (SEM), the following observations were made [10]. In vitreous parts of hard wheat varieties a close starch–protein association was evident, and the matrix appeared to be continuous. No visible air pockets were observed. Nonvitreous parts of hard endosperm were characterized by a network of air spaces. The vitreous part of soft wheat varieties was not nearly as cohesive as the corresponding parts of hard wheats. In the soft wheats, starch granules were dislodged by the sectioning, and the endosperm matrix seemed to be discontinuous.

To further study the relationship between vitreousness and hardness, wheat samples were hand-sorted into vitreous classes [10]. It was found that vitreous grains are harder and have a higher protein content than the corresponding nonvitreous grains of the same variety. Also starch damage was higher for the vitreous grains. It was also found that the average hardness of nonvitreous hard wheats is higher than the average hardness of vitreous soft wheats. The 15-kDa protein, the presence of which is associated with grain softness, was found in vitreous as well as nonvitreous grains from soft wheat varieties, although the vitreous grains are harder.

From the wheat plant's point of view, the embryo is certainly the most important part, for it will grow into the new plant, using the nutrients stored in the endosperm during its development. When wheat is used in foods for human consumption it is necesssary to be aware of the fact that the wheat grain is not homogeneous. A wheat kernel is strongly compartmentalized [11]. It is obvious that the composition must differ between such different physiological parts as the embryo and the endosperm, but there are also composition gradients inside the endosperm. There is, for example, a decrease in protein content from the outer to the inner endosperm [1, 9]. There are also gradients in the lipid composition across the endosperm [6]. In the following subsections, the germ, endosperm, and bran are described in some detail. The approximate chemical analysis of these tissues is given in Table 1.

The differences in composition also result in gradients in properties. The fact that the baking performance decreases from the center to the surface whereas the vitamin and mineral contents decrease in the opposite direction is well known.

A. Germ

The germ, or embryo, is a separate part of the grain, and it could be said to be composed of two major parts, the embryonic axis and the scutellum (see Fig. 1). The embryonic axis is made up of the rudimentary root and shoot. The scutellum serves as a storage, digestive, and absorbing organ [1]. The germ is high in protein and lipid (See Table 1). The level of reducing sugars is also high, 26.3% according to Mattern [4]. The protein present in the embryo is composed of nucleoproteins, albumins, globulins, and proteases [2]. The lipid composition of the germ is characterized by a high level of nonpolar lipids; 79–85% of the total lipids are nonpolar, 14–17% are phospholipids, and there are almost no glycolipids [6, 13]. The fatty acid composition is characterized by a high level of unsaturated fatty acids, about 75% [4]. The most frequent fatty acid is $C_{18:2}$. From a nutritional point of view the high content of tocopherol is of interest [14].

Table 1 Chemical Composition of Endosperm, Bran, and Germ[a]

Component	Whole wheat	Endosperm	Germ	Aleurone	Bran (pericarp)
Protein (%)	8.2–12.1	5.8–16.2	24.3–31.1	18.4–24.3	2.85–7.60
Ash (%)	1.8	0.5–0.8	3.65–9.47	11.1–17.2	1.7–5.1
Fiber (%)	9.0	1.4	8.6	43.0	17.1–73.3
Lipids (%)	1.8	1.6–2.2	5.05–18.8	6.0–9.89	0.0–1.03
Starch (%)	59.2	63.4–72.6	0.0	0.0	0.0

[a]Data are based on 14% moisture.
Source: Data compiled from [4] and [12].

B. Bran and Aleurone Layers

The bran fraction, which constitutes about 15% of the wheat kernel, is composed of layers (Fig. 1) such as the pericarp, seed coat, nucellus, and aleurone cells [2]. The pericarp, which is the outermost part of the kernel, acts as a protective covering. It is built up of several layers (see Figs. 1 and 2) and is about 45–50 μm thick [1]. The cells in the different layers differ in several respects, for example, in size and shape and in cell wall thickness. The composition of the bran (except the aleurone cells) is characterized by high levels of nonstarch carbohydrates such as cellulose and pentosans (Table 1). The lipids in the pericarp are mainly nonpolar (86%), with diglycerides and free fatty acids comprising the most abundant lipid classes [6].

Bran has a detrimental effect on loaf volume. However, the effect is related to the composition of the bran and to the mill it comes from, as the method of separating the bran and endosperm differs among mills. The coarser the bran fraction, the more detrimental its effect will be [15]. The detrimental effect is attributed to a decrease in gas-holding capacity [15]. This effect might be overcome by increasing the amount of water and adding an emulsifier such as sodium stearoyl-2-lactylate [16].

The aleurone layer is a remarkable tissue. Physiologically it belongs to the endosperm; by definition, it is the outer layer of the endosperm. However, during milling, the aleurone separates with the bran and can therefore, practically speaking, be regarded as belonging to the bran [2]. The aleurone layer is usually one cell thick, and the cells are large and rectangular with thick walls. The aleurone cells are easily observed in the fluorescence microscope [11]. The cell walls contain ferulic acid, which is autofluorescent and renders the cell walls an intense blue. The composition of the aleurone cells differs considerably from that of the other bran components. Whereas the outer bran layers are high in cellulose, hemicellulose, and ash, the aleurone is high in protein, lipids, and ash. The lipid content of the aleurone layer is around 8–10% [6]. The composition is very similar to that of the germ lipids and is characterized by being about two-thirds triglycerides (60.3–75.1%), with lower levels of glycolipids (2.2–9.8%) and phospholipids (13.8–17.9%) [6]. The lipids are present as spherosomes, oil droplets bounded by a monolayer membrane [6, 11]. The aleurone cells are metabolically very active during the early stages of development, which explains their special composition.

C. Endosperm

The endosperm, which makes up the largest part of the grain, is mainly composed of starch granules surrounded by a continuous protein matrix (Fig. 3). Its chemical composition is given in Table 1; it is characterized by a very high level

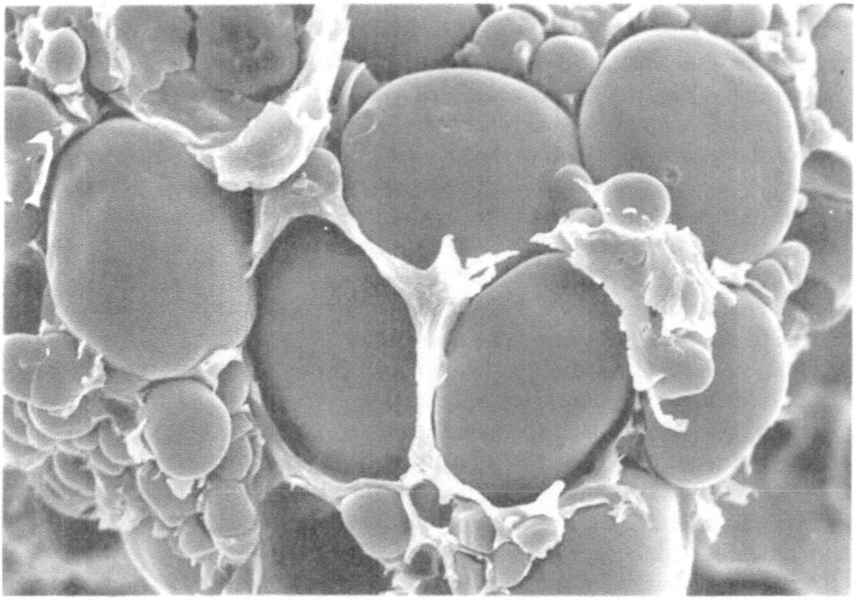

Fig. 3 Starch granules in a protein matrix. (Kerstin Brismar and Waheeb Heenen, EM-lab., SLU, Svalöv, Sweden.)

of starch. The values in Table 1 are average values for the whole endosperm. If the endosperm is separated in subfractions, from outside to center, gradients in composition are evident (cf. [4]).

The outer part of the endosperm, the subaleurone layer, is distinctly different from the main part of the endosperm. In hard wheat endosperm, this layer forms a continuous shell around the endosperm, whereas in soft wheats it is often discontinuous. The cells in the subaleurone layer are small and cubic, whereas those of the inner endosperm are larger and either needle-shaped or, in the central endosperm, polyhedral [1,2]. The cell walls differ in thickness in different parts of the kernel; they are thickest near the aleurone layer and around the crease [1].

The number of starch granules is highest in the central endosperm cells, whereas most of the protein will be found in the outer endosperm cells. In some wheats the protein content of the subaleurone cells can exceed 80%. These cells contain very few starch granules [2,7]. This situation can be expected to result in the kernel having a vitreous appearance. If vitreousness is used to identify hard wheats, then soft wheats of high protein content will be judged as hard. The protein present in the endosperm is mostly storage protein. Protein extracted from the inner endosperm is composed of roughly equal amounts of gliadin and glutenin [2]. In developing kernels the storage protein is present as protein bodies. In the mature endosperm these protein bodies are compressed to a continuous matrix by the developing starch granules [1].

When the content of the endosperm cells is inspected, the benefits of the bimodal partical size distribution of the wheat starch granules (Chapter 2) are evident. The large, lenticular granules are synthesized first, whereas the small, spherical granules are synthesized later and will fill the empty space between the large granules. A large amount of carbohydrate energy is thus stored in a small volume.

As mentioned above, there are lipid aggregates in the endosperm in the form of spherosomes [6]. They are found in the highest concentration in the subaleurone layer, and they constitute about 0.77–1.03% of the endosperm. Their composition is 16.7–34.1% triglycerides, 13.3–19.5% other nonpolar lipids, 20.4–38.3% glycolipids, and 23.6–35.3% phospholipids [6]. A gradient in lipid composition across the starchy endosperm is evident. Differences between varieties in spherosome content in the endosperm have been observed [6].

III. MILLING

The purpose of the milling process is to separate bran and germ from the endosperm, and the endosperm is then delivered as flour from the mill. However, it is not enough just to obtain a clean separation of bran and endosperm; the endosperm should also be reduced to a suitable particle size. Moreover, the yield should be as high as possible; all the endosperm should ideally be in the flour, and the bran and germ fractions should be free of contamination by endosperm cells or particles. To describe how much of the wheat grain is found as flour after the milling the term *extraction rate* is used. An extraction rate of 100% signifies that 100% of the wheat grain is delivered as flour; this flour could therefore also be described as a whole wheat meal. The bulk product from the mill, the straight-grade flour, represents an extraction of around 70% [17].

Another aim of the miller is to deliver flour that does not vary in quality from batch to batch. In order to achieve a stable flour quality over time, the miller can blend different wheat varieties. As a consequence of the milling procedure, the miller has access to a considerable number of milling streams. These streams are blended to give a flour of a specified quality. Not all streams, however, are suitable for blending. The composition as well as the physicochemical properties of these streams differ, and it is thus possible to compose flours of different qualities by blending the streams in different proportions and by excluding or including certain streams.

If we consider the milling process from the point of view of surface chemistry, the fragmentation of the endosperm in an air environment must result in a tendency for each particle to adopt a hydrophobic surface whatever fractures take place. Everywhere throughout the endosperm there are amphiphilic molecules possessing such an ability to adopt an orientation or conformation that will reduce the surface energy. In the mature endosperm the successive drying will "freeze" a hydrophilic surface structure on macromolecules, organelles, and so on in the aqueous environment. A spherosome, for example, will remain unchanged in a dry endosperm, but as soon as it comes into contact with a flour particle and its fragile coat is damaged, the oil will be distributed over the surface of the flour particle as a very effective reducer of surface energy.

When the flour is wetted, the particle surface molecules have to switch over to presenting a hydrophilic structure toward the water.

A. Conditioning

Usually the germ is easily removed from the wheat kernel during milling. The germ could be regarded as a more or less separate part of the kernel connected to the endosperm through the scutellum (Fig. 1). The bran, on the other hand, can be more closely attached to the endosperm. If the bran and endosperm are not completely separated (a "clean" separation), the bran fraction will be contaminated with endosperm material, resulting in a lower yield. Under certain conditions, however, it is desirable to use various proportions of bran up to whole meal in breadmaking, which changes the requirements of separation. Here, though, we discuss only the milling of white flour. To facilitate a clean separation, the procedure known as *conditioning* or *tempering* is used. Water is added to the wheat prior to milling, and the wheat is then held at ambient temperature for 18–24 h. Hard wheats are brought to 15–16% moisture content, and soft wheats to 13–14% moisture [4]. Becauseof the different water-holding capacities of the differnt structures, the endosperm softens and the bran becomes tougher. The toughness and lightness of the bran compared to that of the endosperm makes the milling procedure possible. According to Ziegler and Greer [18], the conditioning is a compromise. During the milling process bran and endosperm should easily separate, and the sifting should easily be performed. These two requirements are best achieved at low moisture contents. At the same time it is required that the bran maintain its toughness (it should not break during milling) and that the endosperm soften. These requirements are best met if the grain has a somewhat higher moisture content.

The wheat kernel initially absorbs water very quickly. This water absorption occurs in the outer layers of the bran. Thereafter the rate of water absorption is slowed down because of the testa (see Fig. 1). The rate of water absorption in the endosperm differs between the different types of cells. The final moisture content is reached last by the cells near the crease. Also, temperature influences the water penetration and the softening of the endosperm. How the conditioning is performed in a milling plant depends on the miller's choice of water or steam and time and temperature conditions.

The conditioning is related both to the kernel hardness and to the initial water content. Water penetration occurs more rapidly through soft grains than hard grains and when the initial water content is high [9]. The influence of the initial water content might be related to wetting; in a dry grain, water has to wet more hydrophobic surfaces, which slows down the water penetration [9]. For a hard wheat of high protein content and low moisture content, therefore, a long conditioning time is required.

B. The Milling Procedure

When the grains arrive at the mill they are first cleaned to remove dust, foreign seeds, and other such contaminants. Then they are conditioned as described above to achieve optimum milling conditions.

Conventional milling uses a gradual reduction system, and most mills use a roller system. The number of rolls and the way in which they are combined with sifters differ from one milling unit to another. The actual setup for a plant is illustrated in a milling diagram such as the one in Fig. 4.

The milling system contains two types of rolls, *break rolls*, which are fluted, and *reduction rolls*, which are smooth. The rolls are mounted in pairs and run at a speed differential. The two types of rolls have different functions during milling; the break rolls are there to open up the structure, whereas the reduction rolls reduce the particle size. The wheat is first passed through a series of break rolls. These are progressively more finely fluted and set closer and closer together. As the break rolls are corrugated, the grains will be held by one of the rolls while the other roll crushes the grain in a shearing action. The breaking of the grain structure requires both shear and pressure. Care should be taken to minimize the scattering of the bran, as small bran particles are more difficult to separate from the endosperm [18]. The broken particles pass from the break rolls to sifters. Particles that are too large for flour are conveyed to the reduction rolls, where they are reduced in size. Reduction is facilitated if the material coming to the reduction rolls is of homogeneous size. The bran is taken care of in certain purifiers.

The yield of flour is limited by the percentage of endosperm in the kernel, which was previously given to be in the range of 74.9–86.5% [1]. In the milling processes used today, however, the aleurone layer travels with the bran, and thus the flour yield is less than the highest theoretical value.

Different methods are used to predict flour yield, such as the determination of kernel size, hectoliter weight, and the structure and composition of the cell walls [2, 8]. The correlation coefficient r between kernel size and flour yield was found to be 0.01 for winter wheats and −0.08 for spring wheats [8]. The correlation between hectoliter weight and flour yield was better—$r = 0.34$ for winter wheats and $r = 0.41$ for spring wheats. Pomeranz [2] concluded that both the chemical composition and the structure of the cell walls seem to be similar between wheat varieties that differ greatly in milling behavior.

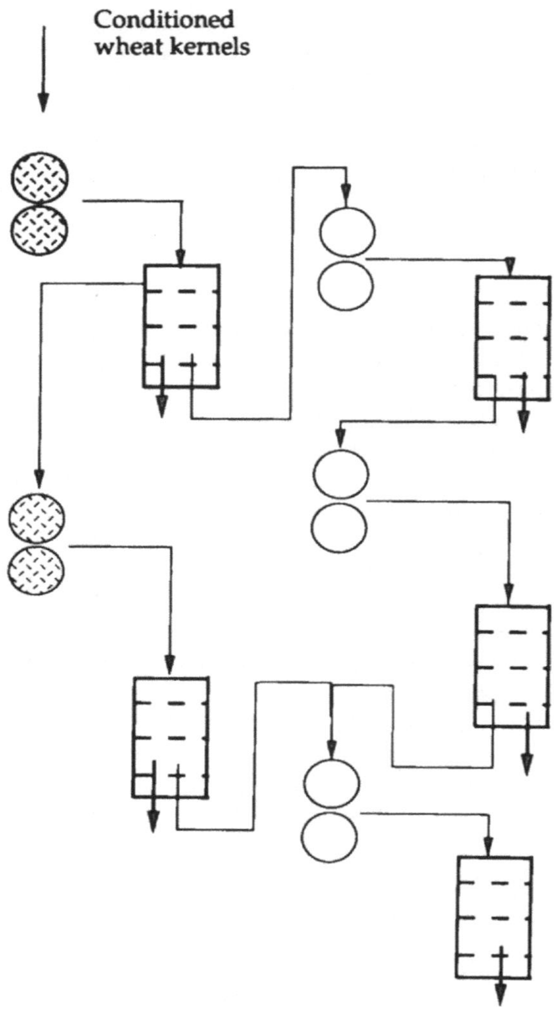

Fig. 4 A milling diagram showing two pairs of break rolls and three pairs of reduction rolls together with sifters.

C. Kernel Hardness

The characteristic hardness of the kernel is apparent during milling, and wheat varieties are divided into hard and soft wheats depending on their milling behavior. Moreover, its hardness determines the end use of a wheat. The hardest wheats, durum, are used for pasta production, and the softest wheats are suitable for biscuits, whereas the wheats most suitable for breadmaking have an intermediate hardness [7]. Depending on the kernel hardness, the flour particles after milling will differ in appearance. The soft wheats tend to be fractured through cells of the endosperm and along individual starch granules, whereas hard wheats tend to fracture along cell walls. When a hard wheat fractures through its cells, the fracture passes through both the starch granules and the protein matrix [7]. The flour particles from a soft wheat will be fluffy, with many starch granules released from the protein matrix, whereas a particle from a hard wheat will consist of a single cell or a few cells. The hard varieties break into relatively large polyhedral flour particles, whereas soft varieties break into very small and irregular particles. The differences in breakage patterns between hard and soft varieties are evident early in the milling process. Flour particles collected above the second set of break rolls show a few large chunks in the case of a hard endosperm whereas a soft endosperm shows much greater disruption [19]. Milling of a hard variety requires more energy, but on the other hand, the hard endosperm is easier to separate from the bran, and the flour is easier to sift and transport [8]. The milling capacity as well as the flour yield will be higher with a hard wheat than with a soft wheat.

Hardness is a characteristic that depends on the variety, but it is influenced by environmental conditions [20]. This means that the hardness may vary from one grain to another in a batch of grains of the same variety. Even grains from the same plant may differ in hardness [9]. To some extent the kernel hardness is influenced by the protein content, but above a certain level the protein does not affect the hardness [8].

Because so much of the behavior of the wheat grain during milling is related to hardness, much effort has been made to measure hardness. The methods can be divided into four groups depending on whether they are based on grinding, crushing, abrasion, or indentation with a stylus [7]. In the particle size index (PSI) test, a weighed sample of grain is ground and sifted under standard conditions, and the weight of sample passing through the sieve is measured [21]. The pearling resistance test measures the ease with which the outer layers of the grain are abraded [22].

Kernel hardness has been attributed to mechanical properties of the individual components (protein and starch), to interactions between them, and to the degree of continuity of the protein matrix. The hardness of protein from different wheat varieties has been found to be very similar [23]. The same is true for starch. The theory that protein–starch interactions, described in detail in Chapter 3, accounts for kernel hardness has thus gained more support. The main hypothesis is that in a hard wheat the interaction between starch and protein is so strong that during milling breakage occurs *through* the starch granules. In a soft wheat, on the other hand, the interaction between protein and starch is much weaker, and, in order to release stress, breakage occurs along the starch granule surface. The reason for the strong adhesion in hard wheat could be the absence of a certain protein termed an "anti-sticky" protein [24]. Such a protein could also influence the development of a continuous protein structure around the starch granules. Scanning electron microscopy has been used to show that in soft varieties the endosperm structure is very open and several air pockets are present [20]. In a hard endosperm the protein matrix is continuous and seems to surround all the starch granules. It has been found that there is a minimum amount of protein necessary to develop a continuous protein matrix, and this amount of protein is influenced by environmental conditions. Over a certain range of protein content, the hardness will depend on the protein content, but when the protein content is higher than what is necessary for the continuous protein matrix, it will no longer influence the hardness.

1. Yield and Particle Size Distribution

Flour yield is related to kernel hardness. This is because of the easier separation of bran and endosperm in hard varieties. However, it is not an easy task to determine the kernel hardness. Svensson [8] used both a sifting method and the pearling resistance test to determine kernel hardness. He found correlation coefficients for flour yield of $r = -0.73$ for the sifting test on spring wheats, $r = -0.61$ for the sifting test on winter wheats, $r = 0.42$ for the pearling resistance test on spring wheats, and $r = 0.52$ for the same test on winter wheats. The sifting test gives a negative correlation because the method measures the material passing through a 75-μm sieve; that is, a low value indicates a hard kernel.

The particle size distribution differs between hard and soft varieties in that a soft variety contains many more small particles. Particles as small as individual starch granules are present. For a hard wheat, 15.8% of the material was found to have diameters below 25 μm, whereas for a soft wheat the corresponding value was 35.9% [9]. Flour can also be air-classified according to particle size, and then the smallest particles have the highest protein content [25].

2. Starch Damage

As described above, milling of a hard wheat frequently causes breakage through the starch granules [23]. The amount of starch damage might then be taken as an indication of the kernel hardness. Moreover, the miller can adjust the amount of damaged starch in the flour by varying the setting of the roller gap. A soft wheat will release the starch granules from the protein matix, and it is therefore difficult to achieve mechanically damaged starch during milling of soft wheats. In fact, the rolls have to be set close enough that the individual starch granules are crushed. This close setting of the roller gap could also damage the protein [8].

The level of starch damage differs between flours, depending on both the varieties used and the milling process. In a winter wheat material, the amount of damaged starch ranged from about 7 to 13 Farrand units (FU) [8].

The presence of mechanically damaged starch in a flour is of great importance to the properties of the flour. First of all, damaged starch affects the water absorption of the dough. The water-holding capacity of normal starch is 0.44 g water/g dry starch, but that of damaged starch is 2.00 g water/g dry starch [26, 27].

The gelatinization behavior of the starch is also affected. Differential scanning calorimetric measurements have shown that the gelatinization enthalpy ΔH decreases in relation to the level of starch damage [28, 29]. A possible explanation is that part of the crystalline domains are destroyed or destabilized by fractions caused by the milling. Then part of the starch will "gelatinize" when the water is added to the flour at ambient temperature. Moreover, the onset temperature of the gelatinization will decrease when the level of starch damage increases. This could also be a result of destabilization of the crystalline domains. In Fig. 5 some typical DSC thermograms are shown for wheat flour–water mixtures in which the level of starch damage differs between the flours. As discussed in Chapter 2 part of the DSC endotherm could be ascribed to the amorphous part of the starch granule. The starch damage evidently results in a release of the coupling between crystalline and amorphous domains and even an increased amount of amorphous regions (the fraction zone will be disordered).

A certain level of starch damage is required to facilitate the gas production by

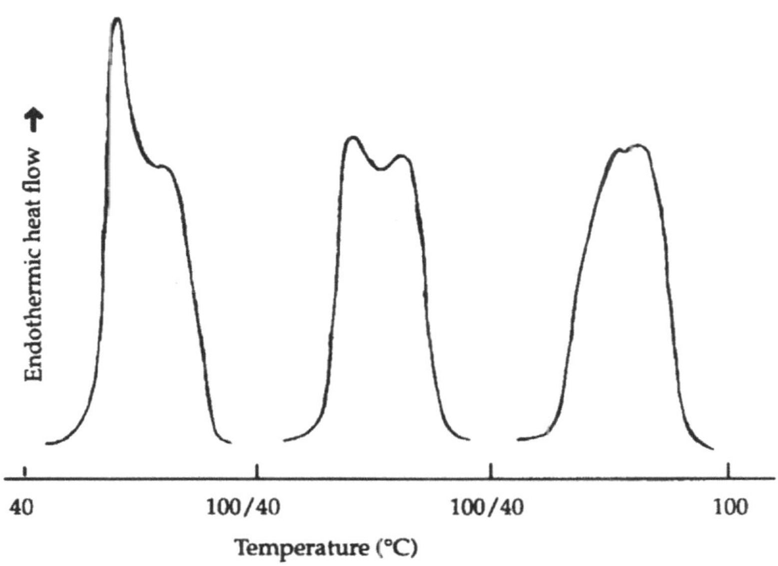

Fig. 5 Differential scanning calorimetric thermograms of flours differing in the level of starch damage (unpublished results). The level of starch damage increases from left to right.

the yeast, but starch that is too heavily damaged will render the dough sticky, an(the crumb will not set properly. There is also a tendency for loaf volumes to b(decreased [30–32]. It has been concluded that the potential loaf volume does no differ between hard and soft varieties but that the level of starch damage is to(low in a soft wheat to bring out its potential loaf volume [33].

IV. FLOUR COMPOSITION

A. Extraction Rate

The chemical composition of a wheat flour depends on the extraction rate. A long as this is below 100% the composition of the flour will differ from th(composition of the grain. As a consequence of the removal of bran and germ, th(flour's nonstarch carbohydrate content will be low. The lipid content of th(whole kernel is in the range 2.8–3.2%, whereas the lipid content of the endo sperm is 0.8–1.2% [6]. Of the lipids in the kernel, 44–57% are nonpolar lipids 8–14% are glycolipids, and 31–42% are phospholipids. In the endosperm, th(nonstarch lipids are composed of 33–47% nonpolar lipids, 20–35% glycolipids and 22–35% phospholipids [6]. The variation in chemical composition witl differences in extraction rate is illustrated in Table 2.

The milling process is also a mixing; the gradients in the endosperm describe(above are lost, and the components are redistributed. In the wheat kernel th(different components are strictly separated, and some of them can never com(into contact. After the milling this situation is changed, and new possibilities fo interaction are opened. Most of the triglycerides in the kernel are located in th(germ, but during the milling these lipids come into contact with the endosperm

Table 2 Chemical Composition of Wheat Flours in Relation to Extraction Rate[a]

	Extraction rate of flour			
Constituent	42–46%	70%	80%	100%
Ash (%)	0.34–0.37	0.41–0.43	0.59–0.71	1.5–1.6
Fiber (%)	trace	trace	0.13–0.20	2.0–2.2
Protein[b] (%)	7.7–12	8.0–12	8.3–13	9.0–14
Lipids (%)	0.76–0.87	1.1–1.2	1.3–1.5	2.0–2.5
Starch (%)	72–77	71–75	70–74	64–68

[a]All data based on 14% moisture.
[b]Protein determined by the Kjeldahl method using the conversion factor 5.7.
Source: Data compiled from [18].

If flour is milled from degermed wheat, the lipid content will be much lower in the resulting flour than in a flour obtained from the whole kernel [34]. The amount of triglycerides in the flour is 407 mg/100 g in the whole grain flour compared to 243 mg/100 g in the degermed flour. The levels of polar lipids are more or less unaffected by the degerming procedure, indicating that it is just the liquid oil from the germ that is distributed on the flour particle surfaces.

Another point of importance is the way in which the components are present in particles of the flour; to some extent the different constituents are in different compartments. It must be expected that polar lipids are present in membrane structures, whereas the storage protein is present as a continuous protein matrix. This has implications for the events taking place when the flour particles are hydrated (the surface aspects of the milling and wetting are discussed in Section III of this chapter).

The difference in composition due to the extraction rate causes variations in the flour properties. From a technological point of view the most evident factor is perhaps the differences in baking performance [35]. One reason for the deterioration of baking performance with increasing extraction rate could be the redistribution of germ lipids. It is known that the occurrence of free fatty acids and too much germ oil can impair baking quality [36]. The distribution of germ oil among the flour particles will also expose the oil to lipase degradation. The presence of more nonpolar lipids in flours of higher extraction rates can furthermore result in a change in phase behavior of the lipids. Some of the polar lipids present can be used to emulsify the germ oil, and thus they are not available to take part in the stabilization of the air/water interface during baking.

Another aspect of baking with whole meal flours is the deterioration in baking performance during storage. This effect is especially evident in the case of whole meal stored under ambient conditions. The deterioration can be explained by changes in the lipid components [37]. The lipase activity present in the bran fraction causes hydrolysis of the triglycerides, and the free fatty acids exert a negative effect during dough mixing. This is partly due to the effect on phase properties (see Chapter 2) and partly due to chemical changes. The lipids thus undergo peroxidation reactions that are catalyzed by lipoxygenase present in the germ [38].

B. Flour Stream

Not only will the composition vary owing to the extraction rate, but it will also differ between flour streams. The proportions of the different flour streams are as follows. At the break rolls, less than 1% flour is obtained in each break from a hard wheat, whereas more is obtained from a soft wheat or from mixed grists. About 2–20% of the total flour is thus obtained from the break rolls. The early reduction rolls each give 10–25% of the flour, so the main part of the flour is

Table 3 Chemical Composition in Relation to Flour Stream[a]

	Break streams		Reduction streams		
Constituent	Early	Late	Early	Middle	Late
Ash (%)	0.62	1.1	0.45	0.40	0.60
Protein (%)	10.6	13.0	10.0	10.4	10.1
Lipids (%)	1.13	2.68	1.16	1.54	2.25

[a]Data based on 14% moisture.
Source: Data compiled from [18] and [39].

obtained from the early reduction system [18]. A flour stream from the early breaks will differ considerably from a flour from the early reduction rolls (Table 3). As already discussed, there are gradients in protein composition. A flour stream of high protein content does not necessarily indicate good baking performance. The late-break flour in Table 3 with a protein content of 13.0% gave a loaf volume of 506 mL, whereas the early-reduction flour with a protein content of 10.0% gave a loaf volume of 729 mL [39].

The different properties of the protein in the flour streams can be illustrated by the rheological behavior of their doughs. The flour streams from the break rolls give low values of the shear modulus G in stress-relaxation measurements, whereas the reduction flours give considerably higher values [40]. The correlation between G and protein content, starch content, and starch damage was calculated. The best correlation coefficients were obtained for starch damage where r^2 values of 0.541–0.775 were obtained, depending on wheat and water content in the dough. For starch or protein, r^2 was 0.273 or below.

Starch damage differs between flour streams, and in the study cited above [40] it was 5.3–25 FU for break streams and 20–33 FU for reduction streams. In another study the amount of starch damage was found to increase with each passage in the break system (from ~8 FU to ~21 FU) and with each passage in the reduction system (from ~10 FU to ~26 FU) [39].

REFERENCES

1. MacMasters, M. M., Hinton, J. J. C., and Bradbury, D. (1971). Microscopic structure and composition of the wheat kernel, in *Wheat Chemistry and Technology*, Y. Pomeranz, Ed., American Association of Cereal Chemists, St. Paul, Minn., p. 51.
2. Pomeranz, Y. (1981). Grain structure and end-use properties, *Food Microstruct.*, 1: 107.

3. Lindblom, H. (1989). Spannmålskärnans byggnad, Stiftelsen Cerealia för Forskning och Utveckling, Stockholm.
4. Mattern, P. J. (1991). Wheat, in *Handbook of Cereal Science and Technology*, K. J. Lorenz and K. Kulp, Eds., Marcel Dekker, New York, p. 1.
5. Bechtel, D. B. (1989). How the structure of the wheat caryopsis should be modified to increase its end-use value, in *Wheat Is Unique*, Y. Pomeranz, Ed., AACC, St. Paul, Minn., p. 71.
6. Hargin, K. D., and Morrison, W. R. (1980). The distribution of acyl lipids in the germ, aleurone, starch and non-starch endosperm of flour wheat varieties. *J. Sci. Food Agric.*, 31: 877.
7. Simmonds, D. H. (1974). Chemical basis of hardness and vitreosity in the wheat kernel, *Baker's Dig.*, 48(5): 16.
8. Svensson, G. (1981). Varietal and environmental effects on wheat milling quality, *Agri Hort. Genet.*, 39: 1.
9. MacRitchie, F. (1980). Physicochemical aspects of some problems in wheat research, in *Advances in Cereal Science and Technology*, Y. Pomeranz, Ed., AACC, St. Paul, Minn., p. 271.
10. Glenn, G. M., and Saunders, R. M. (1990). Physical and structural properties of wheat endosperm associated with grain texture, *Cereal Chem.*, 67: 176.
11. Fulcher, R. G. (1982). Fluorescence microscopy of cereals, *Food Microstruct.*, 1: 167.
12. Pomeranz, Y. (1988). Chemical composition of kernel structures, in *Wheat Chemistry and Technology*, Y. Pomeranz, Ed., AACC, St. Paul, Minn., p. 97.
13. Hargin, K. D., Morrison, W. R., and Fulcher, R. G. (1980). Triglyceride deposits in the starchy ensoperm of wheat, *Cereal Chem.*, 57: 320.
14. Morrison, W. R., Coventry, A. M., and Barnes, P. J. (1982). The distribution of acyl lipids and tocopherols in flour millstreams, *J. Sci. Food Agric.*, 33: 925.
15. Pomeranz, Y., Shogren, M. D., Finney, K. F., and Bechtel, D. B. (1977). Fiber in breadmaking—effects on functional properties, *Cereal Chem.*, 54: 25.
16. Lai, C. S., Hoseney, R. C., and Davis, A. B. (1989). Effects of wheat bran in breadmaking, *Cereal Chem.*, 66: 217.
17. Hoseney, R. C. (1986). *Principles of Cereal Science and Technology*, AACC, St. Paul, Minn., p. 133.
18. Ziegler, E., and Greer, E. N. (1971). Principles of milling, in *Wheat Chemistry and Technology*, Y. Pomeranz, Ed., AACC, St. Paul, Minn., p. 115.
19. Davis, A. B., and Eustace, W. D. (1984). Scanning electron microscope views of material from various stages in the milling of hard red winter, soft red winter, and durum wheat, *Cereal Chem.*, 61: 182.
20. Stenvert, N. L., and Kingswood, K. (1977). The influence of the physical

structure of the protein matrix on wheat hardness. *J. Sci. Food Agric.*, 28
11.

21. Symes, K. J. (1961). Classification of Australian wheat varieties, based o the granularity of their wholemeal, *Aust. J. Exp. Agric. Anim. Husb.*, 1 18.
22. Chesterfield, R. S. (1971). A modified barley pearler for measuring hard ness of Australian wheat, *J. Aust. Inst. Agric. Sci.*, 37: 148.
23. Barlow, K. K., Buttrose, M. S., Simmonds, D. H., and Vesk, M. (1973) The nature of the starch–protein interface in wheat endosperm, *Cerea Chem.*, 50: 443.
24. Greenwell, P., and Schofield, J. D. (1989). The chemical basis of grai hardness and softness, in *Wheat End-Use Properties—Wheat and Flou Characterization for Specific End-Use*, H. Salovaara, Ed., University o Helsinki, p. 59.
25. Sosulski, F. W., and Nowakowski, D. M. (1989). Processing and utiliza tion of air-classified wheat flour fractions, in *Wheat Is Unique*, Y. Pome ranz, Ed., AACC, St. Paul, Minn., p. 431.
26. Greer, E. N., and Stewart, B. A. (1959). The water absorption of whea flour: relative effects of protein and starch, *J. Sci. Food Agric.*, 10: 248.
27. Bushuk, W. (1966). Distribution of water in dough and bread, *Bakers Dig.*, 38.
28. Stevens, D. J., and Elton, G. A. H. (1971). Thermal properties of th starch/water system. I. Measurement of heat of gelatinisation by differen tial scanning calorimetry, *Starch/Stärke*, 23: 8.
29. Eliasson, A.-C. (1989). Some physico-chemical properties of whea starch, in *Wheat End-Use Properties—Wheat and Flour Characterizatio for Specific End-Use*, H. Salovaara, Ed., University of Helsinki, p. 355
30. Shelton, D. R., and D'Appolonia, B. (1985). Carbohydrate functionalit in the baking process, *Cereal Foods World*, 30: 437.
31. Czuchajowska, Z., Pomeranz, Y., and Jeffers, H. C. (1989). Wate activity and moisture content of dough and bread, *Cereal Chem.*, 66: 128.
32. Larsen, N. G., Baruch, D. W., and Humphrey-Taylor, V. J. (1989). Th effect of laboratory and commercial milling on lipids and other physico chemical factors affecting the breadmaking quality of wheat flour, *J Cereal Sci.*, 9: 139.
33. Pomeranz, Y., Bolling, H., and Zwingelberg, H. (1984). Wheat hardnes and baking properties of wheat flours, *J. Cereal Sci.*, 2: 137.
34. Morrison, W. R., and Hargin, K. D. (1981). Distribution of soft wheat kernel lipids into flour milling fractions, *J. Sci. Food Agric.*, 32: 579.
35. Barnes, P. J., and Lowy, G. D. A. (1986). The effect on baking quality o interaction between milling fractions during the storage of wholemeal, *J. Cereal Sci.*, 4: 225.

36. Morrison, W. R. (1989). Recent progress on the chemistry and functionality of flour lipids, in *Wheat End-Use Properties—Wheat and Flour Characterization for Specific End-Uses*, H. Salovaara, Ed., University of Helsinki, p. 131.
37. Tait, S. P. C., and Galliard, T. (1989). Chemical aspects of functional properties of whole meal flour, in *Wheat End-Use Properties—Wheat and Flour Characterization for Specific End-Uses*, H. Salovaara, Ed., University of Helsinki, p. 101.
38. Galliard, T. (1986). Oxygen consumption of aqueous suspensions of wheat wholemeal, bran and germ: involvement of lipase and lipoxygenase, *J. Cereal Sci.*, 4: 33.
39. Olered, R., Johansson, H., and Persson, S. (1978). Undersökningar av olika mjölfraktioner av höst-och vårvete från Skåne-Möllan, *Sver. Utsädesfören. Tidskr.*, 88: 219.
40. Eliasson, A.-C., Silverio, J., and Tjerneld, E. (1991). Surface properties of wheat flour-milling streams and rheological and thermal properties after hydration, *J. Cereal Sci.*, 13: 27.

6
Dough

I. INTRODUCTION

In this chapter we give a description of dough-making from the point of view of colloidal and surface chemistry. Three ingredients constitute the minimum needed to make a loaf of bread—flour, yeast, and water. The proportions among these ingredients may vary, but typical values are 100:2:65 [1]. Our description is based on the simple model in Fig. 1, where we illustrate the dough as an aqueous continuous medium of gluten and starch containing the dispersed gas phase. In this model there are also interfaces such as the air/water interface. The simple model in Fig. 1 will gradually be elaborated as we incorporate the knowledge concerning the components described in Chapter 2. We also describe the role of added ingredients—salt, sugar, redox agents, gluten, enzymes, fats, and emulsifiers. One way to study the behavior of a dough is with the aid of rheological methods, and part of this chapter is devoted to the description of fundamental rheological measurements on wheat flour doughs. Finally, we summarize the model that emerges from our discussion.

II. THE CONTINUOUS GLUTEN PHASE

The continuum of the dough is essentially the gluten gel, as described in Chapter 2. The starch, as will be described later, ultimately also forms a continuous starch–water phase. According to rheological measurements, the continuous gluten phase is a liquid. Furthermore, the continuum of the structure is an aqueous zone, and the importance of a continuous water matrix will be discussed.

To obtain the continuous phase it is not enough to simply add water to the flour, at least not if we wish to obtain the gluten gel within a reasonable period of time. We also have to apply mechanical treatment to the dough. Chemical reactions, such as oxidation reactions, are also important for dough development.

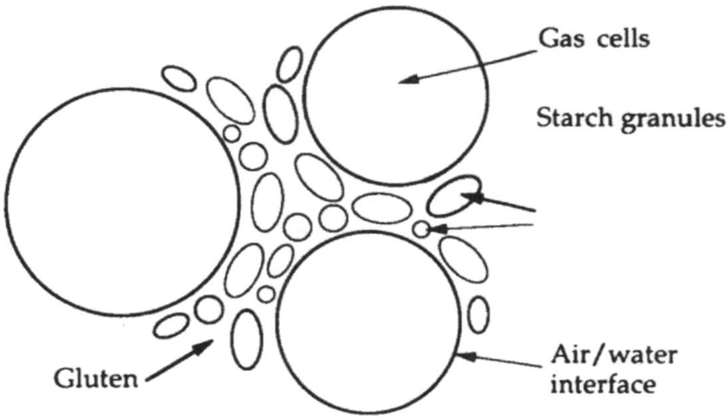

Fig. 1 A simplified model of the dough structure.

A. The Aqueous Environment

It is not possible to make bread without water. Water is necessary for gluten formation, and water is the medium for all types of interactions and reactions that occur during the breadmaking process. Because of differences in water content between crust and crumb, different reactions take place in the crust compared to the crumb [2]. The quantity of water used influences, for example, the rheological behavior of the dough and the extent of starch gelatinization. If the water content is not at the optimum level, the dough is difficult to handle, and the loaf volume is also affected [3]. The quantity of water that should be used is determined by measuring the *water absorption,* a concept that will be described below.

The water content also influences the storage stability of wheat flour. It should be below 14% to prevent microbial growth and chemical changes during storage [3].

1. Quantity of Water and Water Absorption

The water content of a standard bread dough is about 40% [3]. However, the ingredients in the formula are usually expressed as a percentage of the flour by weight, and the water content in a bread dough will then be around 65% [1]. To get the best results in baking there is an optimum level of water. When the water content is lower than the optimum, the mixing time will decrease, and with a higher water level the mixing time will increase. The loaf volume is more affected by a water level that is too low than by one that is too high [4]. The loaf volume will be less with water content below about 45% [5], and at water contents close to 35% there is almost no difference in volume between yeast and non-yeast doughs.

During the subsequent steps in the breadmaking process the total water content in the product (dough, crumb, crust) changes. This could be exemplified with the following figures: mixed dough, 46.4% water; proofed dough, 47.5%; bread crumb after 1 h, 44.0%; and bread crumb after 24 h, 43.9% [4]. The slight increase in water content during proofing is probably a result of the fermentation and absorption of some moisture in the fermentation cabinet [4].

The optimum level of water addition is related to the composition of the flour, and therefore it is necessary to determine this optimum level for each flour. This may, of course, be done in test baking, but it is more common to determine *water absorption* by the use of the Brabender farinograph. The *water absorption* of a flour is described as the amount of water necessary to bring the dough to a specified consistency (normally 500 Brabender units, BU) at the point of optimum development [6]. Both quantity and quality of protein influences water absorption [7–10]. Absorption increases linearly with the amount of protein, but the slope of the regression line depends on the wheat variety. The level of damaged starch also influences the optimum level of water, and the water absorption of the flour increases when the level of damaged starch increases [7, 11, 12]. The content of pentosans and β-glucans will also affect water absorption [13].

2. Distribution of Water in Dough and Bread

From a colloidal point of view, water is important because it constitutes the continuous medium of dough. This continuous gluten phase is not immediately recognized as a liquid phase, as the viscosity is so high. The storage proteins present are not dissolved in the water; rather they are hydrated, as discussed in Chapter 2. That water constitutes the continuous phase in the dough is illustrated by the electrical conductivity of the dough [5]. The conductivity decreases when the water content decreases, and it is almost zero when the water content approaches 35%. This latter result could be interpreted as showing that there must be more than 35% water in the dough to achieve a continuous conducting phase [5]. This water appears to be a second phase of "free" water. This aqueous phase dissolves water-soluble components and is the medium for chemical reactions and for dissolving carbon dioxide up to its saturation point.

The presence of a second phase of "free" water in the dough is evident from the fact that it is possible to separate a liquid phase from the rest of the dough by ultracentrifugation [5]. The apparent water content of the gel phase will be 34.5%; applying higher centrifugal force does not release more of the free liquid phase. This value is in accordance with the water content necessary to obtain electrical conductivity in the dough, as discussed above [5]. The liquid phase released consists of 3.4% protein, 0.3% lipid, 3.0% sodium chloride (NaCl), 7.0% remainder (mainly soluble carbohydrate), and 86.0% water.

When the water content of flour is successively increased without any mixing there is a drastic change when it reaches 23% (w/w), $a_w = 0.95$ [14]. At this concentration the flour loses its free-flowing nature, showing a pronounced interparticle interaction. A very sharp peak of resistance is observed at a water content of 35% (w/w) in the flour [15]. This water content is also in agreement with that giving a minimum in extensibility of the dough. A free water phase starts to separate at 23% (w/w), and when the capillary absorbed water amounts to 12% the mechanical resistance on mixing is at its maximum. This must be due to the lubrication conditions of the colloidal gluten gel structure units (see Chapter 1).

The "free" water present is not evenly distributed throughout the dough at the beginning of dough mixing. Microscopic investigations of the ultrastructure of dough and bread after freeze-fracturing have revealed three different localizations of water [16]. Water may be present in the dough as large regions of bulk water, as small droplets in the protein network, or surrounding the starch granules. The distribution of water among these three types of locales is influenced by additives such as salt and sugar. It is important, however, to keep in mind that the conductivity of the dough shows that the "free" water phase must be continuous. This indicates that the dough is bicontinuous—the gluten gel being one continuous phase interpenetrated by the other "free" water phase. This will be discussed in the final section of this chapter. In the bread no "free" water is observed; presumably the starch absorbs all the water during gelatinization.

Because the components in the dough differ in their affinity for water there is a distribution of water not only between continuous and dispersed phases, but also among the various components. Part of the water is so closely associated with the macromolecules that its properties are changed; for example, its freezing point differs from that of ordinary bulk water. The amount of water that does not freeze at 0°C (or close to this temperature), here called nonfreezeable, has been determined by differential thermal analysis (DTA), and a value of about 0.3 g/g dry flour was obtained [17]. Differential scanning calorimetric measurements gave the same value, 0.33 ± 0.016 g/g dry flour [14], corresponding to a water content of 23–24% [18]. It has been observed that this amount does not depend on the flour strength, and it is not affected by protein content, amount of damaged starch, dough mixing time, or added chemicals such as sodium chloride or ascorbic acid [17]. For a starch–water suspension of about the same starch/water proportions as in a dough, the amount of nonfreezeable water is 0.30 g/g dry matter [19]. This value changes to about 0.38 g/g dry matter after heating to 100°C. The amount of nonfreezeable water is about the same for proteins [20]. However, this water should not be regarded as "bound," as it differs from ordinary bulk water only in its freezing behavior.

The total amount of water present in a dough can be measured, but its distribution is not easily determined. The local water environment determines, for example, the degree of starch swelling and gelatinization. The temperature in the crumb does not exceed 100°C during baking. To completely gelatinize starch, the conclusion temperature of gelatinization (T_c) must therefore be below 100°C. This is the case if the water/starch ratio is greater than 0.8 : 1 [21]. If the water/starch ratio is lower in the dough owing to the distribution of water, partly gelatinized starch granules may be present in the bread crumb. The local water content can also be expected to govern which liquid crystalline lipid phases are to be formed. We do not know exactly the local water content for the flour components, and moreover the distribution may change depending on the ingredients present.

The distribution of water in the dough has been estimated from the water absorption values of the various components. Interactions between the components and the possibility of phase separation are neglected in this approach. The water uptake of the different components was determined to be as follows: starch, 0.44 g/g; damaged starch, 2.00 g/g; gluten, 1.1–2.2 g/g; and pentosans, 15 g/g [22, 23]. From these values the water distribution in dough was calculated, and it was found that 45.5% of the total water is associated with starch, 31.2% with proteins, and 23.4% with pentosans [23]. As argued above, however, we have a two-phase separation into a gluten gel aqueous compartment and a "free" water aqueous compartment (with starch and pentosans), and then the gluten gel contains the fraction 23/40 of the total water (with 40% water in the dough).

If we alternatively start with the structural description in Chapter 2, the gluten gel can accommodate only about 55% (w/w) of water, and any excess will form a separate phase. Taking into account the storage protein content of the flour, the limit of water in the gluten phase of the dough is in good agreement with the 23% observed in the dough. In other words, the two-phase nature of the gluten gel + starch–water in the dough follows directly from the basic structural properties of gluten.

A study of the water distribution in bread determined that 77% of the water was associated with starch, 23% with pentosans, and none with the proteins [23]. If the gluten proteins lose their affinity for water during baking, starch has to take up more of the water. In fact, transport of water from gluten to starch during the gelatinization process has been calculated from the influence of gluten on the gelatinization behavior of starch [24]. Such a redistribution of water has been suggested as one reason for the staling of bread [25]. This will be discussed further in Chapter 7. It seems impossible, however, that the water content of the gluten coagel is zero. Certainly there must at least be some inclusion of capillary water as a space filler. This is also evident from model experiments. A heated gluten gel will still retain some water after transformation to a coagel (see Chapter 2.).

3. Influence of Water Content on the Rheological Behavior of Dough

The reheological properties of a wheat flour dough are extremely sensitive to water content [26–30]. This is illustrated in Fig. 2, and it is evident that a decrease in the amount of water added has a greater effect than an increase, at least within the range of water content illustrated here [29]. The amount of

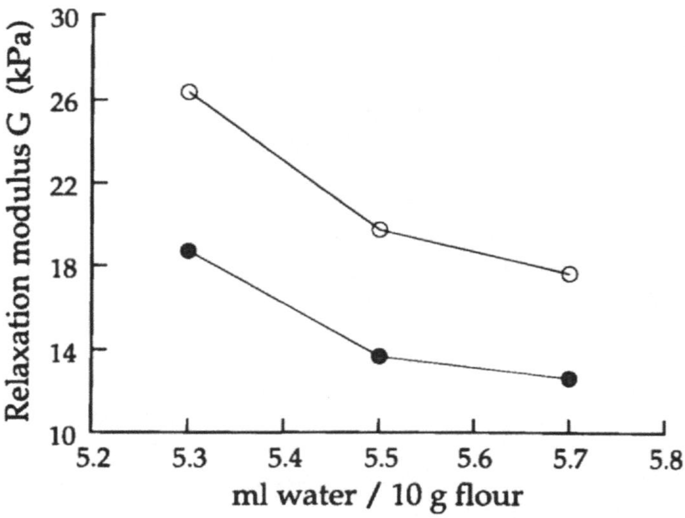

Fig. 2 The influence of water content on the stress-relaxation behavior of dough. (From [29].)

damaged starch has a strong influence on these relationships [28]. It is interesting to note that for two flours with different levels of damaged starch (9.3 and 20.9%, respectively) it was possible to make doughs having the same value of the storage modulus (G') by adding water, but was not possible to achieve the same phase angle (δ) by adjusting the water content. The flour with the low level of damaged starch had a G' value of 12.4 kPa, and δ was 32.9°. The flour with the high level of damaged starch with the same amount of water added (5.5 mL to 10.0 g flour), gave $G' = 46.2$ kPa and $\delta = 21.5°$. When the amount of water was increased to 7.0 mL/10.0 g flour, G' decreased to 14.9 kPa, but δ increased only to 25.3°.

The dependence of G' and G'' (loss modulus) on water content is related to the protein content of the flour [26]. When the protein content increases, G' and G'' both become less sensitive to the water content. However, increasing the water content of the dough will not change the water content of the gluten gel [31], a result that is completely in accord with the swelling behavior of gluten proteins described in Chapter 2.

4. Solutes in the Aqueous Phase

The continuous "free" aqueous phase will affect the behavior of the dough because of the amount and type of soluble substances, such as salts and minerals, as well as organic material.

a. pH

During dough mixing the pH will be determined by the wheat proteins (through dissociation of charged amino acid side chains). The initial pH is about 5.5, and the dough has a considerable buffer capacity. Fermentation will still be able to reduce the pH owing to the saturation of carbon dioxide in the water phase.

At normal temperature and pressure (0°C and 1 atm), saturation is reached at 0.04 M concentration, and the following reaction equilibrium gives a pH of 3.7 (where $pK_a = 6.5$):

$$H_2O + CO_2 \rightleftharpoons H_3O^+ + HCO_3^- \qquad (1)$$

When the temperature is raised, the carbon dioxide will be gradually expelled, and the equilibrium will shift to the left. The temperature during fermentation therefore has a direct effect on the pH development, in addition to the general effect of temperature on microorganism metabolism and reproduction. The lowest pH values are achieved when sourdoughs are produced using various strains of lactic acid bacteria, and the final pH level of about 3.6 will also limit the bacterial metabolism.

The pH of the dough influences the physical properties of the gluten gel and the lipid–water phases. We will first consider the lipids, as they are simpler in this respect. The charged species of the polar lipids are the ones that can induce phase transitions by a pH shift. There are two main components of this type, phosphatidylserine and phosphatidylinositol (with their lysoanalogues), which have a total concentration of about 1% (w/w) in the nonstarch lipids and 2% in the starch lipids. This concentration may seem small, but it is enough to induce lamellar phase → reversed hexagonal phase transitions if divalent ions are present, provided that these lipids have a pH so that they are fully charged. The amount of divalent ions in an "average" water will be enough.

Depending on lipase activity in the flour there can also be considerable amounts of fatty acids among the lipids, a negative quality factor known for a long time. If pH is above 5, most of the fatty acids will be ionized. The reason for their negative effect is certainly that they will be solubilized into the polar lipid bilayer. Calcium ions in the water layer have a strong tendency to contract adjacent charged polar heads. This means an increase of the average wedge shape of the lipid molecules or increase in divergence of the hydrocarbon chains (see Chapter 1), which will induce a phase transition from the lamellar liquid crystalline phase toward the cubic and reversed-hexagonal phases. The charged species of phospholipids (with calcium present) will act in the same direction. This means that a reduction in pH, which will reduce the dissociation, will also be favorable for the lipid effect on gas-holding capacity.

The gluten proteins range widely in their isoelectric point (IP); most of the gliadin-type proteins have IP values above 6.5, whereas the glutenin proteins

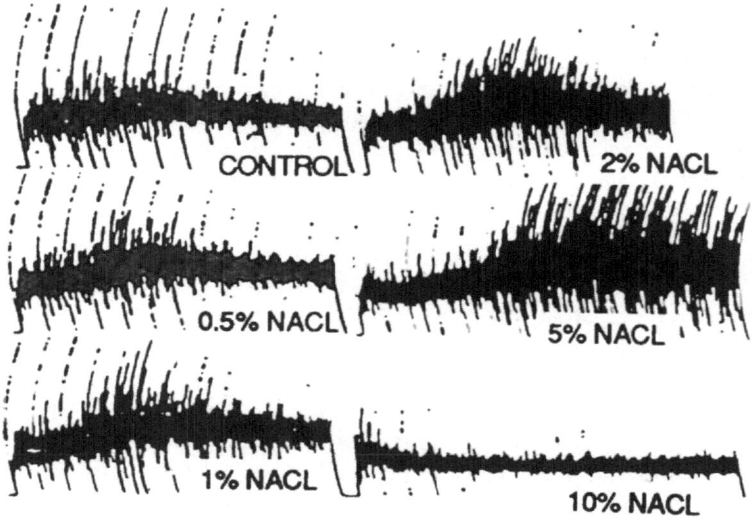

Fig. 3 The effects of salt on mixograph characteristics. (Reprinted with permission from [38].)

because they are dissolved in the added water. From the discussion above it is evident that the hardness of water—its content of calcium and other minerals—must also be of importance. Water of medium hardness is preferred [3]. An extremely soft water causes the dough to be very soft and sticky, whereas extremely hard water retards the fermentation [3]. Ions in the water can also influence the phase behavior of polar lipids.

The effect of salt on lipid–water phase behavior is usually drastic when charged species of the lipids are present. If, for example, some ionic phospholipids are present in a mixture of cereal phospholipids and galactolipids, which form the lamellar phase with water, a minor amount of calcium will switch this structure to a reversed-hexagonal one with quite different functional effects in breadmaking. It has been observed that lecithin forms the lamellar liquid crystalline phase in distilled water, whereas a mixture of several phases is formed in hard tap water [45]. The lamellar liquid crystalline phase was also observed in tap water, but the L2 and hexagonal phases were also present (Fig. 4). As the two latter phases do not have an improving effect in baking, the water used to disperse added emulsifiers could decrease the beneficial effects expected.

The gelatinization of starch is affected by the presence of ions [46–48]. However, for an effect to be observed, the salt concentration has to be so high that it is unrealistic in baking (Table 1). On the other hand, at high concentrations it is possible to gelatinize starch at room temperature [47, 49].

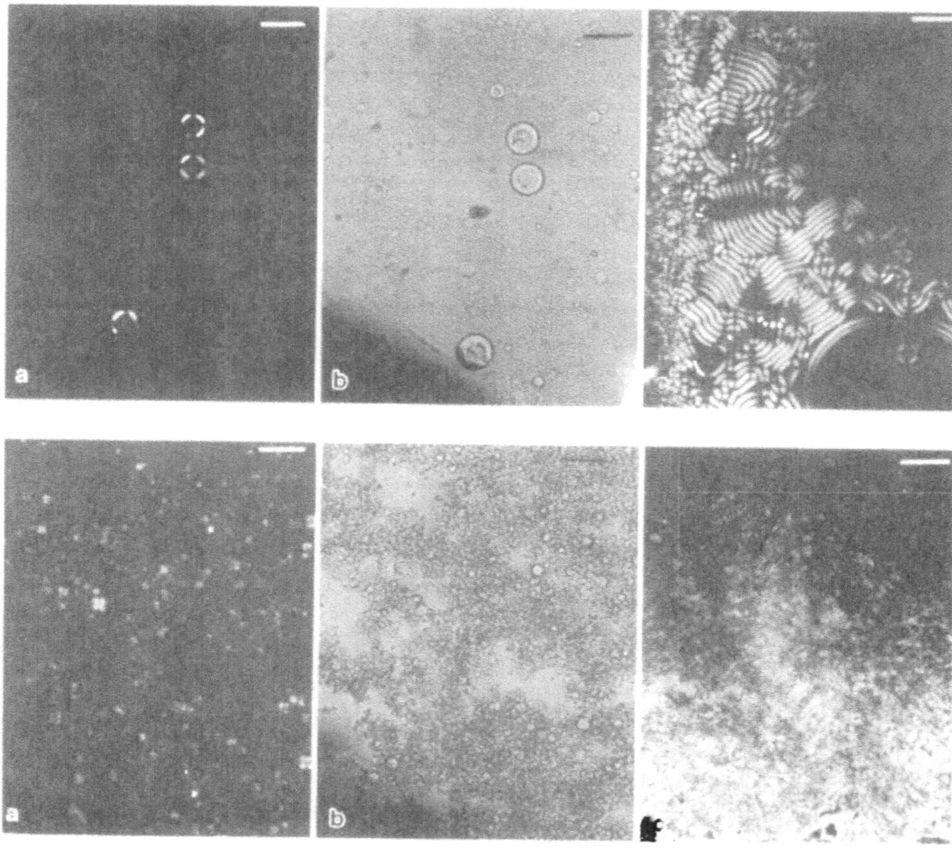

Fig. 4 Phase behavior of lecithin in distilled water (upper row) and hard tap water (lower row). (a) Polarized light; (b) ordinary light; (c) polarized light after drying the sample. Bars indicate 40 μm, (Reprinted with permission from [45].)

c. Sugar

The addition of sugar (usually sucrose) has some beneficial effects in baking. It serves as an easily accessible carbohydrate for the yeast, although the amount added must not be so great that the osmotic effect has a negative influence. It contributes brown color to the crust, through the Maillard reaction (enzymes from the yeast will hydrolyze sucrose to fructose and glucose, which can react with proteins) and through caramelization. Sucrose certainly adds sweetness to the product.

At the low levels of sucrose used in bread formulas, its influence on the

Table 1 Influences of Solutes on the Gelatinization Temperature Range of Starch

Solute	Solute/starch ratio	Gelatinization onset Temp. (°C)	Gelatinization temp. range (°C)
Sucrose[a]	25:100	67	46
	50:100	78	32
	100:100	98	22
NaCl[b]	0	50	36
	3	58	30
	6	64	24
	9	68	20
	12	65	23
	15	65	23
	21	61	29
	30	59	32

[a]Water/starch ratio = 0.70:1.
[b]Percentage of aqueous phase is given. Solvent/starch ratio = 2:1.
Source: Data compiled from [46], [48].

baking process might be restricted to the examples just given. When the sugar/flour ratio is considerably increased, it can also be expected to have an influence on the structure and behavior of doughs as well as that of the products. At a level of 6.0 g sugar/100 g flour, the sugar affects the water distribution in the dough [16]. The water is found to be more evenly distributed, and large regions of bulk water are not detected in the presence of sugar.

Addition of sucrose changes the viscosity of starch–or flour–water suspensions both before and after heating. When wheat starch and water are mixed at a 1:1 ratio, a more or less dry powder results. If one part sucrose (dry) is added (i.e., starch:water:sucrose = 1:1:1), a free-flowing liquid is obtained [48, 50]. The G' value for a wheat starch paste increases at low levels of sucrose in the paste (up to 10%), but further addition of sucrose results in a decrease in G' [51].

The addition of sugar raises the gelatinization temperature of starch (Table 1). All sugars do not have the same effect; disaccharides are more effective than monosaccharides (on a molar basis as well as on a weight basis), and of the disaccharides sucrose is the most effective [47, 52]. At the levels of addition used in breadmaking, the effect on starch gelatinization may not be observable, but in other products with a high sucrose/flour ratio this effect can cause the gelatinization temperature range to be shifted to such high temperatures that complete gelatinization does not occur below 100°C [46]. This effect of sugars on starch gelatinization has been described as an inhibition of the starch gelatinization, for example, due to competition for water [52]. However, the effects cannot be satisfactorily explained in these terms. Instead, it might be argued that the plasticizing effect of water is lessened when sugars are dissolved; the sugar molecules are bigger than the water molecules, and they will not lower the glass transition temperature of starch to the same extent as water alone [53, 54]. Another consequence for starch gelatinization is observed at high sugar/starch ratios and low water/starch ratios. Not only is the gelatinization endotherm shifted toward higher temperatures, it also narrows [48, 50]. This seems to indicate that the gelatinization occurs more easily in the presence of sugar, albeit at a higher temperature. One explanation could be the lower viscosity in the system, described above.

B. Development of the Continuous Phase

To facilitate the formation of the continuous gluten phase, the ingredients are mixed. Mixing accomplishes three things: the ingredients in the bread formula are blended into a homogeneous mass, the proteins are developed into a three-dimensional structure that has the capacity to hold gas, and air cells are included [55]. The significance of mixing for the continuous gluten phase is discussed here, and the inclusion of air is discussed in Section III.

1. Mixing and Dough Development

The appearance of the dough changes during mixing. The lumpy mass obtained after the initial wetting is transformed into a coherent mass that exhibits elasticity. At this stage the dough has a decreasing tendency to stick to the mixer. Eventually the whole piece of dough will wind around the mixer blade. Many bakers use this effect to determine when the optimum development is reached [56]. When mixing has been performed to the peak of development, the appearance of the dough is silky and glossy, and the dough has elastic properties. If mixing is continued beyond the peak of development that is, if it is overmixed, the dough becomes very sticky and extensible.

In the mixing process, mechanical energy is put into the dough. The energy input depends on the type of mixer and the type of flour, and on whether mechanical dough development is used. Some figures could be cited to indicate the magnitudes of energy required. Modern mixers have an energy input of 25 kJ/kg dough in 5–12 min, whereas older mixers give 10–15 kJ/kg dough over 30 min [55]. In processes such as the Chorleywood process, where intense mechanical dough development is used, the energy input as well as the mixing intensity have to be above minimum levels [57]. A typical value is 40 kJ/kg dough, and the mixing time needed is only a few minutes [55].

The quality of the final loaf of bread is strongly dependent on the mixing, and for each combination of flour and mixer it is possible to find an optimum stage of dough development (Fig. 5). Consequently, a dough can be undermixed, mixed to the optimum, or overmixed. To know when the optimum is obtained the baker can observe the appearance of the dough and how it sticks to the machine, as described above. Some information concerning the optimum in dough development can also be gained from recording dough mixers such as the farinograph or mixograph. In this type of mixer the power to mix the dough or, in other words, the resistance to mixing, is recorded. The resulting mixing curve is described with such terms as "dough development" and "breakdown."

An example of a mixing curve is given in Fig. 5. As long as there is only flour in the mixer, no resistance against the mixing is recorded. When water is added, the dough development starts, and this is seen as an ascending part of the mixing curve. Depending on the strength of the flour, the curve will reach a maximum either as a well-defined peak or as a plateau. Immediately after wetting, the glutenin molecules are distributed randomly in the dough [58]. During mixing the glutenin molecules are extended and aligned, resulting in a cooperativity of the secondary forces [58, 59]. The time to "peak development" has been suggested to depend on the time it takes to stretch and extend the glutenin molecules [60]. The more glutenins and the higher their molecular weight, the longer the development time will be. After further mixing, a decrease in the mixing curve is recorded and the breakdown starts. The point of optimum dough development is at the maximum in the mixing curve or slightly after this peak [10]. The rate of breakdown shows the stability of the dough and its sensitivity to

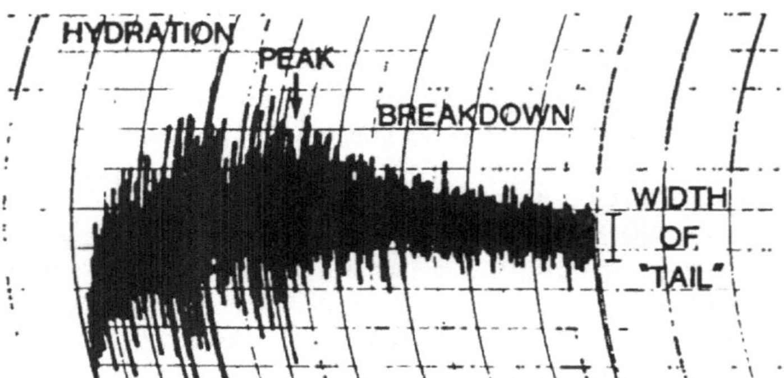

Fig. 5 A mixogram showing hydration, a peak (i.e., a dough with minimum mobility, or an optimally mixed dough), and breakdown. (Reprinted with permission from [160].)

mechanical treatment. The flour of best baking performance has medium to medium-long mixing times. If the flour is difficult to mix, that is, if it takes very long to reach the peak of development, the dough may not develop properly in an ordinary mixer [57].

2. Oxidation and Reduction

Redox processes controlled by enzymes play a central role in all living cells, and wheat has numerous enzymes that have a very good substrate in the storage protein because of the high content of sulfhydryl groups in these proteins. Oxidizing/reducing (redox) agents such as potassium bromate ($KBrO_3$) and ascorbic acid have long been a valuable tool for the baker. Other common agents are oxygen (perhaps used unintentionally by the baker), cysteine, and enzyme-active soy flour [61]. When redox agents are used, the situation is the same as for water, salt, and other ingredients: there is an optimum level of addition, and this optimum level depends on the type of flour used. The amount of oxidant is on the order of 25 mg/kg flour, whereas intense mechanical dough development requires around 100 mg/kg [55]. The oxidation requirements increase with protein content, and, furthermore, the requirements are higher for varieties with short mixing times [10]. Not only is the loaf volume affected by oxidants, but also the mixing behavior is changed when an oxidant is added.

The improving effects of oxidants are especially evident for a newly milled flour; the flour performs poorly but will improve with storage. The thiol content of a newly milled flour decreases by 20–30% over a few hours in air [62]. The flour matures more quickly with the addition of oxidizing agents such as ascorbic acid or $KBrO_3$ [63]. This shows that the effect of maturing of a newly milled flour is due to oxidative changes. Oxidation processes can also cause a bleaching of the flour. However, the two most common oxidants—$KBrO_3$ and ascorbic acid—do not bleach the flour [55]. An important factor is, of course, whether the oxidation agent can act in the dry flour or is activated during dough mixing.

The structure of oxidized wheat flour doughs has been studied by light microscopy [64]. It was found that the gluten network in the oxidized dough was not as homogeneous as in the dough mixed to optimum without oxidants. Moreover, the protein network seemed to be less stretched than in the dough without oxidants. When the doughs were sheeted, the protein network became very much oriented in the direction of the sheeting [64]. Such an orientation was not observed in a reduced dough. These results are consistent with the gluten gel structure shown in Fig. 31 of Chapter 2; the larger the molecules become owing to oxidation, the longer the relaxation times are after shear.

a. Oxidation and Polymerization of Protein Molecules

The influence of oxidizing and reducing agents on dough behavior, especially mixing behavior, has focused interest on the presence of the sulfur-containing amino acid cysteine in the wheat proteins. Oxidizing agents make the dough stiffer, whereas reducing agents weaken the dough [10]. This could be easily explained by the following mechanism:

$$\text{Protein–SH} + \text{HS–protein} \underset{\text{reduction}}{\overset{\text{oxidation}}{\rightleftharpoons}} \text{protein–S–S–protein} \quad (2)$$

This reaction would cause cross-linking of protein molecules and thus increased stiffness of the dough. Reduction causes splitting of the disulfide bonds and thus a reduction in molecular size, causing the dough to become weaker.

The oxidation reaction illustrated by Equation (2) could be due to oxygen in the air. Extensigrams for gluten gels prepared under oxygen and nitrogen have been compared [65]. If different wheat varieties are milled under nitrogen and the gluten gels are also washed out under nitrogen, the resulting gels are all very weak (and very similar to each other). If the gluten gels instead are washed out under air, strong wheat varieties are strongly improved whereas the wheat varieties of poor baking performance are almost uninfluenced (Fig. 6). The improving effect obtained for gluten in air can also be obtained if KBrO3 is added to the gluten isolated under nitrogen. The improving effect of ascorbic acid is greater for gluten isolated in nitrogen than for gluten isolated in oxygen [66]. These results show that suitable proteins must be present for oxidation.

When reaction (2) goes toward the right, the number of thiol groups decreases and the number of disulfide bridges increases. The amount of thiol groups does decrease during mixing, and the effect is most pronounced for wheat varieties of good baking performance [67]. An increase in the number of disulfide bridges is

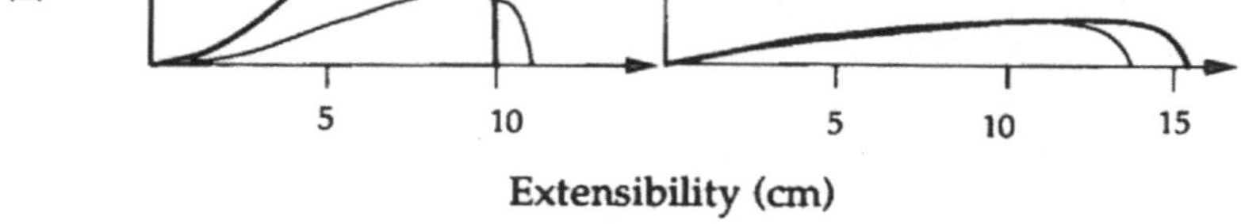

Fig. 6 The strength of gluten gels prepared in oxygen and nitrogen. To the left is shown gluten from the wheat variety Monopol, and to the right gluten from Maris Huntsman. (Adapted from [65].) Thick lines represent oxygen and thin lines nitrogen.

difficult to detect for experimental reasons, but there are several indirect indications of increases in molecular weight due to oxidation. Stress-relaxation measurements on gluten in the presence of oxidants (ascorbic acid, $KBrO_3$) indicate a polymerization or cross-linking [68]. Potassium bromate and ascorbic acid both increase the stress relaxation modulus (G) of gluten. The effect goes in the same direction as that due to treatment of the gluten gel with glutaraldehyde, which is a general cross-linking agent of protein molecules [68]. The relaxation time, $t_{1/2}$, also increases in the presence of $KBrO_3$ and ascorbic acid, indicating a higher degree of elasticity.

A suspension of wheat flour can also consume oxygen because of the oxidation of lipids, particularly in the presence of lipoxygenase [61, 69]. Lipid hydroperoxides formed in this way can in turn oxidize proteins.

b. Thiol–Disulfide Interchange Reactions

If thiol-blocking agents are added to a dough, the effect will be the same as for oxidation—a stiffer dough—although the number of disulfide bonds does not change. The reason for this is not understood. The blocking agent itself may influence protein association. One function of an oxidizing agent may be to remove the thiol groups from the system. Exchange reactions involving S–S and SH groups have been suggested, and the reactions might involve a thiol-containing molecule of low molecular weight such as glutathione (GSH) [10, 70], as illustrated in Fig. 7.

There is about 0.1–1.6 μmol of SH groups per gram of flour [61, 62], but values as high as 4.7–7.2 μmol of SH groups per gram of flour (or 60.4–69.3 μmol/g protein) have been reported [67]. There is no correlation between the amount of thiol groups and baking performance. The number of disulfide groups is higher than that of thiol groups, with reports of 5–17 μmol S–S groups/g flour [61] or about 100 μmol/g protein [62]. There is no correlation between the S–S content and baking performance [67].

Not all thiol and disulfide groups present in the dough take part in these

Fig. 7 Sulfhydryl–disulfide exchange reactions involving low molecular weight sulfur-containing molecules.

reactions; the groups that take part are described as *chemically active* [70]. Moreover, the reaction may or may not influence the viscoelastic behavior of the dough, and we therefore have *rheologically active* SH and S–S groups. It is believed that the rheologically active thiol groups are present in small peptides. Such a molecule would allow many SS/SH exchanges to occur because the peptide is not consumed (see Fig. 7) [70]. Such small molecules present in the dough are glutathione and thioctic acid [71]. An exchange between the gluten protein molecules and a small peptide (as in Fig. 7) is more likely than an exchange between two protein molecules [71].

Small molecules containing SH groups or disulfide bonds—for example, glutathione sulfide (GSSG)—weaken the dough because of the decrease in molecular weight of the proteins. The thiol-containing compounds are the most effective in this respect [71, 72]. When an oxidant is added, the exchange reactions causing softening of the dough stop, because the oxidant reacts with the small thiol-containing compounds [73]. The removal of thiol groups might be beneficial for the handling properties of the dough, as stickiness has been found to correlate with the thiol group content of the protein [63]. The SH/SS exchange reactions taking place during mixing, as well as direct oxidation, may change the molecular weight distribution. The final distribution depends on the initial distribution and on the actions of oxidants and reductants present [74].

The number of thiol groups decreases during mixing; 45% of the initial amount of GSH is lost during the first minute of mixing [61, 67, 75]. If the dough is allowed to rest, the decrease in GSH continues, and the level is virtually zero after about 10 min. The number of SH groups decreases if the dough is mixed in air in the presence of an oxidant (potassium iodate) or a thiol-blocking agent [*N*-ethylmaleimide (NEMI)], but if the dough is mixed in nitrogen the number of SH groups is not affected [75, 76]. It has, however, been suggested that the number of thiol groups may also decrease in N_2 because of the lipids [62]. Although the number of thiol groups decreases, an increase in the number of disulfide bonds has not been observed [67, 75]. This could be due to the possibility that SH/SS exchange reactions occur without a net change in the number of disulfide bonds. However, we must also be aware of the possibility that the analytical tools are not sensitive enough to detect the small increase in disulfide bonds that might occur [77]. Oxidizing and reducing agents influence the mixing curves. If an oxidant is present, the best baking results are obtained when mixing proceeds to the peak optimum. If the level of oxidant is decreased, mixing has to continue beyond the optimum in the mixing curve [10].

When ascorbic acid is used as an improver in the Chorleywood process, the result depends on the mixing atmosphere [78]. The best result is obtained in an oxygen-enriched atmosphere, 60% O_2–40% N_2. No such influence of mixing atmosphere is observed for $KBrO_3$.

c. Reduction Reactions

If cysteine (containing the thiol group) or cystine (containing the disulfide bridge) is added to gluten, the result will be the same: a softer gluten [42]. The addition of glutathione to dough will also have a softening or weakening effect [79]. The softening effect is thought to be due to reduction of the disulfide bonds. An increase in the number of thiol groups might be anticipated because of the increased stickiness of the reduced dough [64]. The stress-relaxation behavior of gluten treated with cysteine also changes in a way that can be interpreted as due to a decrease in the number of cross-links [68]. Both G and $t_{1/2}$ decrease, thus resulting in a softer gluten with a smaller amount of network structure. Also the extensigram is affected in a way that can be interpreted as a weakening when cysteine or glutathione is added [66].

Previously it was claimed that only a very few of the thiol groups and disulfide bonds are rheologically active. It was observed that the major changes in molecular weight of the protein fractions of highest molecular weight occur when 4–5% of the disulfide bonds are broken [65].

There are situations in breadmaking when addition of a reducing agent such as cysteine (usually L-cysteine hydrochloride) would be beneficial. If the flour is too strong for the mixer used, it will not develop even if the mixing time is increased [80]. The addition of cysteine might then bring about the proper dough development. The energy requirements in mechanical dough development can be reduced if a reducing agent is used [81].

d. Mechanisms of Some Common Redox Agents

The most commonly used oxidants are ascorbic acid and potassium bromate, which are both slow-acting oxidants [82]. Examples of fast-acting oxidants are potassium iodate (KIO_3) and azodicarbonamide. The optimum level of addition differs between the oxidants, and there are also synergistic effects [79]. Ascorbic acid alone might not be enough as an oxidant. Moreover, the effect of ascorbic acid seems to vary depending on whether traditional methods or modern intense mixing systems are employed [61]. For a flour of medium to short mixing time and/or high protein content, both 80 ppm ascorbic acid and 10 ppm $KBrO_3$ were required for the best result [9]. When gluten was treated with 8500 ppm $KBrO_3$, the addition of ascorbic acid still had an effect on the rheological properties: it increased G as well as $t_{1/2}$ [68].

The mechanisms for the action of the oxidants are not clear in all details. The situation is perhaps simpler in the case of $KBrO_3$, for which the reaction in Fig. 8 has been suggested. When bromate is added, more disulfide bonds and fewer thiol groups are found in the dough compared with the control without bromate [83]. The effect of $KBrO_3$ is most evident toward the end of proofing [34].

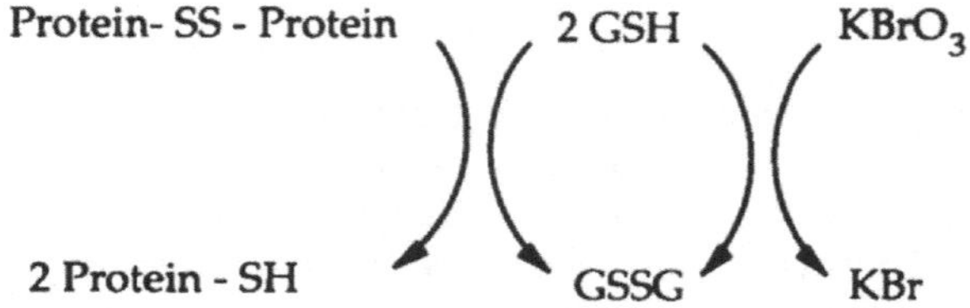

Fig. 8 The mechanism for the function of $KBrO_3$ in the improvement of wheat flou dough. (Adapted from [73].)

In the case of ascorbic acid the mechanisms in Fig. 9 has been suggested L-Ascorbic acid must first be oxidized to L-dehydroascorbic acid (DHAA [61] which is the active compound. Addition of DHAA has the same influence on th rheological behavior of the dough as the addition of ascorbic acid [79]. De hydroascorbic acid will then take part in other reactions involving glutathion and causing the exchange reactions illustrated in Fig. 9 [71, 73, 84]. The enzym glutathione dehydrogenase has been detected in wheat flour [61, 73]. It has bee observed that DHAA reacts very quickly with glutathione in water extracts o flour [79].

The improving effect of ascorbic acid is obtained only with L-ascorbic acid When D-isoascorbic acid is added to dough it is transformed to dehydro-D isoascorbic acid during mixing, but this latter compound has no influence on th dough development [79] (owing to the involvement of enzymatic reactions, th antipode will not have the same effect).

The first step in the reactions involving ascorbic acid, its conversion o DHAA, can be brought about in several ways. There is a specific enzym *ascorbic acid oxidase,* that catalyzes the reaction, but the effect of the D forn mentioned above and other results indicate that the oxidation does not require th enzyme [61]. That other factors, besides the presence of the ascorbic aci oxidase, are involved is illustrated by the following experiments [85]. When flour–water mixture was undermixed (1 min instead of the optimum 3½ min)

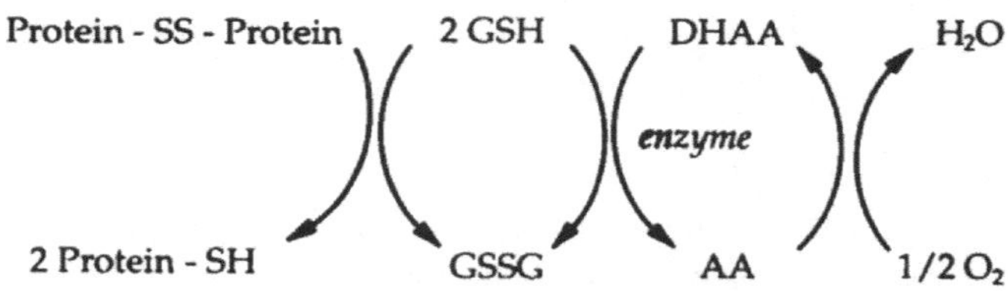

Fig. 9 The mechanism for the function of ascorbic acid (AA) in the improvement o wheat flour dough. (Adapted from [73].) The enzyme indicated is glutathione de hydrogenase. DHAA-dehydroascorbic acid.

78% of the added ascorbic acid was recovered. During the following 30-min resting period, almost no conversion to DHAA took place. If yeast was included in the dough, more conversion to DHAA occurred both during the 1 min of mixing and during the rest period. After the dough rested for 15 min, 33% of the added ascorbic acid was analyzed as ascorbic acid. If the flour–water suspension was mixed to optimum (3½ min), about half of the added ascorbic acid was converted to DHAA during the mixing. And finally, when yeast was added and the dough was mixed to optimum, all the added ascorbic acid was converted to DHAA during the mixing. These results indicate that the oxidation of ascorbic acid to DHAA can be performed by the oxygen in the air or by the action of the yeast.

It has also been speculated that the first oxidation step (ascorbic acid → DHAA) could be brought about as a side reaction in connection with other oxidation reactions occurring in the dough, such as lipid oxidation. Functional aspects of breadmaking related to lipid oxidation have perhaps been neglected, other than regarding lipid hydroperoxides as a transitory stage that can couple to protein oxidation. This oxidation may also involve lipoxygenase [55]. There are possible end products with polar regions in the middle of the acyl chains, such as hydroxy fatty acid groups, and such a chemical change should be expected to stabilize the lamellar liquid crystalline phase of the lipids. The production of trihydroxy-substituted fatty acids in a dough has been demonstrated. Graveland [86] proposed that the hydroxy acids produced by lipoxygenase from linoleic acid will be transformed into hydroxy fatty acids by a coupled reaction involving glutenins. The involvement of the lipids in these reactions is illustrated by the fact that the reaction with oxygen occurs at a slower rate in a delipidized flour [87]. When wheat flour is brought into contact with water, oxygen is consumed in the oxidation of unesterified polyunsaturated fatty acids [69]. The oxygen consumption is much higher in wheat whole meal (0.1–3.0 μmol O_2/g in 10 min at 25°C) than in white flours (0.01–0.03 μmol O_2/g in 10 min at 25°C).

If lipoxygenase and linoleate are incubated with ascorbic acid, the acid will be oxidized [72]. It is well known that addition of lipoxygenase (added as soy flour) increases dough stability during mixing [61, 82]. Lipoxygenase may influence the rheological properties of the dough directly and oxidize the thiol groups in the protein and thus take part in the SH/SS exchange reactions [82].

e. Chlorination

Chlorine treatment of flour has been used for a long time for bleaching and for modifying baking properties. This may be why the process is "generally recognized as safe," although it is clear that we lack information on the chemical characteristics of all substances that can be formed. We will come back to this question at the end of this subject.

Chlorination of the storage proteins in their conformation in the flour should be regarded as a solid-state reaction:

$$Cl_2 + P\text{–}SH \; . \; . \; . \; . \; . \; . \; HS\text{–}P \rightarrow 2HCl + P\text{–}S\text{–}S\text{–}P \qquad (3)$$

The dotted line in the formula indicates that the reacting groups must be close enough to one another to make oxidative cross-linking possible. In this respect there is an obvious difference between this reaction and oxidation reactions in a dough, where water provides a medium for mobility of the peptide chains into conformations favorable for cross-linking.

Oxidation of carotenoids and other lipids will lead to a bleaching of the flour, which can easily be followed analytically. The oxidation of the gluten proteins has been indirectly demonstrated by the observation that after chlorination the oxidative conditioning of fresh wheat flour has no effect. The oxygen effect during the maturing of flour shares the characteristic feature of chlorine oxidation—the reaction takes place in the solid state.

The effect of chlorine on starch was discussed in Chapter 3. Thus there are indications of the transient formation of chlorine–amylose and chlorine–amylopectin inclusion complexes when chlorine is present, which can lead to irreversible surface structure changes on the starch granules.

In addition to oxidation effects on the lipids, there may be a second effect. Recent work on marine lipids [88] has shown that chlorine can react with double bonds to form quite stable dichlorosubstituted acyl chains. If such lipids are formed in wheat they will behave like methyl-branched fatty acid groups in lipids from a physical point of view (the van der Waals radius of chlorine is close to that of a methyl group). Furthermore, such lipids will not form hydroperoxides on contact with lipoxygenase and oxygen.

The possibility of the formation of unknown compounds such as the chlorinated lipids described here gives a reason to question the chlorination of wheat flour until more is known about the toxicology of the chemical reactions involved.

3. Mixing Requirements

We have already discussed the fact that flours differ in their mixing requirements. A flour of good baking performance should have medium to medium-long dough development and mixing time. If the mixing time is very short the flour can easily be overmixed, and if the mixing time is very long it might never reach its optimum [10].

The mixing requirements are related to the composition of the gluten phase. Gliadin imparts extensibility to the dough, whereas glutenin imparts strength and elasticity [10]. The more gliadin there is, the shorter the mixograph peak development time will be. However, with the same gliadin/glutenin ratio, a flour of good baking performance will still have a longer dough development time than a flour of poor baking performance [10].

A wheat variety of good baking performance and medium to long mixing time will have a higher proportion of high molecular weight (HMW) proteins [89]. The oxidative improvement of good wheat varieties is also coupled to an increase in the HMW proteins [65]. These results indicate that the mixing requirements of a flour are related to the molecular weight distribution.

The influence of different protein fractions on mixing behavior is illustrated in Fig. 10. A more detailed analysis was performed with wheat gluten that was fractionated into 10 fractions by successive extractions with dilute HCl [90]. The fractions obtained differed in molecular weight distribution, and sodium dodecyl sulfate polyacrylamide gel electrophoresis (SDS-PAGE) and size-exclusion high-performance liquid chromatography (SE-HPLC) were used to classify the proteins as glutenins or gliadins [90, 91]. The first four fractions extracted contained low levels of glutenin (usually below 50%); they decreased peak development time in the mixograph and had a slightly decreasing effect on the loaf volume when added at the 1% level to a flour [90, 91]. The next five fractions were higher in glutenins (~55–80% of the total protein), and they increased peak development time as well as loaf volume. The last fraction to be extracted was very high in glutenins and contained a considerable amount of protein that was not soluble in the buffer used for SE-HPLC. This fraction had a negative influence on loaf volume and increased the development to a much lesser extent than the previous fractions.

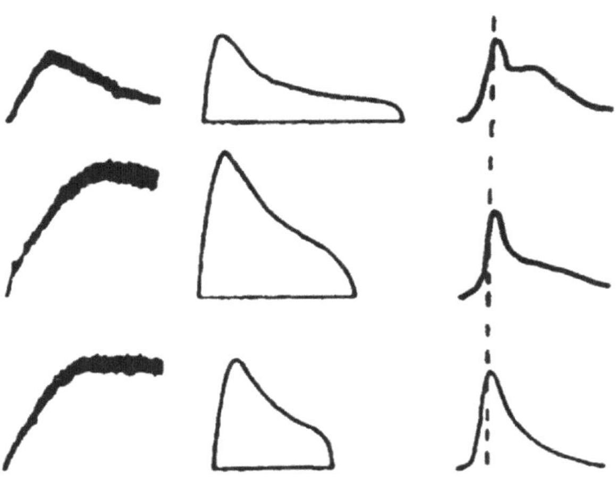

Fig. 10 The influence of different wheat protein fractions on the mixing behavior of wheat flour doughs. The (left) mixograph, (middle) alveograph, and (right) gel filtration patterns for proteins extracted from wheat flour with 2 M urea (top and middle rows) and 0.1 M NaOH (bottom). (Reprinted with permission from [117].)

4. Rheological Characterization of Doughs in Relation to Mixing

The most obvious reason for performing fundamental rheological measurements on dough is to be able to describe the point of optimum development in basic rheological terms. In small-amplitude oscillatory measurements it is possible to detect subtle differences that otherwise might be neglected [92], and this method has also been used to characterize wheat flour doughs after various treatments. Wheat flour doughs mixed in different types of mixers (mixograph, farinograph, a Hobart mixer, and a Krupp mixer) were found to have the same G' and δ values when mixed to optimum (the optimum judged from baking results) [93]. For the flour used in this investigation, G' was 11.1–13.1 kPa and δ was 26.5–28.1 for the four mixer types. The mixing time influences the storage and loss moduli in a way that is related to the wheat flour variety [94]. For one variety, a decrease in G' with mixing time was observed, whereas δ was more or less unaffected [95].

The changes occurring in the dough during the resting period were followed by rheological measurements [30, 94], and it was found that the rheological changes are strongly dependent on the wheat variety [96]. For one variety a considerable increase in G might be measured during dough rest, whereas another variety does not change at all. These varieties also differed in their response to under- and overmixing.

5. Changes in Flour Components due to Mixing

Dough mixing affects the appearance of the dough, its rheological behavior, and its colloidal structure. These changes can mainly be assigned to the protein. Being the continuous phase of the dough, the aqueous gluten gel and the changes it might undergo during mixing are, of course, of utmost importance. The influence of mixing on the protein is discussed in this section. The fate of the lipids is also described, in order to decrease the confusion created by the description of lipids as "bound" or "free." Starch constitutes about 60% of the volume of the gas-free dough [97]. Nevertheless it could be regarded as of minor importance in relation to mixing (except, of course, for the influence of damaged starch on the water absorption discussed in Section II.A). As a result of the mixing, the starch forms a continuous network with the "free" water. The nonstarch polysaccharides may undergo oxidative gelation during mixing. The importance of this oxidation for the mixing process is not known [77], and nonstarch polysaccharides are therefore not described further here.

a. Protein

Protein structure undergoes considerable changes in the chain of events from the flour particle to the loaf of bread. In the wetted flour particle the protein is present as strands that have a diameter in the micrometer range. These strands are easily visible in the light microscope. In a "dough" immediately after addition of water but before mixing or kneading, the individual flour particles can still be seen, although they stick together. After a few seconds of mixing, protein gel aggregates are elongated and arranged in bundles. In the undermixed dough there is a coarse three-dimensional network of these partly aggregated strands of gel-like units. At optimum mixing, these strands form films less than 1 μm thick [98]. Further mixing causes the films to be perforated and eventually completely disrupted. In a complete dough (including yeast, salt, sugar, shortening, and an oxidant), the protein gel phase develops into large sheetlike areas with numerous lipid inclusions.

When a dough is mixed from flour and water, the protein solubility increases. This has been observed for a number of extraction media, including 1% SDS [99], 0.05 N acetic acid [100], aqueous urea solutions [101], and AUC (acetic acid–urea–cetyltrimethylammonium bromide) [102], and some results are given in Table 2. It is also observed that protein extractability increases further with

Table 2 Extractability of Wheat Proteins in Relation to Mixing

		Extractability (% of nitrogen extracted)[a]		
Flour	Mixing conditions	SDS-soluble	2-ME-soluble	Residue
SRA	Flour	72.6	25.9	1.5
	3 min in a 10g mixograph	93.5	4.5	2.0
	15 min in mixograph	95.0	3.1	1.9
KSU	Flour	63.8	33.5	2.7
	4 min in mixograph	89.7	8.0	2.3
	15 min in mixograph	93.9	3.5	2.5
		Percentage of total flour protein extracted in 0.05 N acetic acid using different homogenizing times[b]		
		30 min	60 min	90 min
Red River 68	Flour	54	57	58
	15 min in farinograph	62	62	62
Manitou	Flour	61	63	64
	15 min in farinograph	80	80	80
Talbot	Flour	62	63	64
	15 min in farinograph	84	84	85

[a]Data from [99]. SDS, sodium dodecyl sulfate; 2-ME, 2-mercaptoethanol.
[b]Data from [102].

increased mixing beyond the optimum development. The extractability of protein from flour and dough differs between varieties, and it is greater for a weak flour. Moreover, the increase in extractability due to mixing is higher for a weak flour [102]. The proportion of glutenin in the extract also increases more with mixing time for a weak flour [100].

Two hypotheses were put forward earlier to explain why protein extractability increases with mixing. One is that protein aggregates are broken down, causing the protein molecules to be more accessible to the extraction solvent. In this explanation it is suggested that only noncovalent bonds (hydrogen bonds, ionic linkages, hydrophobic interactions) are broken. Consequently, the size of each individual protein molecule remains the same. The gluten gel structure shown in Chapter 2 seems to be fully consistent with such changes. The aqueous zone between hydrophobic units will be formed during the mixing process, as the flour represents a glassy continuous protein matrix that requires energy to be transformed into the aqueous gel state.

In the other hypothesis, depolymerization is given as the reason for the increase in protein solubility. It is the glutenin molecules in particular that are suggested as being the molecules affected. Depolymerization could be due to at least three different mechanisms: SH/SS exchange, enzymatic degradation, or splitting of S–S bonds due to shearing.

Thiol–disulfide exchange reactions between the proteins and low molecular weight thiol compounds would result in a decrease in molecular weight according to the reaction

$$P_1\text{–SH} + P_2\text{–S–S–}P_3 \rightarrow P_1\text{–S–S–}P_3 + P_2\text{–SH} \tag{4}$$

It is assumed that P_1–SH is of much lower molecular weight than P_2–SH [75]. If so, the solubility should increase when the reaction proceeds in the right direction. The number of low molecular weight (LMW) thiol compounds was suggested to influence the mixing strength of a flour [75]. The importance of LMW thiol compounds for dough breakdown is illustrated by the fact that a dough mixed from flour that has been dialyzed against water does not break down during prolonged mixing [103]. However, if glutathione is added, breakdown occurs. The dialysate from the dough was found to contain cysteine (44.31×10^{-7} mol) and glutathione (17.73×10^{-7} mol) [103]. The number of thiol groups decreases during mixing, and the decrease is greatest for wheat varieties of good baking performance [67]. The amount of thiol groups was determined to be 60.4–69.3 μmol/g protein in the flour, which decreased to 44.0–59.8 μmol/g protein in the dough [67]. The decrease in thiol groups is favorable for the handling properties of the dough, as dough stickiness is related to the amount of these groups [63].

Covalent bonds can be broken by proteases, but in ordinary dough development involving a sound flour, such reactions are not thought to be of great importance [75, 103].

Depolymerization might also occur because of splitting of the disulfide bonds due to excessive shear forces [104]. The splitting of disulfide bonds results in thiol radicals that react with compounds containing activated double bonds [99, 103]. If radical scavengers are added, such as butylhydroxyanisole (BHA) and butylhydroxytoluene (BHT), the mixing curve is stable [99]. A prerequisite for depolymerization to occur due to shearing is that the molecules be large enough. The molecules should be expected to split at the center, as the highest tension builds up there. The strain during mixing in, for example, the farinograph is high enough to break the weakest of the covalent bonds present, the disulfide bonds. A consequence of this type of depolymerization is that there will be a change in molecular size distribution toward smaller molecular weights. The protein molecules of highest molecular weight undergo depolymerization, whereas those below a critical value are not affected at all [104]. It is thus of great importance where in the polymer the disulfide bonds are located [60]. Those in the center of the molecule will be more "rheologically active" if bond splitting occurs.

The most straightforward way to prove whether depolymerization occurs during mixing would, of course, be to determine the protein molecular weight distribution before and after mixing. However, this simple approach does not work because of the high molecular weight of the glutenins; there are no gel filtration media that can show the small changes we are looking for [99]. Another approach would be to measure the changes in the SH and S–S content. However, as discussed above, our methods are again not sensitive enough. There are some indications that a depolymerization does in fact take place. The protein extracted from flour and doughs has been further analyzed concerning protein class (Osborne fractionation) and molecular weight (SDS-PAGE and gel filtration). The albumins and globulins are not affected during dough mixing [75]. Depolymerization should affect the proteins of highest molecular weight, whereas those below a certain value should not be affected at all [104]. This is what is observed. The amount of low molecular weight glutenins in the extract is greater in the dough than in the flour [99, 100, 105]. Consequently, the amount of residual protein is less in the extract [105]. SDS-PAGE showed no changes in molecular weight of the reduced glutenins and residual proteins [105]. If depolymerization occurs, it should not be expected that the glutenin subunits would be affected but rather that the large aggregates would build up from these subunits.

The gliadins also appear to be affected at very excessive dough breakdown [105]. The amount of protein in this solubility class increases, and SDS-PAGE shows new bands in the HMW region. This is probably the result of glutenin splitting; the fragments will in this analysis behave as if they belonged to the gliadin group.

If cysteine is present during mixing, the extractability of protein increases. This effect is not altered by mixing time and is explained by the reduction of intermolecular disulfide bonds [100]. It has not, however, been possible to detect any changes in the molecular size distribution of protein extracted after the mixing of doughs in the presence of cysteine or other rheologically active additives ($KBrO_3$, KIO_3, NEMI, ferulic acid) [99].

Our conclusions from the different investigations is that during the initial stage of mixing, increased protein solubility is due to breakage of noncovalent bonds, when the glassy protein structure from the endosperm forms small units surrounded by an aqueous zone. This gluten gel structure was presented in Chapter 2. Upon further mixing, involving dough breakdown, it seems likely that splitting of disulfide bonds occurs.

b. Lipids

When we discussed the physical properties of the different wheat lipid–water phases in Chapter 2, we described their dispersion behavior. Only the lamellar liquid crystalline phase is spontaneously dispersed, and only this phase is effective in stabilizing hydrophobic/hydrophilic interfaces (such as air/water or oil/water interfaces).

When the dough is mixed, the wheat lipids form aqueous aggregates in the gluten gel; this has been clearly demonstrated by electron microscopic studies [106]. Furthermore, the X-ray diffraction data of the gluten gel under varying water content has shown that the lamellar liquid crystalline phase exists in the gel [107]. What happens to these lipid inclusions during dough mixing? To answer this question we have to consider the mixing effects on dough on two levels.

The segregation of lipids into phases rich in polar lipids (liquid crystals) from those rich in nonpolar lipids (oil phases) takes place spontaneously when the total wheat lipids are exposed to water. In the flour, however, the lipids are not molecularly mixed as in the lipid extract. Mixing in water means that the polar lipids in the flour will start to swell in the water, as shown by the phase diagrams in Chapter 2, whereas the nonpolar lipids in the beginning just form oil droplets. Contacts between these two types of aggregates brought about by further mixing results in exchange of polar and nonpolar lipid species. Thus the flour lipids will tend to reach the same equilibrium between lipid–water phases as is obtained from the extract. However, they will never reach equilibrium in a practical mixing case. The degree of mixing will therefore determine the chemical variation of lipid species in the different lipid aggregates in the gluten gel. It is obvious that this can influence the functionality of the lipids, as different mixtures will exhibit different structures of the phases, as seen in Chapter 2, Fig. 6. In this connection it is also necessary to consider added emulsifiers, which will always interact with the lipids naturally in the dough. Also, the effect of the emulsifiers may therefore vary with mixing conditions.

The second level involves the functional effects on the dough due to the lipid phases once they have reached a kinetically stable composition. Here also the result will vary depending on mixing conditions. Considering the lamellar phase to be the most important, every liposome-like aggregate can be regarded as a depot that will be distributed more uniformly over hydrophobic/hydrophilic interfaces the longer we mix. Thus their contribution to the improvement of baking performance will basically increase with mixing time.

C. Influence of Some Added Ingredients

In the previous sections of this chapter we have described how the main components of flour—proteins, lipids, starch—together form the dough. We have also described how their behavior and their interactions are affected by the environment in the aqueous dough system—how pH, salts, sugars, and redox agents influence the dough development. In many bread formulas, other ingredients also are added, and in the present section we describe a few of these optional ingredients (gluten, enzymes, fats, and emulsifiers) in more detail.

1. Gluten

Gluten is defined as the aqueous gel of storage proteins formed when a wheat flour–water mixture is worked mechanically (see Chapter 2). The term is also used for the dried powder of this gel. As such, gluten is an intrinsic component of the dough. But gluten ("vital gluten") can also be used as an additive to improve the baking performance of a flour. There are several reasons for using gluten as an additive. If the wheat used for baking is not of adequate quality it can be fortified either with other wheat varieties or with gluten. If the flour is going to be used in specialty breads, gluten can be used to strengthen it so that it can carry an extra "load" of, for example, dietary fiber [108]. Gluten can also be used to improve the quality of bread baked from frozen doughs [109].

It has to be realized that there is no such thing as a "gluten," as there are as many different glutens as there are wheats. Wheat varieties differ in baking performance, and it seems obvious that this is reflected in the isolated glutens. Fractionation and reconstitution experiments in which gluten has been exchanged between varieties show that a gluten from a wheat variety of poor baking performance also behaves poorly when it is exchanged for the gluten of another variety [110]. Another factor to be aware of is that a high quality gluten can be severely damaged during the isolation procedure. Laboratory-prepared gluten almost always shows better "vitality" than commercial gluten samples [111]. The most detrimental step during gluten isolation is probably the drying procedure. Excessive heat decreases the vitality of the gluten [112]. When gluten is washed in excess water, substances influencing its rheological behavior are washed away [113].

The user of gluten has to face the problem of how the gluten quality should be analyzed, and the best test of quality is test baking. However, this is a rather tedious procedure, and the results are not always easy to interpret. Alternative methods are therefore used, including SDS sedimentation, hydration time, and water absorption [111, 112, 114].

Even if a gluten is isolated from a wheat of good baking performance and the good baking performance has been kept during the isolation procedure, the baking result may still not be as good as expected. The added gluten must interact with the gluten in the flour to which it is added, and the combination must show the correct balance in properties [115]. This is perhaps a neglected aspect of the behavior of added gluten, but nevertheless it is of great importance. The incorporation of the added gluten into the gluten gel formed during mixing can be more or less complete [116].

If the added gluten is to improve the flour to which it is added, it has to complement the properties of this flour's gluten. If, for example, the gluten in the flour has too high a gliadin/glutenin ratio, the added gluten should be very rich in glutenin. If, on the other hand, the glutenin in the flour is of a very high molecular weight, then the added gluten should perhaps be rich in glutenin of lower molecular weight in order to facilitate dough development. By adding an HMW fraction from gluten it is possible to increase the dough development time of a flour and make it more stable to overmixing [117]. The addition of an LMW fraction has the opposite effect. To convert a weak flour to a strong one it was necessary to add a mixture of whole gluten and HMW gluten [115]. When part of the flour is replaced by gluten, mixing tolerance is increased, the extent of increase depending on the source of the gluten [8]. It has been quantified how gluten increases the stiffness of a dough according to the relaxation modulus G. The increase was found to be greater for flours that already had high G values and could be described as having good baking performance [118].

Commercial gluten samples differ also in their spreading behavior at the air/water interface [119]. A very low spreading rate together with a low final surface pressure indicates heat damage (see Chapter 7).

Because of the way in which it is prepared, gluten contains lipids [about 1–10% (w/w)], and the amount and composition of these lipids can influence its interactions with the various types of flours. Furthermore, these lipids are exposed to oxidation damage during the storage of the gluten powder, as the lipid molecules will tend to cover the particle surfaces. This oxidation is certainly one relevant factor of the gluten quality.

2. Enzymes

Enzymes present in the dough are of fundamental importance throughout the breadmaking process until they are denatured in the oven. Amylases are needed for the fermentation, and other endogenous enzymes influence the storage proteins as well as the lipids during mixing. The addition of enzymes to the flour or the dough makes it possible to standardize and optimize breadmaking.

a. Amylases

α-Amylase, the most important of the amylases, produces α-maltose, dextrins, and glucose. It is an endoenzyme that splits $\alpha(1 \rightarrow 4)$ bonds. A minimum of water content, about 25% in wheat flour, is required to activate cereal α-amylase [120]. This minimum is probably related to the point where "free" water starts to be formed (about 23%), as discussed earlier in this chapter. It is clear that a "free" water phase outside the gluten phase is needed for enzymatic hydrolysis. The α-amylase activity of a flour is an important quality factor in breadmaking. The activity is measured by the falling number using standardized equipment in which a flour–water mixture is gelatinized, and the enzymatic starch degradation is followed by the time it takes for the stirrer to sink through the gel. Low falling numbers, below 100, result in a sticky crumb and poor volume.

α-Amylase is denaturated at a low pH (<4), and a flour with high α-amylase activity can therefore be used in sourdough-based bread [121]. When the α-amylase level is too low, however, the yeast fermentation cannot proceed with acceptable speed. It is therefore necessary to supplement the flour with either an α-amylase or a malt flour. This is usually done in connection with milling. A disadvantage of malt flour is its dark color. Addition of α-amylase is therefore preferred, and fungal amylases have the advantage of being thermolabile enough (some bacterial amylases will not be inactivated even when 100°C is reached during the baking). Fast breadmaking processes, like the Chorleywood process, require higher α-amylase activity than conventional baking, and α-amylase flour supplementation is usually needed.

Flour also contains β-amylase [which can hydrolyze $\alpha(1 \rightarrow 4)$ linkages to produce β-maltose from the nonreducing end of amylose], and its natural level in flour is normally adequate.

b. Proteases

Proteolytic enzymes are sometimes needed to modify the gluten phase of the dough toward weaker mechanical strength. The hard wheat flours with high protein content often produced in the United States may need to be supplemented with a protease, and proteases of fungal origin are usually preferred. When a very weak gluten phase is desired, as in the production of cookies and crackers, proteolytic enzymes can be used, sometimes in combination with cysteine to reduce S-S bonds.

c. Lipoxygenases

Lipoxygenase oxidation of polyunsaturated fatty acids in flour is coupled to the storage protein oxidation (see Section II.B). Perhaps the earliest application of lipoxygenase in breadmaking was as a bleaching agent. During the 1930s, soy flour with active lipoxygenase was added to wheat dough to bleach the carotenoid pigments [122]. Another unusual application of lipoxygenase is in French breadmaking [123], where horse-bean flour has been used to improve bread aroma. The reason for the improvement is the production of hexanal from the decomposition of 13-hydroperoxide.

3. Fats

A few percent by weight of fat is often added to dough to improve the crumb structure and also to obtain a larger volume. The texture is changed toward finer and more uniform pores with thinner walls, which gives an improved softness. The probable role of the fat crystals is to mechanically strengthen the pore walls, whereas the oil will act as a lubricant for the solid particles of the flour during dough mixing. When high-speed mixing is used in a continuous breadmaking process, it is necessary to use fat crystals with a high melting point, and usually about 10% (w/w) of the fat is therefore fully hardened (N. Krog, personal communication). Accordingly, in the Chorleywood process 0.7% of such a fat is used in the dough.

We have done some comparative studies in our department on the effect of different crystal forms of fats [124]. The β and β' crystal forms were obtained from hydrogenated rapeseed oil, and in order to get a stable α form we had to use a rather odd triglyceride, an acetic acid diester of a monoglyceride from hydrogenated lard. The baking experiments were done in a resistance baking chamber [125]. The addition of 2% lipid (calculated on the flour weight) had no effect on the volume of the dough but a strong effect on the volume after oven spring. The values of the height of the bread observed are given in Table 3. It should also be pointed out that the amount of solid to liquid in the fat was adjusted to be similar in the three types of fat.

Table 3 Height of Breads Baked with Different Crystal Forms of Fat

Type of lipid added	Height of bread (cm)
No lipid added	5.95
Soybean oil	7.00
Fat with α-crystals	7.80
Fat with β'-crystals	9.35
Fat with β-crystals	9.60

Source: Data from [124].

These effects may be partly explained by the lubrication during dough mixing improving the uniformity of the dough, and mechanical effects of the fat crystals stabilizing the lamellae between the gas cells. The platelike shape of β-crystals should be expected to be superior as a stabilizer compared to the β' needle shape.

Finally it should be pointed out that the most pronounced effect on bread volume caused by the addition of fat is obtained when the fat crystals do not melt below the starch gelatinization temperature (fully hydrogenated oils). The mechanical stabilization of the structure during oven spring seems to be the obvious explanation.

4. Emulsifiers

Even before World War II, mixtures of mono- and diglycerides were used in breadmaking. These first types of emulsifiers were made by adding glycerol to triglyceride oil to attain reesterification. Emulsifiers are used for various reasons in baking but not for emulsification. The term "emulsifier" is therefore not accurate but is used for historical reasons.

All emulsifiers belong to the group of polar lipids. Some of the additives, like mono- and diglycerides, occur naturally in very small amounts among the wheat lipids. Others are synthetic analogues of polar lipids. It is obvious that the emulsifiers will interact with water to give aqueous phases similar to those of wheat lipids, as described in Chapter 2. Furthermore, they will mix with the naturally present wheat lipids and influence their aqueous phase properties. Such molecular mixing always takes place when oils or liquid crystalline lipid phases are exposed to one another. We will later consider these effects as a basis for understanding the functions of different emulsifiers.

The emulsifiers used most today are distilled monoglycerides. They are mainly added to complex starch ("crumb softener") and as aerating agents in cookies and cakes. Second in importance according to volume of application are the diacetyl tartaric esters of monoglycerides (DATEM), which are used as "dough strengtheners." Their effect in baking is to improve mixing tolerance and uniformity of pore structure and also to increase the bread volume. Next in importance are the sodium or calcium salts of stearoyl lactylates (SSL and CSL, respectively). They are, in fact, poorly defined, consisting of a mixture of free fatty acid, monolactoyl, and di- and even polylactoyl esters of stearic acid. The effect of SSL and CSL in breadmaking is similar to that of DATEM, and their mechanism of action is the same as that of polar wheat lipids.

In principle, there is no difference between the addition of an emulsifier and the addition of a fraction of wheat lipids to a dough. It is known from fractionation and reconstitution studies that the polar fraction of wheat lipids is positive with respect to baking performance in the same way as emulsifiers are (see Chapter 2, Section II). As discussed in connection with wheat lipids, there is a clear relationship between the existence of a lamellar phase as the dominant structure of the lipid–water system and good breadmaking performance. Of the different liquid crystalline phases it is only the lamellar phase that has the possibility of exposing large interfaces. These can be either hydrophilic or hydrophobic, which is ideal for dispersion properties and foam stabilization. It is also striking that both DATEM and SSL form the lamellar liquid crystalline phase under the conditions that exist in the dough.

The monoglycerides that are most used today are the pure type of molecularly distilled ones based on saturated acyl chains. The preparation process will yield an equilibrium mixture of about 90% 1-monoglyceride and 10% of the 2-isomer, due to acyl migration. Monoglycerides from fully hardened lard are common. The phase diagram of this lipid and water has been determined [126] and is shown in Fig. 11. It can be seen that below about 60°C, no liquid crystalline phase is formed.

The monoglyceride is added as a dry powder that consists of crystals (a β-crystal form) or as an aqueous dispersion of the crystals. In exceptional cases the aqueous gel phase is first prepared and added. The gel phase is formed by cooling the lamellar liquid crystalline phase (see Fig. 13 in Section III); the chains crystallize, keeping the lamellar structure unchanged. It is difficult to handle, however, as it is not thermodynamically stable. There will therefore be no interaction with the wheat lipids during dough mixing or fermentation. When the gelatinization starts, however, the monoglyceride will also form an aqueous liquid crystalline phase, which will increase its degree of dispersion in the dough and by that its reactivity. Complexing reactions with amylose and amylopectin will take place during the gelatinization step. Of course, there is the possibility of reaction occurring during dough mixing between damaged starch and free amylose and the monoglyceride monomers dissolved from the crystals into the "free" water phase. The solubility of these monoglycerides is about 10^{-6} mol/L, which means that the solubility is the limited step in the reaction.

Other emulsifiers such as DATEM and SSL will affect the dough by favoring the lamellar liquid crystalline phase, as mentioned above.

The crumb softening and antistaling effect that are provided by all these emulsifiers were discussed in Chapter 3.

There are also other baking emulsifiers in use that are rather far from polar lipids in their chemical constitution. Examples are ethoxylated monoglycerides and polyoxyethylene sorbitan monostearate. These can be regarded as very similar to synthetic surfactants, whereas the ones discussed before are much closer to the naturally present lipids and their metabolites. Monoglycerides, for example, are formed from all fats we eat by enzymatic digestion in the intestine.

In considering the nutritional aspects of baking emulsifiers it is important also to keep in mind their possible surface effects in the gastrointestinal tract. It is not

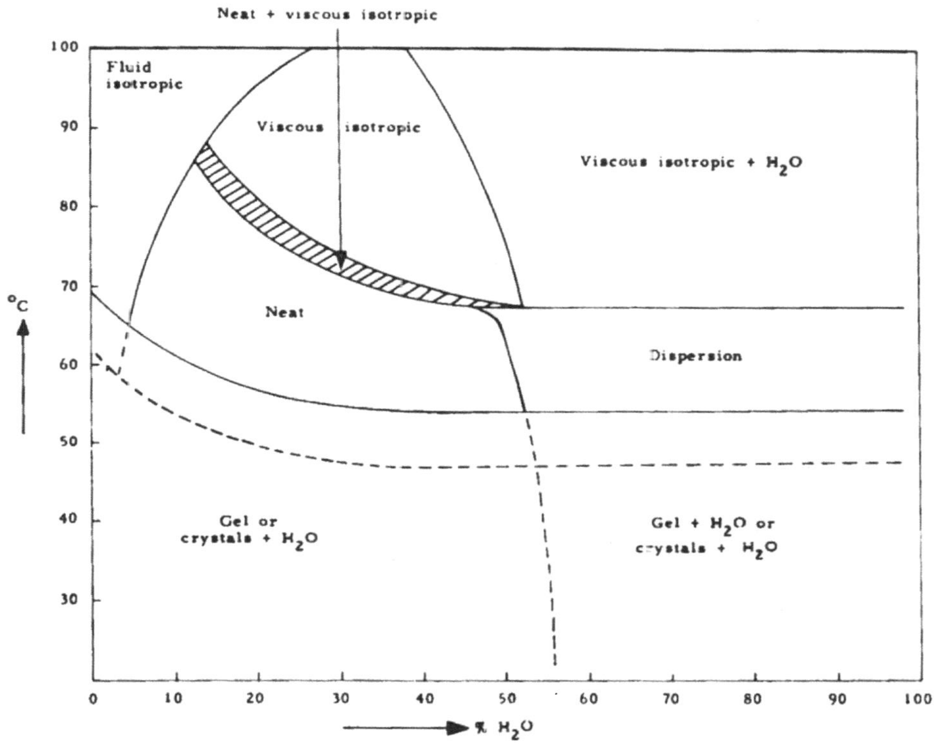

Fig. 11 Aqueous system of fully hardened lard. The different liquid crystalline phases exhibited are the lamellar liquid crystalline phase, at that time termed "neat," and the cubic phase, referred to as viscous isotropic. (Adapted from [126].)

enough to know that they are not absorbed through the intestinal wall; they may still have harmful effects. Such aspects of these types of molecules are discussed further in Chapter 3.

III. THE DISPERSED GAS PHASE

The dispersed phase in the dough consists of the gas cells. The inclusion of gas cells in the aqueous gluten phase is a thermodynamically unstable situation, and with time we should expect the cells to collapse. Their kinetic stability is the reason for the baking performance of wheat.

The gas in the gas cells is obtained from two sources—from air included in the dough during mixing and from the carbon dioxide (CO_2) produced by the yeast.

In this section we describe the yeast and how it produces CO_2 and then discuss the importance of the inclusion of air.

A. Yeast and Fermentation

1. Leavening Agents

The reason for putting yeast, or another leavening agent, in the dough is to leaven the dough—to make it rise—and to produce a porous product. This is achieved by the production of carbon dioxide, and the bread baking process is essentially the production of CO_2 and its entrapment in the continuous gluten matrix of the dough. The first step, the production of CO_2, can be achieved in several ways: by the use of yeast, ammonium bicarbonate, an acidic salt, or sodium bicarbonate. The mechanism for the production of CO_2 varies depending on the leavening agent. In the case of yeast, leavening is caused by fermentation. The yeast ferments sugar, producing carbon dioxide and ethanol. In the case of chemical agents, such as ammonium bicarbonate, decomposition of the molecules causes leavening. Ammonium bicarbonate liberates ammonia, water, and CO_2 in the presence of heat, whereas an acidic salt and sodium bicarbonate react in the presence of moisture and heat to produce salt, water, and CO_2.

The oldest methods of baking a porous bread were based on some kind of "wild" fermentation of the dough and a piece of old sourdough was used to start fermentation. In ancient times, yeast from wine or beer production was also used for the fermentation.

The most common leavening agent in bread is, of course, yeast, *Saccharomyces cerevisiae*. Today we regard baker's yeast as being as necessary in breadmaking as flour or water. Besides its obvious contribution to bread volume by production of CO_2, the role of yeast fermentation is to influence the texture of the dough, to contribute to flavor and aroma, and finally to enhance the nutritional value of the bread. If we compare a yeast-fermented bread with one produced with a chemical leavening agent, the significance of the fermentation becomes obvious even if we consider only the sensory differences. The nutritional effect refers mainly to the phytase activity of the yeast, which improves mineral uptake in the intestines. An alternative is to add phytase directly to the flour. Phytate in the dough (significant when whole wheat flour is used) will decrease the uptake of minerals due to complexing. Phytase will degrade the phytate.

2. Optimum Conditions for the Yeast

Yeast produces carbon dioxide and ethanol from simple sugars according to the formula

$$C_6H_{12}O_6 \rightarrow 2CO_2 + 2C_2H_5OH \quad (5)$$

From 1 g of glucose, 0.49 g of CO_2 is formed, corresponding to 276 mL of CO_2 at 30°C and atmospheric pressure. The carbohydrate source is starch, but the yeast cannot use the starch directly, it has to be hydrolyzed by amylases. To start the fermentation, readily available carbohydrates, that is, simple sugars, have to be present. The yeast digests sucrose and ferments glucose and then fructose. For maltose an adaptation period is required. Lactose and pentoses are not fermented at all [36, 127]. Also, a small amount of damaged starch is beneficial, as the enzymatic hydrolysis starts more quickly at a damaged starch granule. If the amylase activity of the flour is very low (i.e., a very high falling number), extra malt can be added. The same loaf volume is obtained for a loaf baked without added sugar but the optimum malt level as for a loaf baked with 6% sucrose added and an optimum malt level [9]. This volume is higher than for a bread with 6% sucrose but no malt. It should be observed that yeast is sensitive to osmotic pressure, and excessive sugar levels should therefore be avoided [34, 36, 127].

If a bread is baked only from flour and water, the loaf volume will correspond roughly to the dough volume after mixing [9]. The addition of 7.2% yeast to this flour–water suspension causes the loaf volume to increase from 228 mL to 530 mL. The level of yeast is usually 3.5–5.5%, calculated on the amount of wheat flour [9].

The gas-producing power of the yeast is affected by temperature, pH, alcohol concentration, nature of carbohydrate supply, osmotic pressure, and yeast concentration [3]. The time to produce a certain amount of CO_2 (0.4 g CO_2/g yeast) under various conditions is given in Table 4 [128]. The optimum pH for gas production by the yeast is 5.5 [3, 36], but the yeast is quite tolerant to changes in pH, and the gas production is 80% of the optimum value in the pH range 3.7–8.0 [33].

Table 4 Time for Yeast to Produce a Certain Amount of CO_2 Under Various Conditions

Conditions	Time to produce 0.4 g CO_2/g yeast dry matter (min)
2.3% yeast, sugar, 5°C	830
25°C	56
35°C	30
2.3% yeast, no suguar, 25°C	67
35°C	37
4.6% yeast, sugar, 5°C	877
25°C	60
35°C	32

Source: Data from [128].

3. Influence on Dough Behavior

Fermentation has numerous effects on the gluten gel and dough structure. It is obvious that pH is reduced by dissolved carbon dioxide and, when sourdough is used, from lactic acid and short-chain fatty acids also (see Section II.A.4). Furthermore, there are numerous extracellular compounds produced that might also have an effect, such as glutathione in relation to redox reactions of the storage proteins. There is physical work on the dough due to the expansion of the air cells, and there is also the possibility of a slackening effect on the dough due to the yeast reductases acting on substances such as glutathione [129]. Rheological measurements on fermenting doughs have shown that the dough becomes more elastic during fermentation [130].

Ethanol might be expected to influence the two-phase aqueous separation of gliadins as well as lipids owing to specific solubilization. The possible effects on gliadins can be estimated from the phase diagram shown in Fig. 12 [32]. A reasonable ethanol concentration in a dough can amount to about 1% (w/w), which means about 2% in the water phase (or phases). The effect on the phase equilibria, however, can probably be neglected.

A ternary phase diagram involving a phospholipid and ethanol/water was studied in our laboratory [131]. As can be seen from Fig. 13 there are dramatic differences in the effect depending upon the ethanol/water ratio, which in turn is related to the proportion of water in lipid–water phase. According to our X-ray studies of the gluten gel [107], the lipid is fully swollen. That would mean that we can neglect ethanol's effect on the lipid–water phase properties, as clearly shown by Fig. 13.

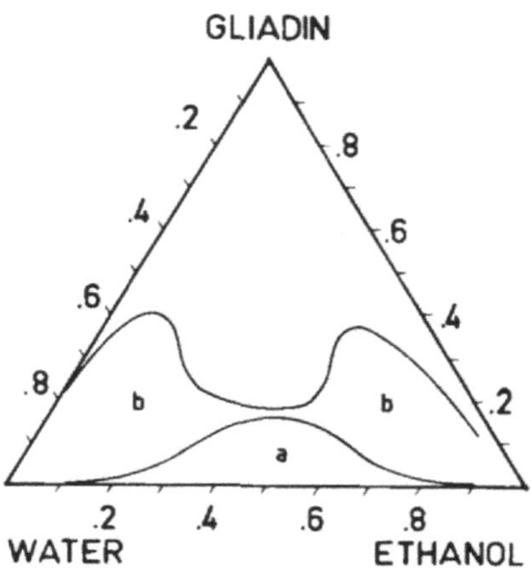

Fig. 12 Ternary phase diagram for ethanol/gliadin/water, (from [32]). The single-phase solution is defined by area b.

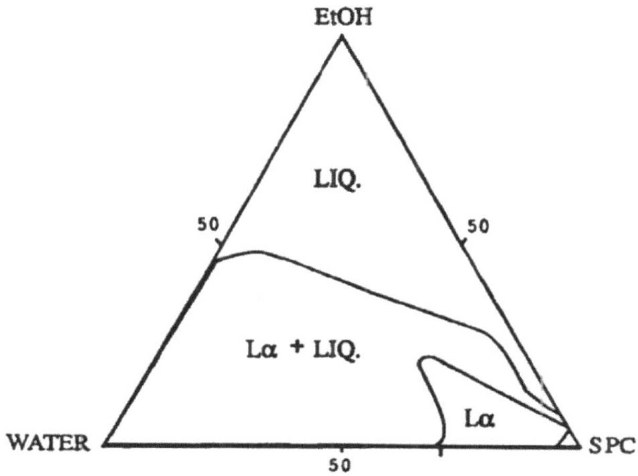

Fig. 13 Ternary phase diagram of phosphatidylcholine/ethanol/water (at room temperature. (Adapted from [131].)

Frozen bread doughs are being increasingly used industrially for later baking. There are two limiting problems in the application of this technique. One is the effect on the yeast and the reduced fermentation performance due to the freeze–thaw cycle. The other is the freeze–thaw tolerance of the gluten gel, particularly with regard to its gas-holding capacity. The mechanical effects of ice crystals are involved in both cases, and in the case of viability of the yeast, osmotic effects are also significant. The technology of frozen doughs and the possibility of improving fermentation were considered in a recent thesis [128].

B. Inclusion of Air Cells

It has already been stated that the reason for putting yeast in a dough is to obtain a leavening effect. The leavening is caused by the gas retention, that is, the dough has the capacity to hold the gas. The increase in pressure due to the production of more carbon dioxide and due to physical changes by brought about oven heat results in an increase in volume. However, a large volume is not enough for a bread of high quality; the porosity must also be acceptable. We want pores of uniform and rather fine size. The porosity is established during mixing and subsequent punching and molding of the dough.

Carbon dioxide is soluble in the aqueous phase of the dough, but to create the leavening effect and the porosity it must be transformed to the gaseous phase. This can occur only if the dissolved carbon dioxide diffuses to the already existing gas cells in the dough. There is no possibility for the creation of new gas cells, and the explanation for this is found in the equation

$$p = 2\gamma/r \tag{6}$$

where p is the pressure, γ is the interfacial tension, and r is the radius of the air cell. If a new gas cell is to be created, its present radius is zero, which means that the pressure has to be infinite. It should be mentioned that an experimental demonstration was reported in 1941 [132] of the fact that the carbon dioxide produced during fermentation cannot form new gas cells; the existing air cells are nuclei for all gas produced. The possible geometries involved in the formation of gas cells are further considered in Section VI.

That the mixing step includes the inclusion of air is evident from the changes in dough density during mixing. This is illustrated in Fig. 14, where the change in density is related to the mixing curve obtained in the mixograph. The initial density value is 1.20 g/mL, and with continued mixing this value decreases to 1.10 g/mL [125]. It should also be noted that the lowest value in density occurs in the breakdown part of the mixing curve. At the optimum in the mixing curve (after 3½ min for this flour) the density is 1.16 g/mL. To include the maximum amount of air, some degree of overmixing is evidently beneficial.

After mixing, the gas cells are spherical, and their diameters are usually between 10 and 100 μm [97, 133]. The changes in shape are discussed in Section VI. The initial value of the density, 1.20 g/mL, corresponds to a gas volume of 3% of the total dough volume [134]. After mixing, about 10% of the dough volume is gas, and the number of gas cells is 10^2–10^5/mm^3 [97].

It should also be noted that the maximum amount of air included depends on the type of mixer. From the initial density of 1.20 g/mL, a decrease to 1.17 g/mL

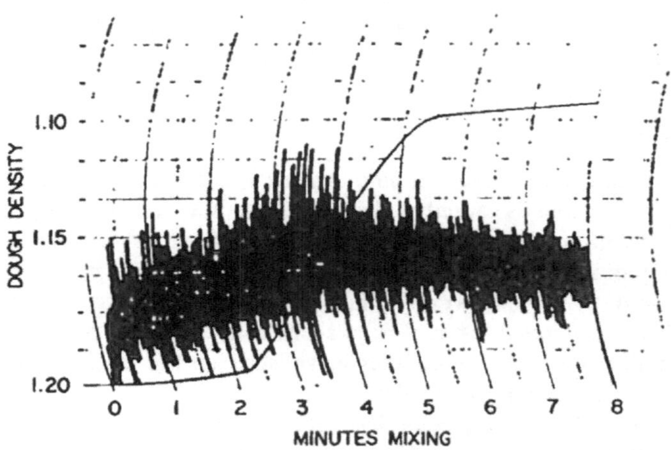

Fig. 14 Changes in dough density (g/mL) during mixing. (Reprinted with permission from [125].)

was observed in the farinograph, to 1.16 g/mL in a Hobart planetary mixer, and to 1.15 g/mL in a Morton twin Z arm mixer [134]. It was also observed that for lower densities to be obtained, the mixing had to be continued to beyond optimum at high intensities. This observation agrees with results obtained in the mixograph, where a density of 1.10 g/mL is obtained with mixing beyond the optimum in the mixing curve [25].

As already indicated, porosity is important to the quality of bread, to produce a desirable fine grain. Porosity can be affected in several ways. The dough can be allowed to ferment, and if it is then punched or remixed the large gas cells are divided into many small ones. Mixing under reduced pressure creates large gas cells that can be divided into many small cells with punching [135]. The addition of emulsifiers influences porosity [125, 135, 136]. An emulsifier such as sodium stearoyl-2-lactylate delays the incorporation of air during mixing; the final density of the dough is the same as without the emulsifier, but the porosity is much finer [125]. The same effect on porosity was observed when lecithin was used to improve the baking performance of wheat flour–rice flour mixtures [136].

The equation relating pressure and gas cell radius [Equation (6)] contains a term for the interfacial tension γ. If the interfacial tension decreases at constant pressure, then the gas cell radius also decreases. Surface-active components in the dough could therefore be expected to affect the porosity in the direction against finer pores. There are surface-active components naturally present in the flour; both the polar lipids and the proteins are surface-active (see Chapter 2). Added ingredients may also be surface-active, especially the emulsifiers. The explanation for the finer grain obtained when emulsifiers are added could be their influence on γ in Equation (6).

C. Mixing Atmosphere

When the dough is mixed in vacuum and immediately baked, a very poor loaf results [132, 137]. The loaf volume is low, and the crumb texture/porosity is characterized by only a few, huge pores with thick cell walls. The dough is very dense, immediately after mixing and almost no gas cells are observed in the microscope [137]. The density of a dough mixed at low pressure was reported to be 1.28 g/mL, whereas the same dough mixed in vacuum was determined to be 1.25 g/mL, and for a similar dough mixed in air, 1.15 g/mL [132]. The difference corresponds to a gas volume of 8% of the total dough. In another study, the density of a dough mixed in air had a density of 1.22 g/mL. However, this work concluded that it was not possible to predict loaf volume from density values [137]. The quality of the final bread was greatly improved when punching steps were included in the process. The loaf volume increased from 590 mL without punching to 680 mL with one punching step, to 845 mL with two steps, and to 933 mL with three steps. The punching might to some extent cause the inclusion of some air, but essentially it causes the large gas cells formed during fermentation to break down into many gas cells of a much smaller size.

Not all gases show the same efficiency for inclusion in gas cells. Both oxygen and carbon dioxide are quite soluble in the aqueous phase of the dough, whereas nitrogen is inert and only slightly soluble [134]. If the dough is mixed in nitrogen followed by oxygen-enriched atmosphere, the final baking result is better than with air. Mixing in nitrogen does not change the dough density compared to mixing in air, but a much finer grain is obtained in the final bread [134]. In other investigations a decrease in density was observed after mixing in N_2, from 1.22 g/mL in air to 1.14 g/mL in N_2 [137]. However, the same good result was obtained when air was blown into the mixing chamber. A general conclusion from these studies is that gases with low solubility are more effective in producing gas cells. There are, however, additional effects due to oxidative changes by oxygen. These are discussed in the following subsection.

D. Oxidation

When the mixing in the farinograph occurs in nitrogen, the farinogram is not very different from the one obtained in air [134]. However, when oxygen is introduced, the mixing curve is shifted to higher values compared to mixing in air or nitrogen. The role of oxygen in mixing is thus not only to be included as gas cells in the dough, but also to act as an oxidizing agent. There must be a flow of oxygen-enriched air above the dough surface to optimize the oxidation unless other oxidizing agents are used. The oxidation reactions that may take place are described in Section II.B.2 in this chapter.

One consequence of the oxidation process is that doughs mixed in vacuum are more yellow than doughs mixed in air, showing that oxygen causes oxidation of carotenoids.

IV. AIR/WATER INTERFACE

When wheat flour particles are placed on the surface of water, some parts will spread at the air/water interface as described in Chapter 3. Differences in loaf volume due to added lipids do not show up until the oven stage, indicating that the air/water interface is stabilized by protein molecules in the dough. Our studies of the surface behavior of wheat storage proteins have revealed that the molecular weight distribution of proteins affect the surface film pressure–area (π–A) and surface area–time (π–t) isotherms. Moreover, the molecular weight distribution can be affected by oxidation, either through repeated compression–expansion cycles of the protein film or by the presence of oxidants in the substrate. As described in Chapter 2, fractions of high molecular weight spread slowly at the air/water interface, whereas fractions of lower molecular weight spread much faster (Fig. 19 in Chapter 2).

Also described in Section III of Chapter 2, an increase in molecular weight distribution as a result of mixing has been observed with the surface balance technique [138]. Glutenin was spread at the air/water interface and allowed to form a protein film. This film was repeatedly compressed and expanded, a treatment that could be regarded as a mixing of the protein film at the air/water interface. For each compression/expansion cycle, an increase was observed in the mechanical strength of the film, indicating an increase in molecular weight. Moreover, SDS-PAGE of collected samples of the film showed an increase in molecular weight, which must have been due to cross-linking (i.e., oxidation). Another way to study this phenomenon is to include an oxidant in the substrate. The spreading behavior of wheat flour on ascorbic acid has been compared to its behavior on distilled water as illustrated in Fig. 15. In this experiment π was recorded after 60 min. It is evident that when the flour is spread on ascorbic acid the π value is much lower than when the same amount of flour is spread on water [139]. The difference between the curves is consistent with a difference in molecular weight as discussed above, as under these conditions the monolayer is in the gaseous state (see Chapter 1).

In this connection it is of interest to consider other reports on the influence of disulfide bonds on foam stability. α_{s1}-Casein is a better foam stabilizer when thiol groups are introduced into the molecule, followed by oxidation (so that disulfide bonds are formed) [40]. The timing of oxidation is also important. The best foam stability was found when the proteins were allowed to absorb at the

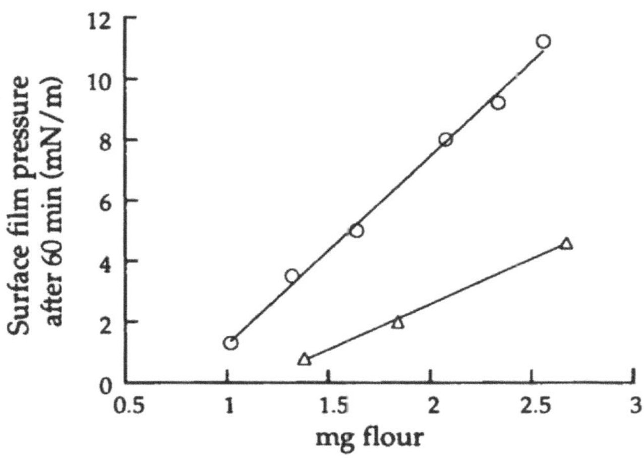

Fig. 15 The spreading of wheat flour on ascorbic acid (△) and on distilled water (○) as shown by surface film pressure. (Unpublished results.)

air/water interface before the oxidation took place. It was further observed that with the thiolated α_{s1}-casein there was an optimum in foam stability at 8.9 thiol groups per molecule of α_{s1}-casein. The decrease in foam stability observed with an increase in the number of thiol groups was attributed to a lower flexibility of the monolayer because of an excess of intramolecular disulfide bonds. These observations are fully consistent with our experience with wheat storage proteins.

V. RHEOLOGICAL BEHAVIOR OF THE DOUGH

The rheological behavior of wheat flour dough is relevant in several respects. First of all, it provides the basis for understanding the handling properties of the dough in the bakery. Furthermore, it may even predict the quality of the final bread. A relationship exists between the mixing curve and the final breadmaking result, and it would be of interest to express the properties of the developed dough in rheological parameters instead of using words such as "silky" or "clean". From a scientific point of view there is also a great interest in rheological properties as they provide a possibility for relating macroscopic properties to microscopic ones, particularly structure. By measuring the rheological properties we can even learn about the structure of the dough, especially if we measure the influence of additives such as oxidants and reducing agents, salts, and emulsifiers.

The aim of many rheological measurements is to find a way to differentiate between wheat varieties according to their baking performance without actually performing the baking test. A rheological test that can correctly predict the baking performance of any wheat variety would save both time and material. However, there is still a long way to go before we have the knowledge necessary to enable us to design such a rheological test. When we decide to measure the rheological properties of dough in order to relate these properties to baking performance (and eventually be able to explain the baking behavior), we are faced with many problems to solve. First, which kind of deformation should be used? The choice is between extension and shear. It has been argued that extension should be preferred because during fermentation the increase in gas volume exerts extensional flow on the protein matrix [55]. On the other hand, because of the high viscosity of the dough there will be laminar flow during mixing, causing shear to be the main type of deformation [60]. Moreover, experiments with shear deformation are easier to perform. Due to the nonlinear behavior of the dough, it cannot be assumed that results obtained in shear are directly related to results in extension [55]. However, data have been presented that show that rheological properties measured in shear are closely related to the properties measured in extension [141].

When the type of deformation has been selected, the next question is the size of the deformation (γ) and the deformation rate ($\dot{\gamma}$). In order to predict the baking performance the strain rate (or strain value) that is most critical for the baking result should be chosen. This choice is not trivial; the dough will meet strain rates in the range of 10^{-4}–10^{2} s^{-1} during the baking process, and the strains range from 100% during sheeting up to 500,000% during mixing [55, 141]. It must be decided whether the low strains and strain rates during fermentation, the somewhat higher strain rates during oven rise, or the very high strain rates during mixing will best differentiate among wheat varieties. Another complication is that gluten and dough are viscoelastic, which means that they can show both elastic behavior and viscous flow. Which property dominates depends on the size of the applied deformation and on the time of measurement. At short times and/or low strain values, gluten and dough behave like a solid, whereas at long times and/or high strain values they behave like a liquid. This behavior is typical of all viscoelastic materials [142].

Although the rheological properties are difficult to measure, rheological methods have been used for a considerable time in cereal science and technology. The methods are usually classified as empirical or fundamental. The empirical methods, or the physical dough testing instruments, were recently reviewed [55]. The fundamental methods differ from the empirical in that the strains and stresses used are known and well characterized in the whole test piece. Moreover, the results are obtained in fundamental units, making it possible to compare the results obtained with different instruments. Some of these fundamental methods were described in Chapter 2, Section III.

The dough structure illustrated in Fig. 1 could be described, from a rheological point of view, as composed of a continuous gluten phase and a dispersed starch phase. In Chapter 2 we described the rheological behavior of the gluten gel, and we now focus attention also on the starch and how it affects the rheological behavior of the gluten phase.

A. Reasons for the Viscoelastic Behavior of Dough and Gluten

The viscoelastic behavior of dough is often attributed to the proteins (see Chapter 2, Section III). For a material to show elasticity it must have the potential to store the energy applied during deformation. This is effectively done in a three-dimensional network [143, 144]. The statistical theory of rubber elasticity has been used to calculate the number of cross-links in dough as well as in gluten. The first rheological model for gluten was based on the idea that the disulfide bonds in the gluten proteins created a permanently cross-linked three-dimensional network. Such a network shows elasticity, and the influence of oxidants and reductants is easily explained by their influence on the number of cross-links in the network. The more disulfide bonds present, the stiffer the network will be. In fact, it has been found that oxidants and heat treatment both cause the number of cross-links to increase [143, 144].

The above model has, however, a clear disadvantage. It cannot explain the viscous flow observed for gluten as well as for dough. The concept of SH/SS exchange reactions was therefore introduced [145]. Reactions such as those illustrated in Fig. 7 are used to explain the viscous flow. If SH/SS exchange reactions take place, there will still be a three-dimensional network in the dough, but this network is not permanent.

It is not necessary to postulate SH/SS exchange reactions in order to explain viscous flow. And it is not necessary to postulate a covalently cross-linked network in order to explain the elastic behavior of the dough. The viscoelastic behavior could also be due to the size and shape of the protein molecules. It is a common phenomenon encountered with many macromolecules (e.g., synthetic polymers) that a polymer solution shows elastic properties without the presence of permanent cross-links. The elasticity is explained with the concept of entanglement [146, 147]. The molecules are of such a size that when deformation occurs they cannot orient themselves quickly enough to release the tension because they are entangled in each other. The result is that a three-dimensional network is formed, and the sample shows an elastic response to deformation. Biological macromolecules such as polysaccharides and proteins form gels in which the network is formed by noncovalent forces. Such networks usually have a transient existence. Examples are gelatin, where the cross-links (or junction zones) are triple helices. In many polysaccharides (carrageenan, agar, alginate), single or double helics are associated, sometimes by the actions of ions, to form junction zones.

For entanglement coupling to occur, the molecular weight has to exceed the critical entanglement molecular weight, and the concentration has to exceed a critical value, the overlapping concentration C^* [146]. The entanglement theory was developed to account for the solution behavior of flexible polymers, and in general it is not possible to apply it to proteins in solution. An important aspect of entanglement coupling is that it should be possible to destroy the network by dilution; that is, when the concentration goes below C^* the network should be lost. This is not possible in the case of gluten. Instead, an aqueous gluten gel in equilibrium with excess water is obtained. The elasticity of gluten, therefore, cannot be explained by the entanglement concept. The β-turns, however, are proposed to be of importance for the elasticity of the gluten gel as described in Chapter 2.

Stress-relaxation measurements of gluten show that the stress approaches zero after sufficient time, and the gluten should then be described as a liquid [68, 148]. The relaxation time for gluten is about 1.5 s [148]. However, strictly speaking, the relaxation cannot be described with a single $t_{1/2}$ value; instead, a spectrum has to be used [55]. Moreover, the stress-relaxation curves of both dough and gluten show two relaxation processes. This is evident from the plots of stress rate versus stress [148]. The first relaxation process occurs over a short time (0.1–10 s), whereas the second process occurs over a much longer time (10–10^4 s). The stress-relaxation behavior has been analyzed according to the theory of cooperative flow, and from this analysis a coordination number (z) is obtained [149]. For the first relaxation process, z was found to be 4, whereas for the second process $z = 2$. The coordination number indicates the number of units that rearrange cooperatively during flow. However, the theory does not give information about the nature of these flow units. For liquid crystalline lipid phases, a coordination number of 2 was obtained for the lamellar phase and one of 6 for the hexagonal phase [150]. In these cases the coordination number of the flow is the same as the coordination number of the structure. A correlation between flow units and structure units is not as evident in the case of gluten or dough. By treating gluten with different chemicals (glutaraldehyde, cysteine, ascorbic acid, $KBrO_3$), some indications of the structure of the flow units were obtained [68]. It was suggested that the primary relaxation process occurs in a partly oriented fibrillar structure, whereas the secondary relaxation process occurs in a lamellar superstructure. The addition of $KBrO_3$ or ascorbic acid to gluten resulted in an increase in G and $t_{1/2}$ [68]. Cysteine, on the other hand, caused a decrease in G and $t_{1/2}$ and a decrease in G' together with an increase in δ [68, 95].

B. The Influence of Starch on the Rheological Behavior of Gluten

Starch makes up about 60% of the gas free volume of the dough [97]. At the dough stage the temperature never approaches the gelatinization temperature of the starch, and therefore we do not need to be concerned with the gel-forming behavior of starch. Starch affects the rheological properties of the dough because of its volume. According to the equation of Einstein, the following relationship exists between viscosity and volume of a dispersed phase [151]:

$$\eta_{rel} = 1 + \alpha_1 \times \Phi \tag{7}$$

In this equation η_{rel} is the viscosity of the suspension, Φ the phase volume of the suspended material, and α_1 a constant. This equation holds for dilute suspensions; higher terms have to be introduced for more concentrated suspensions. Moreover, an assumption for Equation (7) is that the suspended material consists of spherical rigid particles. For other types of particles, α_1 depends on the shape of the particle. Starch granules can be regarded as rigid in the dough but not as spherical. When filler particles are added to a polymer solution, a strengthening effect is expected [152]. The bimodal particle size distribution of starch granules can also be expected to affect their influence on the rheological behavior of the dough. The minimum in the relative viscosity depends on the filler fraction and the blend ratios in the filler fractions [153]. It is also possible for a concentrated dispersion to show elasticity [154].

The strengthening of the gluten gel when starch is introduced has been analyzed [95, 155]. The storage modulus is related to the protein/starch ratio, and G' increases as the protein/starch ratio decreases (Fig. 16). The phase angle δ was also found to depend on the protein/starch ratio, decreasing as that ratio decreased. Another conclusion drawn is that starch is not simply an inert filler particle in the gluten matrix [155, 156]. A region of linear viscoelastic behavior of gluten is quite easily found, but when starch is introduced the linear behavior is lost or at least changed to lower strain values [26].

An interesting observation is that the rheological behavior of a flour reconstituted from its starch and gluten fractions is not the same as that of the unfractionated flour. The reconstituted flour shows higher G' and lower δ values than the orginal flour [95]. If the water-soluble fraction is included, G' decreases and δ increases, but the fully reconstituted dough still differs from the original flour. One explanation could be that lyophilization used in the fractionation procedure affected the components so that the starch–gluten interaction will be changed.

What about the significance of these viscoelastic properties in breadmaking? The same rheological behavior can be achieved with any wheat flour by adjusting the water content [157]. In a recent report, Bloksma [74, 97] concluded that the rheological properties required for the baking process are extensibility and a sufficiently large viscosity. The stress on the importance of elasticity, he concluded, is not valid.

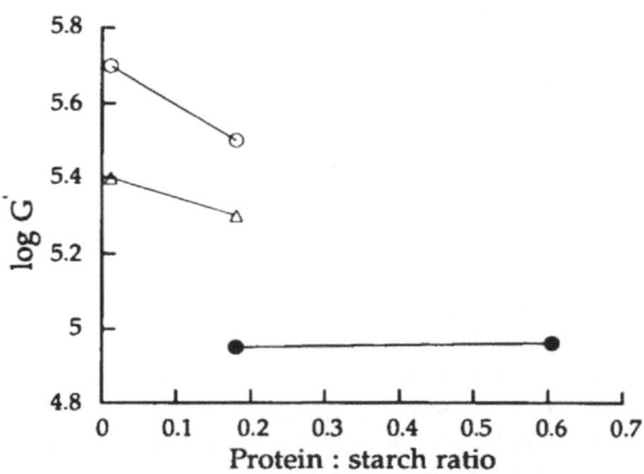

Fig. 16 The influence of the protein/starch ratio on the rheological behavior of dough. (Adapted from [155].) Results were obtained at water levels of (○) 45%, (△) 46%, and (●) 49% water.

C. Flour Quality

Different flours give doughs with different rheological properties. The relaxation modulus obtained immediately after loading (at time $t = 0$) has been used to characterize wheat flour doughs made from different wheat varieties, and for 23 varieties G was in the range 7.8–14.7 kPa [158]. There was no simple relation between G and the protein content. For different wheat flour milling streams from a commercial mixture of spring wheats, G was in the range 12–25 kPa, and for the corresponding flour streams from a winter wheat mixture G was in the range 10–18 kPa [29]. Linear regression analysis was performed to find correlations between the G values and protein content, starch content, and amount of damaged starch. Only in the case of starch damage were correlation coefficients (r^2) greater than 0.500 obtained. The correlation coefficient obtained for the protein content was $r^2 = 0.02$ for spring wheat and 0.245 for winter wheat.

At the same water content and frequency, a hard red spring wheat flour dough had a much higher G' value than a soft red winter wheat [27]. The differences decreased when the water content decreased. In another investigation the wheat flour that gave the best baking performance had the highest G' value and the lowest δ value [30]. It has also been observed that G' depends on the protein content for a single wheat variety, but when different wheats are compared, the G' values do not necessarily reflect their protein contents [26]. Generally, the flours claimed to give better baking performance have the higher G', G'', and G values and the lower δ values. The extensibility, especially during oven rise, differs between varieties [74].

The differences in rheological behavior may be due to differences in molecular weight distribution. A fractionation procedure involving successive extractions in dilute HCl was used to prepare fractions for further studies of their influence on the rheological behavior of dough [159]. The results obtained with small-amplitude oscillatory measurements on doughs after the addition of 1% of the different fractions were in agreement with the influence of the fractions on dough development previously described [90]. The HMW fractions increased G' and decreased δ, so that a stiffer and more structured network was developed in these doughs. Fractions of a lower molecular weight distribution had the opposite effect.

D. Added Ingredients

Most of the ingredients added to a wheat flour dough influence its rheological behavior; in fact, some ingredients, for example, oxidants and reductants as well as emulsifiers, are added with this aim. However, fundamental rheological measurements have been used in only a few cases to study the effect of added ingredients on the rheological behavior of dough. Cysteine decreases G' and increased δ; iodate increases G' but has no significant effect on δ [95]. Lecithin has been shown to increase G in stress-relaxation measurements [45].

VI. THE STRUCTURE OF THE DOUGH

In this final section of this chapter we summarize our discussion and give a somewhat more detailed picture of the dough than the one given in Fig. 1.

A. Formation of Two Interpenetrating Aqueous Phases

As will be argued below, the structural result of dough mixing is the formation of bicontinuous networks of two water-containing phases, the gluten phase and the "free" water phase. The relative amount of water in the "free" water phase (see Section II) is low; with 40% (w/w) water in the dough, about 17% will be "free." How can such a small volume form a continuous medium (the continuous character being obvious from the electrical conductivity discussed in Section II)? The location of the starch is the explanation.

If starch granules are added to a gluten gel, the interface will preferably be the aqueous zone between the protein units (see Chapter 2, Section III). There are no planes of large enough area, however, with such an interface in the gluten gel, and the interfacial energy will certainly be reduced if starch enters the water phase. If the water content is increased above the limit of swelling of the gluten gel phase, it seems obvious that the "free" water will wet the starch granules. The two phases "free" water + starch/gluten gel will have a volume ratio of about 1:1. It is therefore not surprising that they can form interpenetrating networks. In this context we can neglect the other components, the water-soluble proteins and pentosans distributed in the "free" water and the lipid–water phases forming separate aggregates in the gluten phase. Electron microscopic evidence of a "separate" water phase covering starch granules as well as separate water droplets in mixed doughs has in fact been clearly demonstrated [16].

If the surface of starch granules are covered with water, and this phase takes up about half of the total volume of the dough, they will tend to fuse into a continuous network. The reason is that the interfacial energy will be reduced in this way. Thus the contact area toward the gluten gel will be reduced. This is shown schematically in Fig. 17.

The fact that we can easily wash out starch from the gluten gel is consistent with such a phase separation. If the starch granules were attached directly to the gluten gel it would not be possible to "dilute" the granules away in the excess water without disintegrating the gel.

B. The Dispersed Gas Cells

The inclusion of air cells was discussed in detail earlier. Here we will be concerned with the structural details of the changes of the dispersed gas cells brought about through the hydration of the flour, dough mixing, and fermentation.

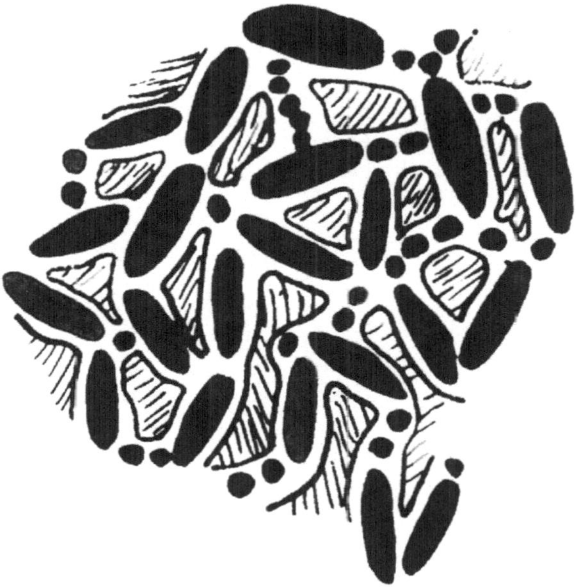

Fig. 17 Two phase systems in the aqueous region of the dough consisting of two interpenetrating water phases. Starch granules with a surface coat of "free" water, are indicated in the figure. These granules are associated into a network interpenetrating the continuous gluten gel. The gluten gel is indicated by the shaded area. The gluten gel is thus considered to fill the space between the water-fused starch granules.

The hydration of the flour means that many air cells are entrapped into the heterogeneous water–flour mix. This is due to the rough surface character of the flour particles. These first air cells are normally not spherical. To understand this we must again start with the Laplace–Young equation [Equation (6)] written in a more general form:

$$\Delta p = \gamma(1/R_1 + 1/R_2) \tag{8}$$

This is the pressure gradient over any surface, and R_1 and R_2 are the principal radii of curvature, defined and discussed in Chapter 1.

A cavity in a starch granule can form a stable gas cell toward outside water if the interface is planar, as shown in Fig. 18. In this context it is of interest to notice that the large starch fraction of wheat can sometimes adopt a biconcave shape. Thus a planar surface toward water gives a stable air cell with zero pressure gradient. This could, by the way, be one explanation for the fact that

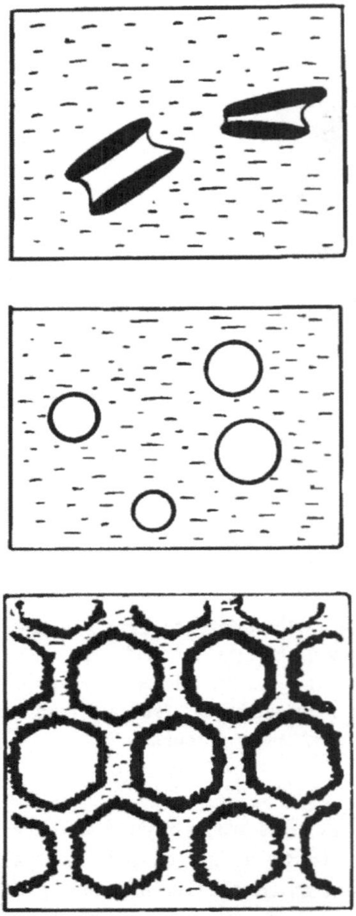

Fig. 18 Development of gas cells from the first stage of mixing (zero average curvature, top), through the end of mixing (spherical cells center), until the foam structure has developed in the final bread (bottom). Only air cells and starch granules or protein molecules at their interfaces are indicated.

wheat starch cannot be replaced by other starches without influencing baking performance.

Another example to illustrate this is the formation of a catenoid, as shown in Chapter 1. Instead of the two rings there may be two circular surfaces on two solid particles. The average curvature can thus adopt the value zero, and therefore the cells can form spontaneously on hydration. As soon as these cells have been formed, surface-active molecules can adsorb at the interface and reduce the interfacial energy.

During mechanical mixing, however, these cells will be transferred to the bulk of the dough and transformed into spherical cells, as mentioned in Section III (see Fig. 18). The reason is the mass transfer taking place and the pressure effects induced mechanically. At this stage the water-soluble proteins and perhaps even glycoproteins are the major candidates for formation of the monolayer at the new air/water interface.

If we cut a fermented wheat flour dough with a razor blade, it is possible to see a structure resembling foam. The size of the gas cells will vary considerably if the dough is not kneaded. With mechanical working, the deformability will tend to become uniform over the dough, which will result in a narrower size distribution of air cells. After the mixing, fermentation will start to close-pack gas cells. The gluten gel proteins will now begin to compete for the gas/water interface. In this process the gliadins and the LMW fraction of the glutenins will first replace the water-soluble proteins, and then they will successively be replaced by MMW proteins and HMW proteins. These changes toward the formation of larger proteins are a general phenomenon of protein adsorption at interfaces (known as the Vroman effect after L. Vroman, who first described these competitive phenomena in blood proteins). The formation of intermolecular disulfide bridges between the interfacial protein molecules, particularly the HMW fraction, will increase the mechanical strength of the film. It will also, however, decrease the compressibility/extensibility of the film as evident from the π–A isotherms.

What about the two aqueous phases and their relationship to the initially formed gas cells and the final foam cells? This question has not been addressed before, and our discussion below—although based on available evidence—is therefore somewhat speculative.

The first entrapped air cells during hydration must be expected to be localized at the surface of the solid starch and storage protein particles. At this stage it is doubtful whether there is more than one aqueous phase, the "free" water. The water-soluble surface-active molecules in this water phase will start to adsorb at the air cell surface, and the gluten gel is formed by absorption of "free" water during the mixing.

After the regular foam structure has been built up in the late stages of fermentation, it is rather obvious that the storage proteins form the interface, and judging by the interfacial properties of these proteins (Chapter 2, Section III) we should assume that the gluten gel itself forms a surface zone and the "free" water-starch the core of the foam lamellae. These structures are indicated in Fig. 19. Lipid aggregates are known to associate with the gluten gel.

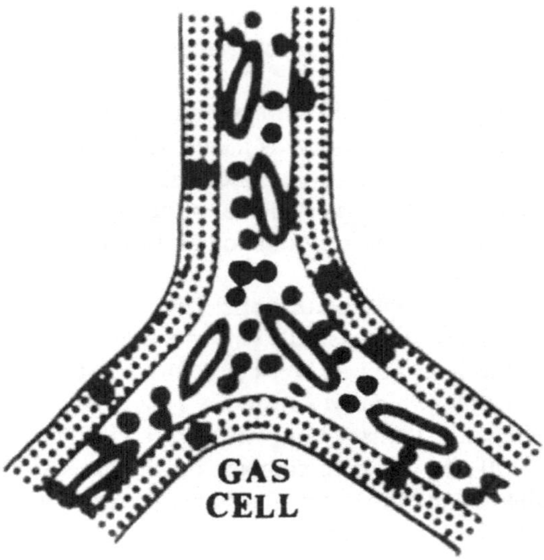

Fig. 19 Final structure of the foam lamellae in the dough. The outer gluten gel phase is shown by the dotted area; larger black regions denote lipid aggregates.

REFERENCES

1. Kulp, K. (1991). Breads and yeast-leavened bakery foods, in *Handbook of Cereal Science and Technology*, K. J. Lorenz and K. Kulp, Eds, Marcel Dekker, New York, p. 639.
2. Westerlund, E., Theander, O., and Åman, P. (1989). Effects of baking on protein and aqueous ethanol-extractable carbohydrate in white bread fractions, *J. Cereal Sci.*, 10: 139.
3. Pyler, E. J. (1971). *Baking Science and Technology*, Siebel, Chicago, p. 546.
4. Czuchajowska, Z., Pomeranz, Y., and Jeffers, H. C. (1989). Water activity and moisture content of dough and bread, *Cereal Chem.*, 66: 128.
5. MacRitchie, F. (1976). The liquid phase of dough and its role in baking, *Cereal Chem.*, 53: 318.
6. AACC (1976). *Approved Methods of the American Association of Cereal Chemists*, AACC 54-21, American Association of Cereal Chemists, St. Paul, Minn.
7. Holas, J., and Tipples, K. H. (1978). Factors affecting farinograph and baking absorption. I. Quality characteristics of flour streams, *Cereal Chem.*, 55: 637.

8. Butaki, R. C., and Dronzek, B. (1979). Comparison of gluten properties of flour wheat varieties, *Cereal Chem.*, 56: 159.
9. Finney, K. F. (1984). An optimized, straight-dough, bread-making method after 44 years, *Cereal Chem.*, 61: 20.
10. MacRitchie, F. (1984). Baking quality of wheat flours, *Adv. Food Res.*, 3: 201.
11. Tipples, K. H. (1969). The relation of starch damage to the baking performance of flour, *Baker's Dig.*, 43(6): 28.
12. Farrand, E. A. (1972). Controlled levels of starch damage in a commercial United Kingdom bread flour and effects on absorption, sedimentation value, and loaf quality, *Cereal Chem*, 49: 479.
13. Lineback, D. R., and Rasper, V. F. (1988). Wheat carbohydrates, in *Wheat Chemistry and Technology*, Y. Pomeranz, Ed., AACC, St. Paul, Minn., p. 277.
14. Davies, R. J., and Webb, T. (1969). Methods, apparatus: new product research, process development and design, *Chem. Ind.*, 1969: 1138.
15. Wood, P. S., Daniels, N. W. R., and Greenshields, R. N. (1972). *J. Food Technol.*, 7: 183.
16. Fretzdorff, B., Bechtel, D. B., and Pomeranz, Y. (1982). Freeze-fracture ultrastructure of wheat flour ingredients, dough and bread, *Cereal Chem.*, 59: 113.
17. Bushuk, W., and Mehrotra, V. K. (1977). Studies of water binding by differential thermal analysis. II. Dough studies using melting mode, *Cereal Chem.*, 54: 320.
18. Daniels, N. W. R. (1975). Some effects of water in wheat flour doughs, in *Water Relations of Foods*, R. B. Duckworth, Ed., Academic, London, p. 573.
19. Eliasson, A.-C. (1985). Retrogradation of starch as measured by differential scanning calorimetry, in *New Approaches to Research on Cereal Carbohydrates*, R. D. Hill and L. Munck, Eds., Elsevier, Amsterdam, p. 93.
20. Levine, H., and Slade, L. (1990). Cryostabilization technology: thermoanalytical evaluation of food ingredients and systems, in *Thermal Analysis of Food*, V. R. Harwalkar and C.-Y. Ma, Eds., Elsevier, New York, p. 221.
21. Eliasson, A.-C. (1980). Effect of water content on the gelatinization of wheat starch, *Starch/Stärke*, 32: 270.
22. Greer, E. N., and Stewart, B. A. (1959). The water absorption of wheat flour: relative effects of protein and starch, *J. Sci. Food Agric.*, 10: 248.
23. Bushuk, W. (1966). Distribution of water in dough and bread, *Bakers' Dig.*, 40(5): 38.
24. Eliasson, A.-C. (1983). Differential scanning calorimetry studies on

wheat starch–gluten mixtures. I. Effect of gluten on the gelatinization of wheat starch, *J. Cereal Sci.*, 1: 199.
25. Willhoft, E. M. A. (1971). Bread staling. I. Experimental study, *J. Sci. Food Agric.*, 22: 176.
26. Smith, J. R., Smith, T. L., and Tschoegl, N. W. (1970). Rheological properties of wheat flour doughs, *Rheol. Acta*, 9: 239.
27. Navickis, L. L., Anderson, R. A., Bagley, E. B., and Jasberg, B. K. (1982). Viscoelastic properties of wheat flour doughs: variation of dynamic moduli with water and protein content, *J. Text. Stud.*, 13: 249.
28. Lindahl, L. (1990). *Rheological properties in wheat flour systems. Molecular and interfacial basis*, thesis, Lund University, Lund.
29. Eliasson, A.-C., Silverio, J., and Tjerneld, E. (1991). Surface properties of wheat flour-milling streams and rheological and thermal properties after hydration, *J. Cereal Sci.*, 13: 27.
30. Abdelrahman, A., and Spies, R. (1986). Dynamic rheological studies of dough systems, in *Fundamentals of Dough Rheology*, H. Faridi and J. M. Faubion, Eds., AACC, St. Paul, Minn., p. 87.
31. Willhoft, E. M. A. (1973). Recent developments on the bread staling problem, *Bakers' Dig.*, 47(6): 14.
32. Carlson, T. L.-G. (1981). Law and order in wheat flour dough. Colloidal aspects of the wheat flour dough and its lipid and protein constituents in aqueous media, thesis, Lund University, Lund.
33. Holmes, J. T., and Hoseney, R. C. (1987). Chemical leavening: effect of pH and certain ions on breadmaking properties, *Cereal Chem.*, 64: 343.
34. Marek, C. J., and Bushuk, W. (1967). Study of gas production and retention in doughs with a modified Brabender oven-rise recorder, *Cereal Chem.*, 44: 300.
35. Niman, C. E. (1981). Salt in bakery products, *Cereal Foods World*, 26: 116.
36. van Dam, H. W. (1986). The biotechnology of baker's yeast: old or new business?, in *Chemistry and Physics of Baking*, J. M. V. Blanshard, P. J. Frazier, and T. Galliard, Eds., The Royal Society of Chemistry, London, p. 117.
37. Galal, A. M., Varriano-Marston, E., and Johnson, J. A. (1978). Rheological dough properties as affected by organic acids and salt, *Cereal Chem.*, 55: 683.
38. Danno, G., and Hoseney, R. C. (1982). Effect of sodium chloride and sodium dodecyl sulfate on mixograph properties, *Cereal Chem.*, 59: 202.
39. Linko, P., Härkönen, H., and Linko, Y.-Y. (1984). Effects of sodium chloride in the processing of bread baked from wheat, rye and barley flours, *J. Cereal Sci.*, 2: 53.
40. Weegels, P. L., Marseille, J. P., de Jager, A. M., and Hamer, R. J.

(1990). Structure–function relationships of gluten proteins, in *Gluten Proteins 1990*, W. Bushuk and R. Tkachuk, Eds., AACC, St. Paul, Minn., p. 98.
41. Preston, K. R. (1989). Effects of neutral salts of the lyotropic series on the physical dough properties of a Canadian red spring wheat flour, *Cereal Chem.*, 66: 144.
42. Kieffer, R., Kim, J.-J., and Belitz, H.-D. (1983). Einfluss niedermolekularer ionischer Verbindungen auf die Löslichkeit und die rheologischen Eigenschaften von Weizenkleber, *Z. Lebensm. Unters. Forsch.*, 176: 176.
43. Kinsella, J. E., and Hale, M. L. (1984). Hydrophobic association and gluten consistency: effects of specific anions, *J. Agric. Food. Chem.*, 32: 1054.
44. Salovaara, H. (1982). Effect of partial sodium chloride replacement by other salts on wheat dough rheology and breadmaking, *Cereal Chem.*, 59: 422.
45. Eliasson, A.-C., and Tjerneld, E. (1990). On the influence of added lipid phases on the rheological properties of wheat flour doughs, *Z. Lebensm. Unters. Forsch.*, 191: 35.
46. Wootton, M., and Bamunuarachchi, A. (1980). Application of differential scanning calorimetry to starch gelatinization. III. Effect of sucrose and sodium chloride, *Starch/Stärke*, 32: 126.
47. Evans, I. D., and Haisman, D. R. (1982). The effects of solutes on the gelatinization temperature of potato starch, *Starch/Stärke*, 34: 224.
48. Ghiasi, K., Hoseney, R. C., and Varriano-Marston, E. (1983). Effect of flour components and dough ingredients on starch gelatinization, *Cereal Chem.*, 60: 58.
49. Lindqvist, I. (1979). Cold gelatinization of starch, *Starch/Stärke*, 31: 195.
50. Eliasson, A.-C. (1992). *Carbohydr. Polym.*, (to be published).
51. Cheer, R. L., and Lelievre, J. (1983). Effects of sucrose on the rheological behavior of wheat starch pastes, *J. Appl. Polym. Sci.*, 28: 1829.
52. Spies, R. D., and Hoseney, R. C. (1982). Effect of sugars on starch gelatinization, *Cereal Chem.*, 59: 128.
53. Slade, L., and Levine, H. (1987). Recent advances in starch retrogradation, in *Industrial Polysaccharides—The Impact of Biotechnology and Advanced Methodologies*, S. S. Stivala, V. Crescenzi, and I. C. M. Dea, Eds., Gordon and Breach, New York, p. 387.
54. Slade, L., and Levine, H. (1988). Non-equilibrium behavior of small carbohydrate–water systems, *Pure Appl. Chem.*, 60: 1841.
55. Bloksma, A. H., and Bushuk, W. (1988). Rheology and chemistry of dough, in *Wheat Chemistry and Technology*, Y. Pomeranz, Ed., AACC, St. Paul, Minn., p. 131.

56. Kulp, K. (1988). Bread industry and processes, in *Wheat Chemistry and Technology*, Y. Pomeranz, Ed., AACC, St. Paul, Minn., p. 371.
57. Kilborn, R. H., and Tipples, K. H. (1972). Factors affecting mechanical dough development. I. Effect of mixing intensity and work input, *Cereal Chem.*, 49: 34.
58. Schofield, D. (1986). Flour proteins: structure and functionality in baked products, in *Chemistry and Physics of Baking*, J. M. V. Blanshard, P. J. Frazier, and T. Galliard, Eds., The Royal Society of Chemistry, London, p. 14.
59. Ewart, J. A. D. (1968). A hypothesis for the structure and rheology of glutenin, *J. Sci. Food Agric.*, 19: 617.
60. MacRitchie, F. (1986). Physicochemical processes in mixing, in *Chemistry and Physics of Baking*, J. M. V. Blanshard, P. J. Frazier, and T. Galliard, Eds., The Royal Society of Chemistry, London, p. 132.
61. Grosch, W. (1986). Redox systems in dough, in *Chemistry and Physics of Baking*, J. M. V. Blanshard, P. J. Frazier, and T. Galliard, Eds., The Royal Society of Chemistry, London, p. 155.
62. Mecham, D. K. (1968). The sulphydryl and disulfide contents of wheat flours, doughs, and proteins, *Baker's Dig.*, 42(1): 26.
63. Noguchi, G., Shinya, K., Tanaka, K., and Yoneyama, T. (1976). Correlation of dough stickiness with texturemeter reading and with various quality parameters, *Cereal Chem.*, 53: 72.
64. Moss, R. (1974). Dough microstructure as affected by the addition of cysteine, potassium bromate, and ascorbic acid, *Cereal Sci. Today*, 19: 557.
65. Belitz, H.-D., Kieffer, R., Seilmeier, W., and Weiser, H. (1986). Structure and function of gluten proteins, *Cereal Chem.*, 63: 336.
66. Kieffer, R., Kim, J.-J., Walther, C., Laskawy, G., and Grosch, W. (1990). Influence of glutathione and cysteine on the improver effect of ascorbic acid stereoisomers, *J. Cereal Sci.*, 11: 143.
67. Graveland, A., Bosveld, P., and Marseille, J. P. (1978). Determination of thiol groups and disulphide bonds in wheat flour and dough, *J. Sci. Food Agric.*, 29: 53.
68. Mita, T., and Bohlin, L. (1983). Shear stress relaxation of chemically modified gluten, *Cereal Chem.*, 60: 93.
69. Galliard, T. (1986). Oxygen consumption of aqueous suspensions of wheat wholemeal, bran and germ: involvement of lipase and lipoxygenase, *J. Cereal Sci.*, 4: 33.
70. Bloksma, A. H. (1975). Thiol and disulfide groups in dough rheology, *Cereal Chem.*, 52: 170.
71. Kaufman, S. P., Hoseney, R. C., and Fennema, O. (1986). Dough rheology—a review of structural models and the role of disulfide interchange reactions, *Cereal Foods World*, 31: 820.

72. Kaufman, S. P., and Fennema, O. (1987). Evaluation of sulfhydryl oxidase as a strengthening agent for wheat flour dough, *Cereal Chem.*, 64: 172.
73. Mair, G., and Grosch, W. (1979). Changes in glutathione content (reduced and oxidised form) and the effect of ascorbic acid and potassium bromate on glutathione oxidation during dough mixing, *J. Sci. Food Agric.*, 30: 914.
74. Bloksma, A. H. (1990). Rheology of the breadmaking process, *Cereal Foods World*, 35: 228.
75. Tanaka, K., and Bushuk, W. (1973). Changes in four proteins during dough-mixing. III. Analytical results and mechanisms, *Cereal Chem.*, 50: 605.
76. Bloksma, A. H. (1963). Oxidation by molecular oxygen of thiol groups in unleavened doughs from normal and defatted wheat flours, *J. Sci. Food Agric.*, 14: 529.
77. Pomeranz, Y. (1988). Wheat flour components, in *Wheat Chemistry and Technology*, Y. Pomeranz, Ed., AACC, St. Paul, Minn., p. 219.
78. Chamberlain, N., and Collins, T. H. (1979). The Chorleywood bread process: the roles of oxygen and nitrogen, *Baker's Dig.*, 53(1): 18.
79. Elkassabany, M., and Hoseney, R. C. (1980). Ascorbic acid as an oxidant in wheat flour dough. II. Rheological effects, *Cereal Chem.*, 57: 88.
80. Stenvert, N. L., Moss, R., and Murray, L. (1981). The role of dry wheat gluten in breadmaking. I. Quality assessment and mixer interaction, *Baker's Dig.*, 55(2): 6.
81. Kilborn, R. H., and Tipples, K. H. (1973). Factors affecting mechanical dough development. IV. Effect of cysteine, *Cereal Chem.*, 50: 70.
82. Faubion, J. M., and Hoseney, R. C. (1981). Lipoxygenase: its biochemistry and role in breadmaking, *Cereal Chem.*, 58: 175.
83. Endo, S., Okada, K., and Nagao, S. (1987). Studies on dough development. III. Mixing characteristics of flour streams and their changes during dough mixing in the presence of chemicals, *Cereal Chem.*, 64: 110.
84. Cherdkiatgumchai, P., and Grant, D. R. (1986). Enzymes that contribute to the oxidation of L-ascorbic acid in flour/water systems, *Cereal Chem.*, 63: 197.
85. Elkassabany, M., Hoseney, R. C., and Seib, P. A. (1980). Ascorbic acid as an oxidant in wheat flour dough. I. Conversion to dehydroascorbic acid, *Cereal Chem.*, 57: 85.
86. Graveland, A. (1971). Thesis, University of Utrecht.
87. Chung, O. K., Pomeranz, Y., Shogren, M. D., Finney, K. F., and Howard, B. G. (1980). Defatted and reconstituted wheat flours. VI. Response to shortening addition and lipid removal in flours that vary in bread-making quality, *Cereal Chem.*, 57: 111.

88. Wesén, C., Carlberg, C. E., and Martinsen, K. (1990). On the identity of different chlorinated organic substances in aquatic organisms and sediments, *Ambio*, 19: 36.
89. Huebner, F. R., and Wall, J. S. (1976). Fractionation and quantitative differences of glutenin from wheat varieties varying in baking quality, *Cereal Chem.*, 53: 258.
90. MacRitchie, F. (1987). Evaluation of contributions from wheat protein fractions to dough mixing and breadmaking, *J. Cereal Sci.*, 6: 259.
91. Lundh, G., and MacRitchie, F. (1989). Size exclusion HPLC characterisation of gluten protein fractions varying in breadmaking potential, *J. Cereal Sci.*, 10: 247.
92. Faubion, J. M., and Hoseney, R. C. (1990). The viscoelastic properties of wheat flour dough, in *Dough Rheology and Baked Product Texture*, H. Faridi and J. M. Faubion, Eds., Van Nostrand Reinhold, New York, p. 29.
93. Mani, K., Eliasson, A.-C., Lindahl, L., and Trägårdh, C. (1992). Rheological properties and breadmaking quality of wheat flour doughs made by different dough mixers, *Cereal Chem.* (to be published).
94. Bohlin, L., and Carlson, T. L.-G. (1980). Dynamic viscoelastic properties of wheat flour dough: dependence on mixing time, *Cereal Chem.*, 57: 174.
95. Dreese, P. C., Faubion, J. M., and Hoseney, R. C. (1988). Dynamic rheological properties of flour, gluten and gluten-starch doughs. II. Effect of various processing and ingredient changes, *Cereal Chem.*, 65: 354.
96. Lindahl, L., and Svensson, G. (1988). Characterization of wheat varieties based on rheological behaviour during stress-relaxation, *Acta Agric. Scand.*, 38: 363.
97. Bloksma, A. H. (1990). Dough structure, dough rheology, and baking quality, *Cereal Foods World*, 35: 237.
98. Amend, T., and Belitz, H.-D. (1990). The formation of dough and gluten—a study by scanning electron microscopy, *Z. Lebensm. Unters. Forsch.*, 190: 401.
99. Danno, G., and Hoseney, R. C. (1982). Changes in flour proteins during dough mixing, *Cereal Chem.*, 59: 249.
100. Tsen, C. C. (1969). Effects of oxidizing and reducing agents on changes of flour proteins during dough mixing, *Cereal Chem.*, 46: 435.
101. Mamaril, F. P., and Pomeranz, Y. (1966). Isolation and characterization of wheat flour proteins. IV. Effects on wheat flour proteins of dough mixing and oxidizing agents, *J. Sci. Food Agric.*, 17: 339.
102. Tanaka, K., and Bushuk, W. (1973). Changes in flour proteins during dough-mixing. I. Solubility results, *Cereal Chem.*, 50: 590.

103. Okada, K., Negshi, Y., and Nagao, S. (1987). Factors affecting dough breakdown during overmixing, *Cereal Chem.*, 64: 428.
104. MacRitchie, F. (1975). Mechanical degradation of gluten proteins during high-speed mixing of doughs, *J. Polym. Sci.*, 49: 85.
105. Tanaka, K., and Bushuk, W. (1973). Changes in flour proteins during dough-mixing. II. Gel filtration and electrophoresis results, *Cereal Chem.*, 50: 597.
106. Marion, D., Le Roux, C., Akora, S., Tellier, C., and Gallant, D. (1987). Lipid–protein interactions in wheat gluten: a phosphorus nuclear magnetic resonance spectroscopy and freeze-fracture election microscopy study, *J. Cereal Sci.*, 5: 101.
107. Hermansson, A.-M., and Larsson, K. (1986). The structure of gluten gels, *Food Microstruct.*, 5: 233.
108. Czuchajowska, Z., and Pomeranz, Y. (1990). Quest for a universal test of commercial gluten quality for breadmaking, *Cereal Foods World*, 35: 458.
109. Neyreneuf, O., and van der Plaat, J. B. (1991). Preparation of frozen French bread dough with improved stability, *Cereal Chem.*, 68: 60.
110. MacRitchie, F. (1978). Differences in baking quality between wheat flours, *J. Food Technol.*, 13: 187.
111. Bushuk, W., and Wadhawan, C. (1989). Wheat gluten is good not only for breadmaking, in *Wheat Is Unique*, Y. Pomeranz, Ed., AACC, St. Paul, Minn., p. 263.
112. Weegels, P. L., and Hamer, R. J. (1989). Predicting the baking quality of gluten, *Cereal Foods World*, 34: 210.
113. Dreese, P. C., and Hoseney, R. C. (1990). The effects of water-extracted solubles from gluten on its baking and rheological properties, *Cereal Chem.*, 67: 400.
114. McDermott, E. E. (1985). The properties of commercial glutens, *Cereal Foods World*, 32: 169.
115. MacRitchie, F. (1973). Conversion of a weak flour to a strong one by increasing the proportion of its high molecular weight gluten proteins, *J. Sci. Food Agric.*, 24: 1325.
116. Stenvert, N. L., Moss, R., and Murray, L. (1981). The role of dry vital wheat gluten in breadmaking. II. A microscopic method for locating dry vital gluten in doughs and determining the interaction with flour gluten and mixer type, *Baker's Dig.*, 55(3): 27.
117. Lee, J. W., and MacRitchie, F. (1971). The effect of gluten protein fractions on dough properties, *Cereal Chem.*, 48: 620.
118. Eliasson, A.-C. (1990). On the possibility to improve baking performance with the addition of gluten or polar lipids, 24:e Nordiska Cerealistkongressen "Spannmål i den framtida kosten," Stockholm.

119. Eliasson, A.-C., and Silverio, J. (1992). Interfacial and rheological behaviour of some commercial gluten samples, *Acta Agric. Scand.* (to be published).
120. Linko, Y.-Y., and Linko, P. (1986). Enzymes in baking, in *Chemistry and Physics of Baking*, (J. M. V. Blanshard, P. J. Frazier, and T. Gaillard, Eds., The Royal Society of Chemistry, London, p. 105.
121. Siljeström, M., Björck, I., Eliasson, A.-C., Lönner, C., Nyman, M., and Asp, N.-G. (1988). Effects on polysaccharides during baking and storage of bread—in vitro and in vivo studies, *Cereal Chem.*, 65: 1.
122. Haas, L.W., and Bohn, R. M. (1934). Flour whitener, U.S. Patent 1,937,333.
123. Nicolas, J., and Drapron, R. (1983). Lipoxygenase and some related enzymes in breadmaking, in *Lipids in Cereal Technology*, P. Barnes, Ed., Academic, London, p. 213.
124. Leissner, O. (1986). A comparison of the effect of different polymorphic forms of lipids in breadmaking, *Cereal Chem.*, 65: 202.
125. Junge, R. C., Hoseney, R. C., and Varriano-Marston, E. (1981). Effect of surfactants on air incorporation in dough and the crumb grain of bread, *Cereal Chem.*, 58: 338.
126. Krog, N., and Larsson, K. (1967). Phase behaviour and rheological properties of aqueous systems of industrial distilled monoglycerides, *Chem. Phys. Lipids*, 2: 129.
127. Shelton, D. R., and D'Appolonia, B. (1985). Carbohydrate functionality in the baking process, *Cereal Foods World*, 30: 437.
128. Dunås, R. (1991). Yeast in frozen bread doughs, thesis, Lund University, Lund.
129. Cooper, E. J., and Reed, G. (1968). Yeast fermentation—effects of temperature, pH, ethanol, sugars, salt, and osmotic pressure, *Baker's Dig.*, 42(6): 22.
130. Cullen-Refai, A., Faubion, J. M., and Hoseney, R. C. (1988). Lubricated uniaxial compression of fermenting doughs, *Cereal Chem.*, 65: 401.
131. Söderberg, I. (1990). Structural properties of monoglycerides, phospholipids, and fats in aqueous systems, thesis, Lund University, Lund.
132. Baker, J. C., and Mize, M. D. (1941). The origin of the gas cell in bread dough, *Cereal Chem.*, 18: 19.
133. Carlson, T., and Bohlin, L. (1978). Free surface energy in the elasticity of wheat flour dough, *Cereal Chem.*, 55: 539.
134. Marston, P. E. (1986). Dough development for breadmaking under controlled atmospheres, *J. Cereal Sci.*, 4: 335.
135. Hoseney, R. C. (1984). Gas retention in bread doughs, *Cereal Foods World*, 29: 305.

136. Rajapaksa, D., Eliasson, A.-C., and Larsson, K. (1983). Bread baked from wheat/rice mixed flours using liquid-crystalline lipid phases in order to improve bread volume, *J. Cereal Sci.*, 1: 53.
137. Mahdi, J. G., Varriano-Marston, E., and Hoseney, R. C. (1981). The effect of mixing atmosphere and fat crystal size on dough structure and bread quality, *Baker's Dig.*, 28.
138. Lundh, G., Eliasson, A.-C., and Larsson, K. (1988). Cross-linking of wheat storage protein monolayers by compression/expansion cycles at the air/water interface, *J. Cereal Sci.*, 7: 1.
139. Eliasson, A.-C., and Miezis, Y. (to be published).
140. Okumara, K., Miyake, Y., Taguchi, H., and Shimabayashi, Y. (1990). Formation of stable protein foam by intermolecular disulfide cross-linkages in thiolated alfa-*s*1-casein as a model, *J. Agric. Food Chem.*, 38: 1303.
141. Menjivar, J. A. (1990). Fundamental aspects of dough rheology, in *Dough Rheology and Baked Product Texture*, H. Faridi and J. M. Faubion, Eds., Van Nostrand Reinhold, New York, p. 1.
142. Mitchell, J. R. (1980). The rheology of gels, *J. Text. Stud.*, 11: 315.
143. Muller, H. G. (1969). Application of the statistical theory of rubber elasticity to gluten and dough, *Cereal Chem.*, 46: 443.
144. Bale, R., and Muller, H. G. (1970). Application of the statistical theory of rubber elasticity to the effect of heat on wheat gluten, *J. Food Technol.*, 5: 295.
145. Goldstein, S. (1957). Sulfhydryl- und Disulfidgruppen der Klebereiweisse und ihre Beziehung zur Backfähigkeit der Brotmehle, *Mitt. Geb. Lebensmittelunters. Hyg.*, 48: 87.
146. Ferry, J. D. (1970). *Viscoelastic Properties of Polymers*, Wiley, New York.
147. Graessly, W. N. (1974). *Adv. Polym. Sci.*, 16: 1.
148. Bohlin, L., and Carlson, T. L.-G. (1981). Shear stress relaxation of wheat flour dough and gluten, *Colloids Surfaces*, 2: 59.
149. Bohlin, L. (1980). A theory of flow as a cooperative phenomenon, *J. Colloid Interface Sci.*, 74: 423.
150. Bohlin, L., and Fontell, K. (1978). Flow properties of lamellar liquid crystalline lipid–water systems, *J. Colloid Interface Sci.*, 67: 272.
151. Frisch, H. L., and Simha, R. (1956). The viscosity of colloidal suspensions and macromolecular solutions, in *Rheology—Theory and Applications*, F. R. Eirich, Ed., Academic, New York, p. 525.
152. Nielsen, L.-E. (1974). *Mechanical Properties of Polymers and Composites*, Marcel Dekker, New York.
153. Farris, R. J. (1968). Prediction of the viscosity of multimodal suspensions from unimodal viscosity data, *Trans. Soc. Rheol.*, 12: 281.

154. Mewis, J. (1976). Rheology of concentrated dispersions, *Adv. Colloid Interface Sci.*, 6: 173
155. Hibberd, G. E. (1970). Dynamic viscoelastic behaviour of wheat flour doughs. III. The influence of the starch granules, *Rheol. Acta*, 9: 501.
156. Szczesniak, A. A., Loh, J., and Wesley, R. (1983). Effect of moisture transfer on dynamic viscoelastic parameters of wheat flour/water systems, *J. Rheol.*, 27: 537.
157. Lindahl, L., and Eliasson, A.-C. (1992). A comparison of some rheological properties of durum and wheat flour doughs, *Cereal Chem.* (to be published).
158. Lindahl, L., and Svensson, G. (1987). Stress-relaxation—a method to characterize the rheology of wheat varieties, in *Cereal Science and Technology*, L. Munck, Ed., The Danish Cereal Society, Copenhagen, p. 111.
159. Eliasson, A.-C., and Lundh, G. (1989). Rheological and interfacial behaviour of some wheat protein fractions, *J. Text. Stud.*, 20: 431.
160. Hoseney, R. C. (1986). *Principles of Cereal Science and Technology*, AACC, St. Paul, Minn., p. 133.

7
Bread

I. EFFECT OF HEAT

When the dough is put in the oven after fermentation, a considerable increase in volume occurs—the oven spring. The increase in volume induces considerable stress on the air/water interfaces in the dough. This is the most critical point in the breadmaking process. All mistakes made earlier in the process will now be revealed. Differences in baking performance between wheat varieties will show up [1,2]. The doughs may have fermented to the same size, but during oven spring the extent of their ability to produce a bread of satisfactory loaf volume is displayed. Furthermore, the influence of some additives such as fat and emulsifiers is not seen until this stage.

The volume increase observed during oven spring is a consequence of the increase in temperature, which induces changes in both the dispersed and continuous phases of the dough. These changes result in the volume expansion and the setting of the crumb. The rate of temperature increase and the final temperature reached differ in different parts of the dough, and this is especially true for the center of the dough compared with the surface. These time–temperature relations are illustrated in Fig. 1.

The increase in temperature directly influences the dough in several ways, including yeast activity, solubility of gases, and volume expansion of evaporated gases. The flour constituents are also affected—the starch gelatinizes, and the protein undergoes certain changes. As a result of the physical changes related to the included gas, an enormous increase in volume occurs, and this increase in volume has some indirect effects on the dough.

The meaning of gas-holding capacity of the dough at this stage of the baking process is discussed in this chapter. The influence of heat on the main components of the dough (lipids, protein, starch) is also described.

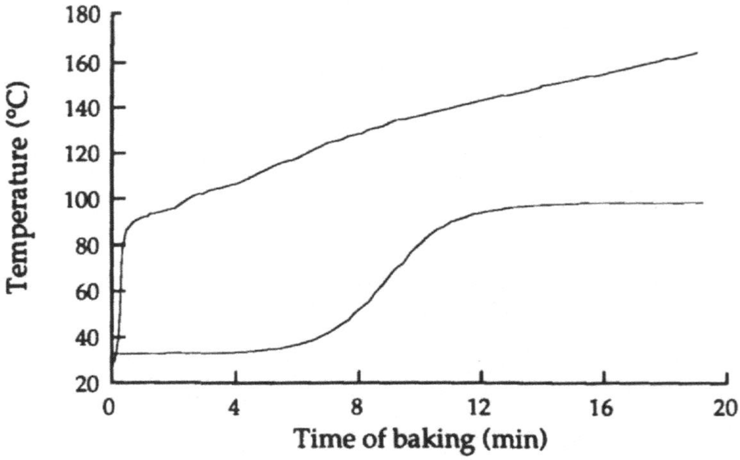

Fig. 1. The temperature increase in the center of the dough (lower curve) compared with the surface (upper curve) during baking. The temperature in the oven was 230°C. (Adapted from [3].)

A. Oven Spring

1. Heat and Mass Transfer

When the dough is put into the oven, heat is transferred by several mechanisms, including convection, radiation, conduction, the condensation of steam, and the evaporation of water. Heat transfer inside the dough is thought to occur through the mechanisms of heat conduction and evaporation–condensation [4].

For some types of bread, steaming is used for a short period at the start of baking. Steaming shortens baking time, but it also maintains an elastic dough surface and creates special crust properties [4]. The changes that occur in the oven can be divided into three periods. During the steaming period the volume of the product is constant, but there is a strong increase in surface temperature (see Fig. 2). The period of steaming is followed by a period of volume increase. The surface temperature continues to increase during this period. Then follows a period when crumb and crust are completed, the volume does not change, but the surface temperature still increases [4].

The addition of steam increases the relative humidity of the air to 0.7–0.9. The dough surface is cold compared to the surroundings, and moisture condenses onto it from the air. The condensation makes the surface temperature rise to the dew point very quickly (see Fig. 2). The amount of water condensed on the surface amounts to about 1% of the weight of the bread [4]. The steaming thus initially causes an increase in the weight of the product. However, as soon as the

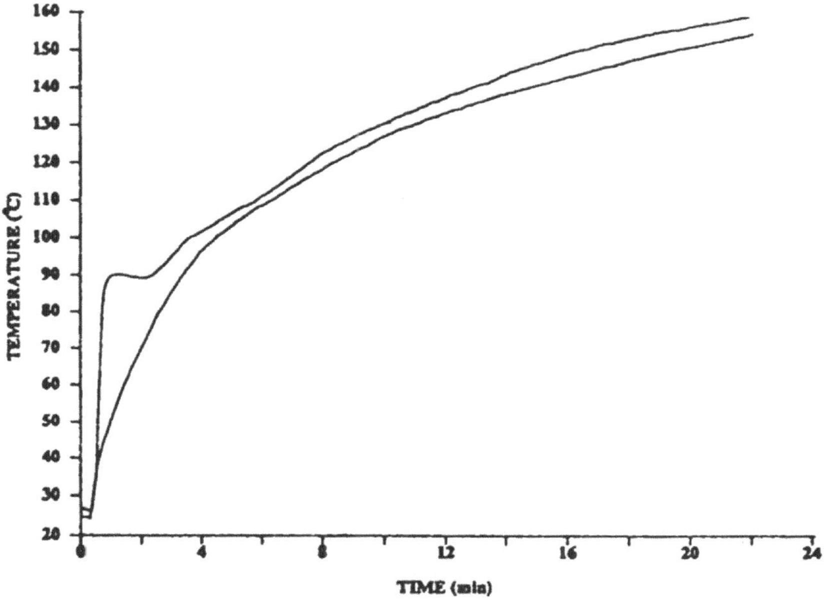

Fig. 2. The temperature increase of bread surface without (lower curve) and with (upper curve) steaming. (Adapted from [4].)

surface has reached the dew point temperature, evaporation begins, causing a loss in weight (Fig. 3). The weight loss during baking depends on the maximum surface temperature; the higher the maximum surface temperature, the higher the weight loss. In one study a weight loss of about 13% at 150°C and 17% at 200°C was recorded [3].

The increase in surface temperature causes the evaporation of water from a zone just below the crust. Part of the water vapor passes through the crust into the surrounding air, but part of it moves toward the cooler regions inside the crumb. Here, the water vapor condenses and causes an increase in the crumb temperature [5]. It has geen suggested that evaporation–condensation is the major mechanism of heat transport in bread. Although conduction is thought to play a minor role, it is nevertheless essential. If conduction was the only or the main mechanism for heat transport in the dough, heat transport in a gas-free piece of dough would be as rapid as the heat transport in a fermented piece of dough. This is not the case, however; the temperature increases much faster in a fermented dough than in a gas-free dough. Heat transport in the dough is thus the result of a combination of conduction in the continuous liquid phase and evaporation–condensation in the gas phase. The mechanism is illustrated in Fig. 4.

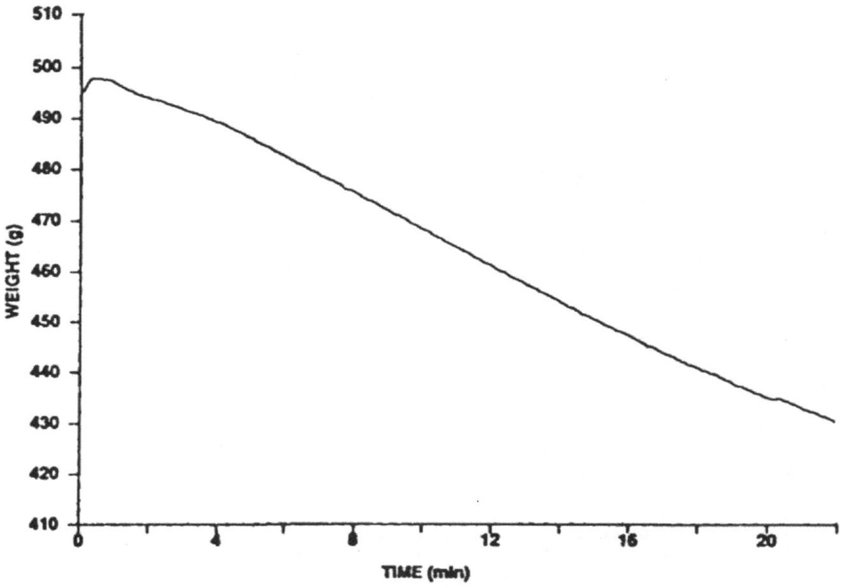

Fig. 3. The changes in loaf weight during baking at an oven temperature of 230°C. (Adapted from [4].)

Evaporation–condensation is assumed to be the only mechanism of mass (water) transport. The transport of water during baking is illustrated by the gradients in the water content of the loaf. Differences occur in water content between crumb and crust, with 44.2% water in the crumb and 15.5% in the crust 1 h after baking as one example [6]. The water content in the center of the crumb was measured at 45.7%, whereas at the edge of the same loaf it was 37.2% [7]. The water content of the center increases during baking, from 0.776 kg/kg dry matter in the initial dough to 0.894 kg/kg dry matter in the bread immediately after baking. After cooling to ambient temperature, the water content of the bread was 0.779 kg/kg dry matter [5]. The water content in the final bread reflects the water addition to the dough [7].

The gas volume of a dough increases as a function of temperature until a certain temperature just above the gelatinization onset temperature of starch. The temperature at which the volume expansion stops has been given to be in the range of 68–80°C [8, 9]. A slowing down of the rate of volume expansion has been observed at temperatures as low as 55°C in some cases [9]. Not only does the total gas volume change, but its composition also changes with temperature. In the early stages of baking, the gas volume is composed almost entirely of CO_2, whereas the fraction of water increases with temperature. The heat trans-

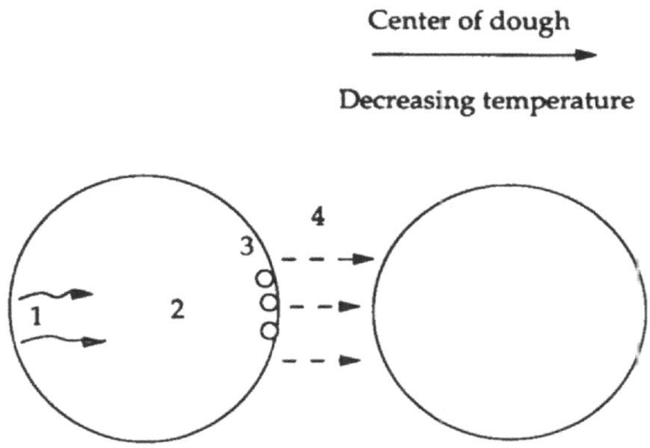

Fig. 4. The four steps involved in the heat transport in dough according to de Vries et al. [5]. (1) Water evaporates at the warmer side of a gas cell, absorbing latent heat of vaporization. (2) It migrates through the gas phase. (3) The water condenses at the colder side of the gas cell, releasing its latent heat. (4) Heat and water are transported by conduction and diffusion through the gluten gel to the warmer side of the next cell. The process then begins all over again.

port is related to the gas volume of the initial dough—the more gas included in the dough, the shorter the baking time, that is, the faster the temperature in the center approaches 100°C [5].

2. Production, Solubility, and Thermal Expansion of Gas

During the early stages in the oven, high yeast activity is observed. In the temperature interval 20–40°C the yeast activity increases by a factor of 2 for each 10°C [10]. At about 60°C, however, the activity stops because the yeast cells are killed by the heat [11]. To support the yeast activity, fermentable carbohydrates have to be present. The inclusion of sugars in the recipe ensures this. However, the activity of the amylases also depends on the temperature. Of the amylases, β-amylases are more heat-labile than α-amylases; α-amylases can be purified from β-amylases by heat treatment at 70°C for 15 min [12]. The optimum temperatures for α-amylases are 60–66°C, and for β-amylases, 48–51°C [13]. α-Amylases thus have the opportunity to act upon the gelatinizing starch during the time period when the temperature increases from 60°C to 70°C. The time available for starch hydrolysis depends on the type of bread and the size of the loaf [14]. For a loaf of 500 g, the time is 2–3 min [13].

The end of oven spring has been estimated from experiments using a resistance oven, and the increase in dough height ceased at about 80°C [9]. However, the temperature at which the volume expansion stops depends on the type of bread and is further discussed in relation to the effect of heat on starch in Section I.B.3. The gas-holding capacity of the dough is further discussed in Section I.C.

B. Influence of Heat on Flour Components

1. Lipids

We have observed no thermal transitions of the wheat lipid aqueous phases between room temperature and 80°C. The nonpolar lipids are all in the liquid state in this temperature interval, and the polar lipids give water phases that are insensitive to temperature when charged species are present.

Oxidative effects are favored by temperature, but still we consider them to be irrelevant during the baking process.

2. Proteins

a. Changes in Conformation

When proteins are heated they usually undergo thermal denaturation (see Chapter 1). The structure of the protein unfolds and becomes less ordered, a random coil. Denaturation is followed by aggregation and, in some proteins, gel formation. The denaturation of proteins has been extensively studied by differential scanning calorimetry (DSC) [15]. Denaturation is observed as an endothermic peak on the DSC thermogram. However, if gluten is analyzed by DSC, a thermogram similar to the one in Fig. 5 is obtained. Four peaks can be observed, but they are all very small [16]. The enthalpy of the most distinct peak, the one at about 65°C,

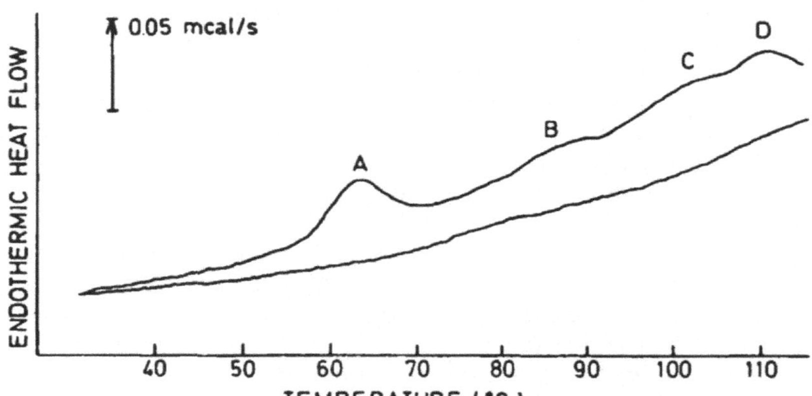

Fig. 5. A DSC thermogram of gluten. The upper curve is the first scan, and the lower curve a second scan. Peaks A and D were assigned to starch, whereas peaks B and C may be due to proteins. (Reprinted with permission from [16].)

was calculated to be 1.5 J/g. Moreover, this transition is attributed to starch, which to some extent will always remain in the prepared gluten [16, 17]. Gluten proteins are not the only proteins that fail to produce denaturation endotherms when they are heated in the calorimeter; the caseins in milk are another example.

At the moment there is no adequate explanation for the lack of gluten denaturation endotherms. The DSC thermograms could be interpreted as showing that no ordered structure exists in the gluten proteins. However, gluten proteins in solution are known to have a considerable degree of secondary structure (Chapter 2, Section III) [18, 19]. Instead, the explanation may be the variations in the gluten proteins. The DSC technique requires considerable cooperativity to produce detectable heat flow, which may not be possible with gluten proteins—the regions of *identically* ordered structure may be too small. Another explanation could be that the gluten proteins are unusually stable with respect to heat.

The thermal denaturation of wheat proteins in solution has been studied by circular dichroism [18, 19]. An increase in temperature from 20°C to 80°C resulted in monotonic but small decreases in the regular structure content of a γ-gliadin. The structural changes, which involved a helix-coil transition, were reversible, and the loss of α-helical content was small, below 10%. Native α- and β-gliadin showed very similar behavior. The marked stability with respect to heat was attributed partly to the disulfide bonds but also to extensive hydrogen bonding. ω-Gliadin underwent a conformational change that reached completion at about 60–70°C. The changes on heating were attributed to an increase in hydrophobic interactions and a decrease in hydrogen bonding. In addition, the heat treatment of low molecular weight (LMW) subunits of glutenin and aggregating gliadins results in some loss of α-helix and β-sheet structure. The losses were greater for the reduced LMW glutenins, an indication that the disulfide bonds are important for the stability of the structure. Carbon-13 NMR spectra recorded before and after retaining of hydrated gluten to 80°C appeared to be unaffected by the heating—peak position and intensities were the same [20]. Pulse NMR showed that the mobility of the protein molecules increased with increasing temperature, indicating that the molecules have a more open structure at higher temperatures [20]. When the samples were cooled, the mobility decreased to the same values as before the heating.

The thermal effects on the gluten gel were also discussed in Chapter 2, with the conclusion from electron microscopic studies that the thermally induced change from gel to coagel is mainly the result of an increase in disulfide bridging between the globular units, keeping the gross structure unchanged. These successive cross-linking reactions are consistent with the results reported above.

b. Changes in Functional Properties

The heating of gluten causes dramatic changes in its physiochemical and functional properties. Baking experiments with reconstituted flours have shown that heating gluten prior to test baking deprives it of its functionality. Heating the gluten to only 55°C has a detrimental effect; the volume decreases by about 10% [21]. Heating to 75°C or higher caused a total loss of functionality.

Heating gluten changes its rheological properties [22]. These changes are obvious from even a visual inspection. A piece of a gluten gel kept for a few minutes at 90°C after cooling will behave like a piece of rubber, with complete loss of cohesiveness. The viscous deformation has been found to decrease with increasing temperature [23], and when gluten is treated in boiling water, its compressibility is found to decrease with increasing time of treatment [24]. By using the statistical theory of rubber elasticity, the molecular weight between cross-links was calculated together with the number of cross-links (23). The molecular weight between cross-links was found to decrease, and consequently the number of cross-links increased. The changes were evident for preheating temperatures above 50°C. The changes in rheological behavior show that irreversible changes take place during heating. If this was not the case, a decrease in viscosity with increasing temperature would be expected, as for any other liquid. These changes are also consistent with the structure shown in Chapter 2.

The surface activity of gluten proteins changes with heat treatment (Fig. 6). Gluten samples heated to 70°C or above spread much more slowly than an untreated control. Also the π value after a long spreading time decreased compared with the control. Another observation concerns the π–A isotherms.

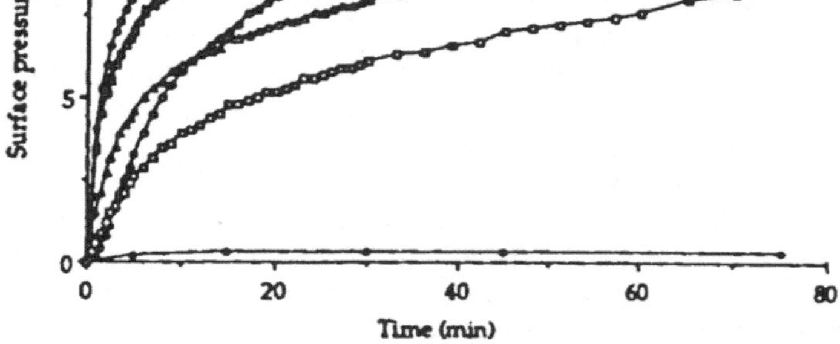

Fig. 6. Spreading behavior of gluten at the air/water interface after heat treatment. The gluten samples were heated to the following temperatures (from top to bottom): 55, 60, 65°C, control without heat treatment, 70, 75, and 100°C. (Reprinted with permission from [25].)

For the control gluten and the mildly heat-treated glutens, the compression cycle showed that a loss of material occurred during compression; that is, some material was squeezed out from the protein film. For the more severely heat-treated gluten, no such loss of material occurred [25].

The solubility of heat-treated gluten has been checked in various solvents. In 0.5 M NaCl the solubility was found to decrease with heating time in a boiling water bath [24]. The remaining solubility after 5 min of boiling was found to be related to the wheat variety. The extractability of gluten proteins in sodium dodecyl sulfate (SDS)–Tris–HCl buffer decreased with increasing temperature [21]. However, a higher temperature was needed to affect the extractability than to affect the loaf volume and crumb texture. The solubility of glutenin in 1.5% SDS was found to decrease after heating, whereas the solubility of gliadin in water was less affected [20]. The solubility of ω-gliadins is not changed at all by heat, whereas the solubility of α-, β-, and γ-gliadins decreases [21].

All the data presented above demonstrate that polymerization occurs during heat treatment of gluten proteins. Presumably this could be achieved by the formation of new disulfide bonds at the expense of thiol groups. However, the total number of SH groups does not change during heat treatment of gluten [21]. Most of the SH groups were found in the SDS-insoluble residue, and it was suggested that the mechanism involves SH/SS interchange reactions among the glutenin proteins. The gliadins are affected only at temperatures above 75°C.

As the heat-induced changes result in an increase in molecular weight of the storage proteins, they should perhaps at first consideration be expected to improve the baking performance, as the molecular weight of the glutenin could be one factor that determines wheat flour quality (see Chapter 2, Section III and Reference 26). In our model of the gluten gel (Fig. 31 in Chapter 2), however, all cross-linking should be expected to take place in the aqueous zone. If we start with such a gel, with an aqueous zone that is not already partly blocked with respect to S—S linkages, we will obtain increased mechanical strength of the gel owing to the formation of S—S linkages in situ during dough mixing. If instead we start with a gluten gel in which almost all possibilities to form S—S linkages have been blocked, as after heating, this gluten will tend to behave like an inert filler.

3. Starch

Gelatinization of starch has been described in detail in Chapter 2. Oven temperature eventually reaches the onset temperature of starch gelatinization (T_o). The temperature range of starch gelatinization is shown in the temperature profile of the center of the crumb during baking in Fig. 7.

a. Gelatinization Temperature Range

The gelatinization process in the dough is affected by the water content and by other ingredients present in the dough. T_o is not affected by the water content, but the conclusion temperature T_c is. This means that the gelatinization tempera-

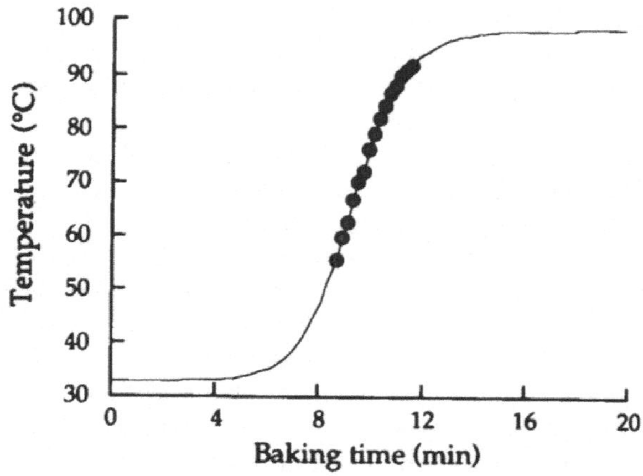

Fig. 7. The temperature interval for the gelatinization of starch (water starch ratio 1 : 1) is indicated by the bullets on the temperature profile of the crumb. (See Fig. 1.)

ture range is greatly affected by the water content. Other ingredients present—for example, sucrose and salt—cause an increase in T_o. The net effect may very well be that the upper part of the gelatinization temperature range is moved into temperatures above 100°C. As the temperature in the crumb never exceeds 100°C, there may be ungelatinized starch granules present in the crumb. In white bread the degree of gelatinization is around 100%, whereas in a pie crust it is about 50% [27]. In sugar cookies it is even lower, 20% or below. The stages of granular dispersion in baked goods, then, can differ considerably. In bread, all types of granules are present—gelatinized, disrupted, dispersed, and enzymatically degraded—whereas in biscuits the granules are swollen, gelatinized, and disrupted [28].

Gluten also affects the gelatinization temperature of starch. An addition of 0.1 g gluten/g starch increases T_o by 7.5°C [29]. As discussed in the previous subsection, gluten proteins do not undergo thermal transitions that are detected by DSC. The DSC thermogram of a wheat flour–water mixture essentially shows the gelatinization of starch, albeit modified by the other flour components. The DSC thermograms of starch and of the flour from which it is prepared are quite similar, but the gelatinization temperature range is shifted toward higher temperatures in the flour compared to the starch [30].

b. Relation Between Changes in Rheological Properties and the End of Oven Spring

The gelatinization of starch puts an end to the oven spring, but at the same time it inhibits the collapse of the crumb structure. Both these effects are explained by the rheological changes following gelatinization and the change from a foam structure to a pore system discussed later in this chapter. During gelatinization the starch granules absorb water and swell to a degree that depends on the availability of water (as the "free" water in insufficient) [31]. There is clearly a redistribution of water from gluten to starch during baking [29, 32]. Such a redistribution has implications for the staling of bread (see Section IV.B).

The gelatinization of starch causes an enormous change in the rheological properties of the dough. At first, when the temperature of the dough starts to increase, the viscosity decreases [8], but at about 60°C the viscosity begins to increase. Dynamic rheological measurements on flour–water doughs show that the storage modulus G' increases and the phase angle δ decreases when the dough is heated to above 55°C [33]. These changes continue to about 75°C; thereafter, the values are more or less constant. The effects on G' and δ are proportional to the starch content in the dough—the more starch, the higher G' is and the lower δ is. The decrease in δ with increasing starch levels shows that starch contributes to a continuous structure in the final bread [33].

The increase in viscosity has two effects. There is an increase in tensile stress at the air/water interface, which finally causes the rupture of the weakest foam cell walls. The increased viscosity also helps to "set" the structure. If the bread is taken out of the oven just at the point where the volume expansion stops, then the loaf collapses [1]. This is a direct consequence of the foam structure not having changed to a pore network as discussed in Section II.A. Unless there is an open system, the individual gas cells of the foam will contract upon cooling, and with that, the entire loaf collapses.

The gelatinization onset temperature of wheat flour measured by DSC is in the range 51–57°C, and the temperature at the maximum in the gelatinization endotherm (T_m) is in the range 60–66°C [30]. Most changes in rheological properties occur at higher temperatures, starting at about 60°C [8]. The leaching of amylose is very low at temperatures below 60°C when the starch concentration is low [34]. Also there is little swelling below 60°C. Increases in amylose leaching and starch gel volume are quite small at temperatures of 60–80°C, but considerable increases in both parameters occur at temperatures above 80°C [34].

The gelatinization of starch is the most obvious prerequisite for the formation of bread crumb. Experiments have even been performed to show that starch is not just an inert filler that dilutes the gluten to a suitable consistency [35]. Breads were baked in which glass beads of the same size distribution as wheat starch were used in place of the starch. The result was a bread that collapsed completely. Neither is it possible to use just any starch for baking. Waxy maize starch, for example, results in a dough that ferments properly, but the loaf's crumb structure collapes after baking [35]. If rice starch is used, differences in volume are observed during proofing [35]. The best result is always obtained with wheat starch, but rye and barley starches perform almost as well [36].

The temperature ranges for the various changes that occur in starch during heating were compared in Chapter 2, Section IV (Fig. 52). Most of the data available were obtained for systems of high water/starch ratios compared to the situation in the dough. During heating the large wheat starch granules undergo certain characteristic morphological changes that are also typical of rye and barley starches [31, 37, 38]. When the swelling of starch granules is studied by scanning electron microscopy (SEM), it is found that for a 10% suspension the typical heat-induced changes observed for wheat starch occur at a higher temperature than for a 1% suspension [31]. The appearance of the starch granules or the stage of swelling could be used to estimate the degree of heat treamtnet [39, 40]. The wheat starch granules in the crumb of the baked bread can be recognized as individual granules [39]. They are oriented parallel with the protein sheets in the dough, and this orientation is even more evident in the bread. The gelatinization of the granules may render them more flexible [35]. The starch granules expand during heating, but the planes of the granules are still parallel to the surrounding protein sheets [41].

It could, then, be expected that the temperature ranges are shifted toward higher values in the dough. T_o and T_m, measured by DSC, could give some information concerning the differences between wheat varieties, as the primary gelatinization process has to occur before the secondary processes such as amylose leaching and gel swelling can take place.

c. Influence of Heat on the Starch Granule Surface

Heating starch changes its surface properties, which may influence the interactions between starch and other components. It has been reported that heating starch makes it more hydrophobic [42]. When gluten proteins were allowed to absorb on heat-treated wheat starch granules, an increase in the amount of protein adsorbed was measured [43]. The effect was greatest when the temperature was increased from 50°C to 60°C, when the amount adsorbed increased from about 6 mg protein/g starch to about 12 mg protein/g starch. When the protein was heated prior to adsorption, no significant change in the amount adsorbed was measured. An increase in the interaction between starch and protein could certainly contribute to the stopping of oven spring. More protein would be required for enveloping the starch granules, and less would remain for stabilizing the increasing air/water interface.

d. The Amylose–Lipid Complex During Baking

The temperature in the crumb does not exceed 100°C, and in an ordinary wheat flour dough heating to just below 100°C causes complete gelatinization of the starch (see Fig. 7). At the water/starch ratio corresponding to the conditions in the dough, the thermal transition due to the amylose–lipid complex is not detected until a temperature above 100°. The temperature at the maximum of the endotherm (T_{cx}) has been measured to be 112.5°C in a wheat flour–water mixture [30]. This is somewhat lower than the value obtained for starch prepared from this flour (114.9°C). When the enthalpy values (ΔH_{cx}) were compared, a higher value was obtained for the flour, 1.84 compared to 1.51 J/g dry matter for the starch [30]. The differences between starch and flour can probably be attributed to the formation of new complexes during heating.

There are possibilities for the formation of new complexes, as the nonstarch lipids of wheat consist of a minor amount of monoacyl lipids (see Chapter 2, Section II), and there may also be a possibility of complex reactions of some diacyl lipids, as shown by the effect of lecithin [44]. Such a complex could be expected to form during gelatinization of the starch. If emulsifiers are added to the dough they may also form complexes with the starch during the gelatinization period in the oven. In fact, when monoglycerides are added to the dough, a larger endotherm is measured in the bread with monoglycerides than in the control without added monoglycerides [45]. The distribution of some crumb softeners [monoglycerides, sodium stearoyl-2-lactylate (SSL), and succinylated monoglycerides] between starch and protein has been measured by extraction of the emulsifiers during various stages of breadmaking [46]. The results need to be reinterpreted, however, in view of the two-phase aqueous separation between the gluten phase and starch in the dough and the loss of gluten gel during baking.

e. The Influence of Added Ingredients on the Gelatinization of Starch

As already described, the gelatinization temperature measured for a flour is somewhat higher than the corresponding temperatures measured for the starch isolated from the flour [30]. The onset of gelatinization could be regarded as the beginning of the loss of gas retention. It could then be argued that if the starch gelatinization is delayed, then the period for volume expansion in the oven will be prolonged. In fact, there are indications that there is such a relation between gelatinization temperature and the baking performance of a flour—wheat varieties of good baking performances have higher T_o and T_m values [30, 47].

The addition of substances that delay the gelatinization of starch could be expected to improve baking performance. Emulsifiers are examples of such substances, but their effects on the temperature of gelatinization are quite small. The addition of 5% (calculated on the starch weight) of SSL caused an increase in T_o of about 2°C [30]. Another emulsifier, lecithin, caused no change in T_o at all. The effects of added emulsifiers on the gelatinization temperature seem to be negligible. Nevertheless, emulsifiers and fats prolong the period of time for volume expansion at least as measured in a resistance oven [2, 48]. For a dough without shortening this period was 5 min, which increased to 6½ min when 3% shortening was added. The addition of 0.5% SSL had a similar effect. In another experiment the set temperature was increased from 55°C in the control to 80°C when 3% shortening was added [48]. It was also shown that a delipidized flour had a set temperature of 77°C. If the total extracted lipids were added back, the set temperature was 55°C; if the nonpolar lipids were added back, the set temperature was 74°C; and if the polar lipids were added back, the set temperature was 55°C.

The effects of added emulsifiers or fats are probably not related to their influence on the primary process, that is, the gelatinization process as measured by DSC or the X-ray diffraction technique. The additives should instead be expected to affect the secondary processes, amylose leaching and granule swelling [31]. Then a delay in the increase in viscosity of the dough should be expected.

The addition of NaCl and sucrose was found to delay the increase in G' when the dynamical rheological properties were measured during heating of flour–water doughs [33]. The decrease in δ started at a lower temperature when NaCl and sucrose were added, but the initial value was somewhat higher than in the control. The net effect was that δ was the same above 55°C in both types of doughs, but G' was lower for the dough containing sucrose and NaCl. When glycerol monostearate (GMS) was added to a 10% wheat starch suspension, a minimum in δ occurred at about 70–75°C compared to about 65°C in the control without GMS. The values of δ were higher in the presence of emulsifiers whereas G' was lower [34]. These results indicate that the addition of emulsifiers changes the rheological properties of the gelatinizing starch in a way that is in accordance with the hypothesis that added emulsifiers improve the baking performance by delaying the setting of the starch gel.

C. Gas-Holding Capacity

In the discussion of dough structure in Chapter 6, the dough at the end of fermentation was modeled as a foam. The gas-holding capacity of the dough is one of the most central functions in breadmaking and is related to the surface structure of the gas cells. This capacity of the dough will also manifest itself during the baking in the fast volume expansion that takes place during heating—the oven spring. The most striking feature of this expansion of gas volume due mainly to evaporation is its irreversibility. This is explained by the structural changes described below.

An extensive analysis of the role of different components in the gas expansion that takes place in the oven was recently reported by Bloksma [49]. The relative volume increases by a factor of about 5. According to Bloksma, the evaporation of water is the dominant factor behind this expansion. The carbon dioxide and ethanol will also evaporate. Bloksma assumed that the partial pressure of the carbon dioxide is in equilibrium with its concentration in the liquid dough and that all ethanol evaporates during baking. As the gas retention is lost soon after starch gelatinization, however, it can be expected that a major part of the ethanol will not contribute to the oven spring as it will evaporate in the neighborhood of its boiling point. In Bloksma's analysis, however, it was assumed that the partial pressure of water equals the saturated vapor pressure.

Moore and Hoseney [9] also estimated the contributions of various components to the gas expansion. They assumed that water evaporation can be neglected. It seems reasonable to expect some evaporation of water from the gluten gel zone adjacent to the gas cells. This aqueous zone (see Chapter 6) may even transport water from the "free" water phase before the starch gelatinizes. The interaction between starch and other components with the "free" water phase, as well as between storage proteins and water in the gas cell surface zone, on the other hand, makes it unlikely that the water vapor pressure can be equal to the saturated vapor pressure. An intermediate value between that of Bloksma and that of Moore and Hoseney is therefore probable, particularly as their total estimates in relation to experimental values gave lager and smaller values, respectively.

The volume change during heating has been well demonstrated with the resistance oven technique [50]. The fast expansion must be accounted for in the surface structure of the gas cells, in order to keep the gas-holding properties. Reconstitution studies by MacRitchie [51] demonstrated that the polar lipids play a critical role at the gas cell interface during this expansion.

We have discussed in connection with the gluten structure (Chapter 2) how heat successively changes the gluten gel to a coagel by polymerization. It is also clear from the monolayer work on gluten that this cross-linking takes place far more effectively at the gas/water interface. It should therefore be expected that the gluten at the interface will lose its stabilizing property early during the heating process. As polar lipids are present in the appropriate state (dispersed as lamellar aggregates) at the interfacial zone, they will first be able to "plug" the opening produced during oven spring and then gradually take over the interface, as further explained below.

It is important in this context to remember that no molecularly mixed films are formed (see Chapter 2, Section III). The proteins and lipids therefore form separate phases in the monolayer. The protein film consisting of the high molecular weight (HMW) fraction will develop reduced extensibility owing to the thermal cross-linking. Therefore it should be expected that a lipid monolayer will spread from the lipid aggregates over the interface during the gas cell expansion, and ultimately there will be a continuous monolayer phase with protein monolayer units dispersed within it. There will probably be many direct contacts between bulk lipids (dispersed as L_α aggregates) and monolayers spread from these aggregates. The film pressure should therefore be expected to approach the equilibrium spreading pressure (ESP). As polar lipids have ESP values of about 40 mN/m, the proteins will gradually be squeezed out from the film.

The localization of the lipids assembled in the gluten gel [52] means that the transition to the coagel will set these lipids free, some of them just at the interface. In the gluten gel they are entrapped and cannot express their surface properties unless an interface happens to be opened up at their position. This is the most probable explanation for the fact that the lipids do not play any surface-stabilizing role until the heating in the oven.

Finally it should be pointed out that a second function of polar wheat lipids is to emulsify the nonpolar oil. Without the presence of polar lipids "passivating" the oil surface, the proteins, which are needed to stabilize the gas cells, will be inactivated by interfacial anchoring to oil drops. The polar lipids have the advantage over proteins in the competition for this interface in their ability to reduce the interfacial energy to a greater extent.

The loss of expansion during heating after starch gelatinization is clearly related to the water transport from the gluten phase to the starch–water compartment. This volume change in combination with the transition from gel to coagel results in the rupture of the gluten gel surface zone that we saw in the dough. The gas cells will fuse into an open pore system during these changes; the detailed structure is described in the next section.

II. BREAD STRUCTURE

When the bread loaf is ready to be taken out of the oven, it consists of crust and crumb. There is, of course, no clear-cut division between crust and crumb, but it is still possible to divide the bread into these parts with good reproducibility [53, 54]. White bread has been divided even further, into crumb and inner and outer crusts. The amount of crust formed depends, of course, on the time in the oven (Table 1) and could amount to 59% or more of the total weight of a bread of "normal" shape (calculated on dry matter) [53].

Table 1 The Proportion of Crust and Crumb Formed with Different Baking Times at 210°C

Experimental conditions	Bread fractions (% of total fresh weight)		
	Crumb	Inner crust	Outer crust
210°C, 22 min	57.8	33.6	8.6
210°C, 35 min	53.0	37.5	9.5

Source: Data extracted from [54].

A. Crumb

We finished the last section by describing the gross structural changes during gelatinization and the transition from gel to coagel. These processes induce the crucial steps in the metamorphosis from dough to bread. The most striking changes from a macroscopic point of view are the opening of the foam type of gas cells to give pores and the solidification of the aqueous bulk medium. This solidification is due partly to the loss of cohesiveness when the gluten gel is transformed into a coagel and partly due to the gelatinization in a regime in which there is no excess water outside the starch granules.

The significance of curvature in porous structures was discussed in Chapter 1. There are two functional aspects of pore curvature besides the basic effect of pores on bread texture. One is the mechanism of heat transfer discussed above. The evaporation and condensation of water vapor within the pore system is directly related to curvature. A second aspect is the adsorption of the compounds responsible for crumb flavor that are formed during the heating process. The relation between adsorption and curvature is discussed in Andersson et al. [55]. When the bread has a high fat content, the oil formed at elevated temperatures will tend to cover the surface of the pore system. This process is driven by the reduction in surface energy.

We have tried to relate the crumb pore system to different types of periodic surfaces with constant average curvature. The arguments for this type of pore geometry were given in Chapter 1. Thus a symmetric pore wall with a liquidlike core and pressure equilibrium should be expected to give a pore geometry that, idealized, is equivalent to a minimal surface. One surface appears to approximate the disordered pore network of the crumb. This surface, called a *gyroid* (Fig. 8), represents an effective close-packing of helical pores in three directions in space.

We now have enough information to show how molecules and colloidal units are arranged in the pore wall of the crumb.

Figure 9 demonstrates schematically the structure of the crumb on different levels. The pore surface is considered to consist of a monomolecular lipid film with only a few patches of polymerized HMW storage protein units dispersed within it. The pore wall consists of only one continuous phase—the dried aqueous phase of partly gelatinized starch.

A complication in the formation of pore structure during oven spring is the temperature gradient from the center of the crumb outward toward the surface. Temperature equilibrium is not achieved until the entire crumb has reached about 100°C, when the bread is ready. Before an open pore network is completed, the temperature gradient will result in a pressure gradient that can influence the gas cell structure. Such a gradient in the gas cell expansion will cause the lattice to dilate in one dimension, from the center of a hypothetical cylindrical bread toward the surface, as shown in Fig. 10. If such an expanded foam lattice is

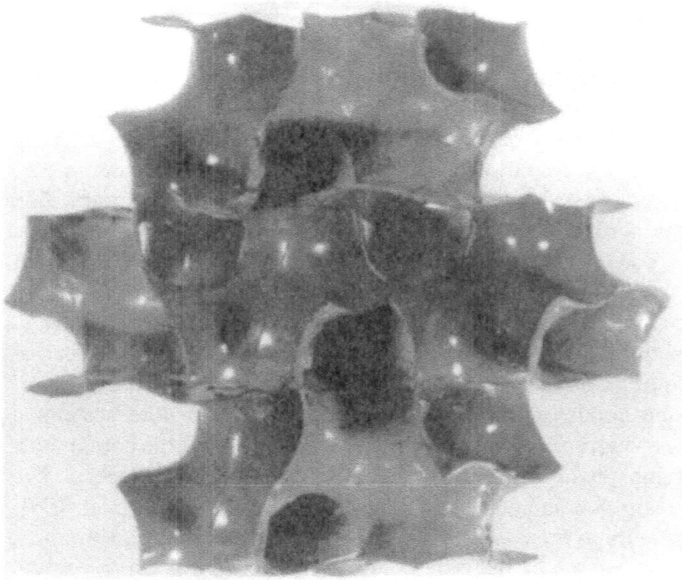

Fig. 8. A plastic model of the gyroid surface.

transformed into a corresponding pore system (see Chapter 1), the pore structure will show the same dilatation. Such a bread is easily fractured along the inside of the crust.

The heat gradient in the oven must be adapted to the strength of the gluten film forming the interface of the gas cells, in order to avoid the dilatation effect just described. An example of such a bread defect is shown in Fig. 11.

B. Crust

The main difference between the crumb and the crust is, of course, the difference in the temperatures they attain during baking (see Fig. 2). The high temperature in the crust causes the evaporation of water, so the water content of the crust is very low compared to that of the crumb. The dry matter content of the outer crust (Table 1) has been determined to be 89.9% after 22 min at 210°C and 93.7% after 35 min [53]. The corresponding values for the crumb were 56.7% after 22 min and 57.3% after 35 min. For another type of bread, the water content of the crust decreased from 18.03% after optimum baking time (24 min) to 15.50% after prolonged baking (27 min). After a shorter period of baking (21 min) the water content was 21.60% [6]. The corresponding values for the crumb were 44.03% water after optimum baking, 44.08% after 27 min, and 44.28% after 21 min. These figures illustrate that the surface of the bread goes through a drying

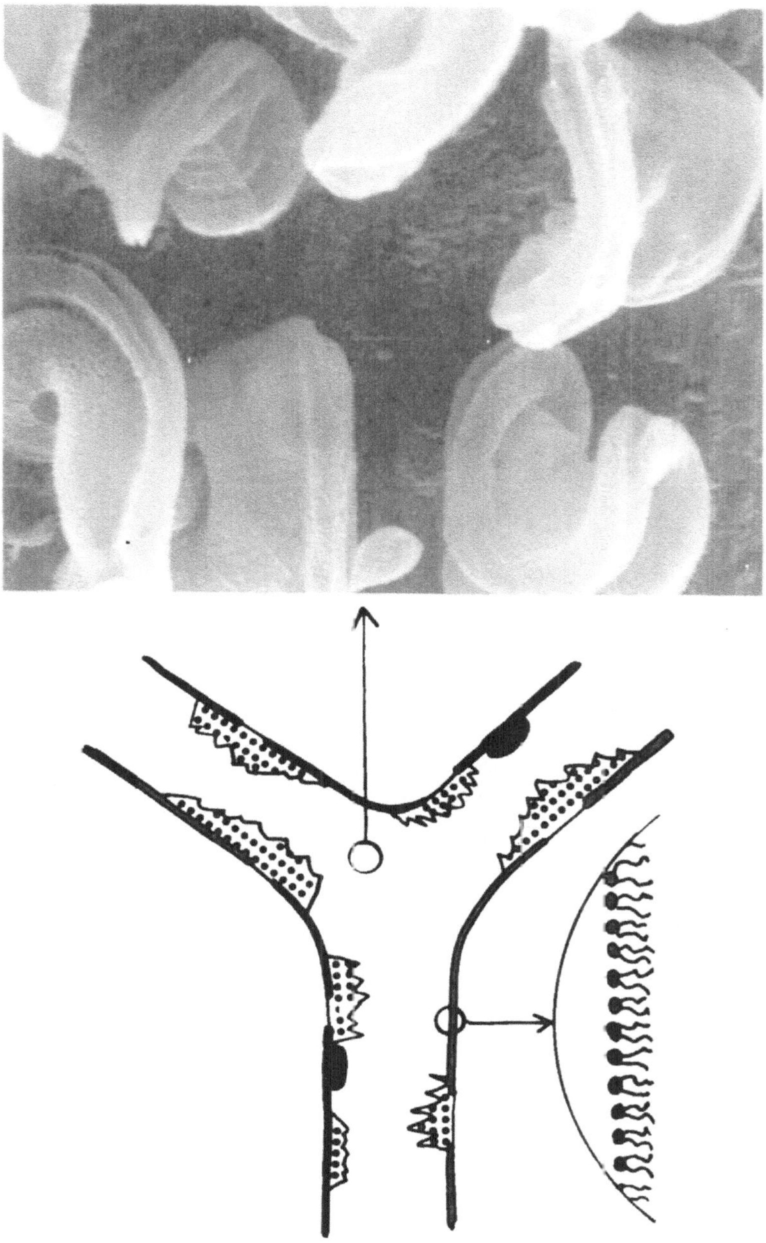

Fig. 9. The structural elements of the pore walls that constitute the bread crumb. The starch gel continuum is illustrated by a scanning electron micrograph.

Fig. 10. Dilatation toward the crust as seen in bread and in a model of foam cells, illustrated by filled circles.

process [4]. Starch granules with X-ray diffraction patterns of the A and V types are found in the crust [39].

The water content of the crust is affected by the baking absorption [6]. If less than an optimum amount of water is added, then the water content of the crust is less, and when more water is added the water content is greater. However, the variation is not large—17.78% for the bread with less water than optimum (the

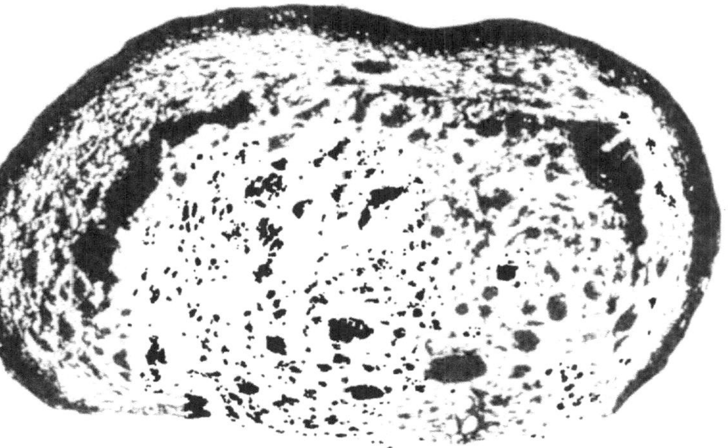

Fig. 11. A bread crumb that has a tendency to be fractured so that the crust separates from the crumb.

baking absorption –3%) and 18.47% for the bread with more water than optimum (baking absorption +3%) [6]. The water content of the crust is also affected by other ingredients; it will increase if salt, sugar, or shortening is excluded from the formula. Furthermore, it increases considerably if the level of damaged starch is increased [6].

The high temperature in the crust also causes different types of chemical reactions to occur, reactions that do not occur in the crumb. These reactions, which include the Maillard reaction as well as caramelization reactions, give the color to the crust [56]. The Maillard reaction requires the presence of a reducing sugar (e.g., glucose, fructose, maltose) and an amino group (e.g., lysine). As the lysine content is low in wheat proteins, the browning is improved by including milk in the recipe. Caramelization reactions involve the polymerization of sugars. The high temperature in the crust causes dextrinization of the starch, and the dextrins formed may then take part in other reactions. Starch will be consumed in various reactions, and this is observed as a lower starch content in the outer crust (74.4%) compared with the crumb (75.1%). The difference is even more pronounced when the baking time is increased to 35 min. Then the starch content in the outer crust is 72.8%, whereas it is 75.8% in the crumb [54]. It has been suggested that fragmentation and transglycosidation reactions of starch cause the observed decrease in the amount of starch analyzed, especially in the crust [54].

The baking of bread causes the formation of resistant starch [57, 58]. However, the formation of resistant starch (see Section IV of Chapter 2) does not contribute to the decrease in starch content, as it requires a higher water content. It is highest in the crumb (0.95–1.02%) and lowest in the outer crust (0.30%), and the amount in the outer crust is not affected by the baking time [54].

From a structural point of view, the crust is a hard, vitreous surface layer formed of collapsed crumb pore walls. It is a continuum of dried starch gel with dispersed protein and lipid aggregates.

The significance of the heating gradient was discussed earlier in relation to pore structure. It is also a factor controlling crust formation. If the crust is formed and "fixed" by drying too early, it will not be able to allow the expansion induced by the crumb. It will then break, and the bread will show cracks along the crust.

An additional function of the crust is to provide a barrier against the loss of water as well as flavor.

III. BAKING BEHAVIOR OF WHOLE WHEAT FLOUR AND NON-WHEAT CEREALS

If wheat has such a superior baking performance it does not seem very reasonable to use other cereals in breadmaking. However, there may be very good reasons for doing exactly this. The incorporation of other cereal flours gives the bread a new taste and may improve its keeping qualities. It is often claimed that rye bread has better keeping qualities than ordinary wheat bread. It may be convenient to use locally produced crops instead of imported wheat. The nutritional value of the bread could be increased if, for example, a lysine-rich flour such as oats or whole kernels from other cereals could be incorporated into the bread formula. Finally, many people suffer from coeliac disease and cannot tolerate wheat, rye, barley, or oats.

Consumer demand during the last two decades for bread with a high dietary fiber content has resulted in increased production of bread baked with whole wheat flour (whole meal). Gaillard and coworkers [59] reported extensive studies of the mechanisms involved in the poorer baking performance of whole meal compared to that of conventional "white" wheat flour. The decrease in performance was found to be caused mainly by hydrolysis of triglycerides owing to lipase activity in the bran fraction.

In this section we summarize some features of breadmaking when cereals other than wheat are used. Usually such breads are made from wheat flour mixed with flours from other cereals, but there are also similar kinds of bread that contain no wheat.

A. Bread Baked from Composite Flours

Wheat dominates breadmaking all over the world. In central Europe, eastern Europe, and the Scandinavian countries, however, there are also a wide variety of other types of bread on the market. Most of them are based on composite flours, often dominated by wheat, with rye being the second most common component. In Sweden, for example, a common household flour for baking is a mixture of the four cereals cultivated there—wheat, rye, oats, and barley. Rye is most common in various sourdough-based breads.

In general, the breadmaking process using a composite flour is the same as that described for wheat, including mixing, mechanisms of gas-holding capacity of the dough, oven spring, and structural aspects. If a standard formula is used for producing an ordinary type of white bread, it is possible to include an increasing proportion of oat flour with only minor effects on the bread up to an oats/wheat ratio of about 20:80 (w/w).

The differences in gas retention between doughs made from corn, rice, and rye flours compared to wheat flour have been studied by He and Hoseney [60]. During proofing the loss of carbon dioxide was greatest in the corn dough, followed by the rice and rye doughs. Gas loss during baking followed the same order. He and Hoseney also showed that diluting the wheat flour with starch to 7% protein still gave a larger loaf volume than rye flour.

In a research project aimed at substituting as much rice flour as possible for wheat flour (for countries that cannot produce wheat), we were able to substitute half of the wheat flour and still get an acceptable bread [5]. This work was also an application of lipid functionality in wheat flour dough. In other words, it demonstrates the significance of lipids in breadmaking, particularly the relevance of the structural state of the lipid–water phase (see Chapter 2). Without this lipid phase, the volume is reduced to an unacceptable level when there is more than about 15% rice flour in the composite flour. The addition of the lipid phase increased the gas-holding capacity, particularly during oven spring, and also gave the bread a uniform pore structure. A typical example of such a bread compared to a bread made with wheat flour is shown in Fig. 12.

The same principle of stabilizing the foam pore structure was applied in the special case of simulating the bread structure described.

B. Gluten-Free Bread

A considerable proportion of the world's population suffers from coeliac disease or gluten intolerance. Trace amounts of proteins from wheat, rye, oats, and barley in their food will induce a severe inflammatory conditon in the intestinal system of these individuals. The proportion of a population group suffering from this disease varies from an average of about 0.1% in the Western world up to about 0.5% in Ireland.

Different substitutes for these cereals have been tried, including cellulose derivatives used as the dough-forming agent in a starch–water mixture. We obtained a bread with good acceptance using mainly rice flour and lipids in the right liquid crystalline form as a gas cell stabilizer. The formula we used is given

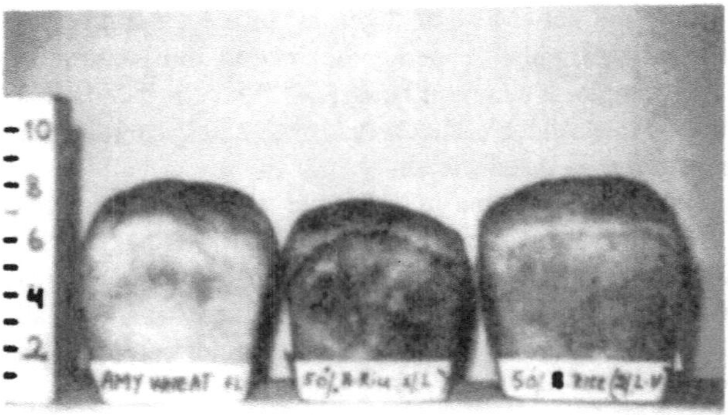

Fig. 12. Bread baked of a rice–wheat 1 : 1 composite flour compared to a wheat bread baked in the same way (to the left). A lamellar liquid crystalline phase of lecithin (2% calculated on the dough weight) was added to the rice–wheat bread (middle) in the form of a liposomal dispersion and (right) as vesicles (unilamellar).

in Table 2. The rice flour and caseinate were mixed dry. The rest of the ingredients were first mixed with water and then added to the rice–caseinate mixture. The lipid was first dispersed in water for a couple of hours (by shaking) to give the lamellar L_α phase. After 30 min of fermentation at 32°C and 60% relative humidity, the dough was put in baking tins and allowed to ferment another 30 min. The baking time was 20 min at 230°C.

C. Rye Bread

As mentioned earlier, rye is frequently used in breadmaking in central and eastern Europe and Scandinavia. It is often fermented with lactic acid bacteria,

Table 2 Formula Used for Gluten-Free Bread

Ingredient	Amount (g)
Rice flour	100
Caseinate	5
Sugar	2.5
Salt	1
Yeast	5
Lecithin	2
Water	65

Source: Eliasson et al. [61].

by adding a sourdough starter, as this gives an attractive aroma and also improves the texture. Yeast is also used, and often a combination of yeast and lactic acid bacteria is used in the sourdough starter. Various strains of lactic acid bacteria are used; some also produce acetic acid, and others produce propionic acid in addition to lactic acid. The acid content provides efficient antimicrobial protection.

Crispbread based on rye is baked from a rye flour dough that is mixed with more water than an ordinary dough for soft bread (usually with about equal amounts of rye flour and water). Yeast and salt are also added, and after mixing, the dough is allowed to ferment a couple of hours before being rolled out into a thin layer. After a brief second fermentation (about 20 min at 38°C) the dough goes into the oven. The temperature is high (about 350°C), and this gives the crispbread the brown color, special flavor, and crispiness. After cooling, the bread must be further dried, and it is kept in special drying rooms until the water content is below 7%, after which it is packed. This low water content makes it stable against microbial spoilage, and it can be stored for months without losing quality, provided that the humidity in the environment is "normal." The adsorption behavior is illustrated by the middle curve of Fig. 18 in Chapter 1. The water content of the crust corresponds to the start of the plateaulike region of the curve. Crispbread can also be baked with skim milk instead of water and it can be mixed into a very porous dough, which gives a grey-white bread.

D. Barley Bread

In northern countries where not even rye can be grown, barley is used in breadmaking. Barley can be grown even above the Arctic Circle. Such bread must be very thin, or it becomes too hard to chew. Like rye crispbread, it is a crisp bread with a similar water content and can also be stored without microbial damage. In the old days, each household in the Nordic countries used to bake their bread only once or twice a year. Milk was used in the recipe (almost as much milk by weight was used as barley flour). Yeast and salt were also added.

IV. STALING

When the loaf of bread is removed from the oven, a series of changes starts that eventually leads to deterioration of the quality. These changes are collectively termed "staling" and include all the processes that occur during storage except microbial spoilage. The consumer detects the staling by the changes in taste as well as in texture. The typical aroma of fresh bread is lost, and a "stale" flavor develops with time. The crumb becomes "dry" and hard, whereas the crust becomes soft and leathery. These changes are described in some detail in this section. The description is based on results obtained from firmness measurements and from thermal analysis (DSC and DTA). Some mechanisms suggested to explain staling are described, and finally some ways in which staling might be avoided, or at least delayed, are discussed.

A. How Staling Manifests Itself

Both crumb and crust are subject to deterioration over time, but the reasons are different and the deterioration is expressed in different ways. As already said, the crumb becomes hard whereas the crust becomes soft during staling. The staling of the crust seems to be the result of simple water uptake, and there are not very many reports about this process in the literature. The staling of the crumb, on the other hand, is not so easily explained and has resulted in a vast number of investigations, including several reviews [62–66].

1. Staling of the Crust

During storage of bread the water content of the crust increases as a result of water transport from the crumb to the crust [67]. During a storage period of 100 h the water content of the crust was found to increase from about 15% to 28%, but the decrease in water content of the crumb was not at all as pronounced, from about 45% to 43.5%. In a zone near the crust the decrease was much more pronounced, from about 45% to 32%.

The increase in the water content of the crust changes its rheological behavior. In fresh bread the crust is crisp and brittle, whereas in stale bread the crust is soft and leathery, or rubbery. The change from brittle to rubbery is the typical change in rheological behavior when the glass temperature (T_g) of a polymer material is passed [68, 69]. The change in the crust is certainly brought about without any change in temperature, but it is known that T_g of starch gels, for example, is very sensitive to the water content [70]. Water acts as a plasticizer and decreases the T_g of the material [71]. For the crust the situation could then be that in the fresh bread T_g is well above room temperature, whereas at a somewhat higher water content T_g is depressed to below room temperature and then the change in texture follows.

During oven baking many flavor compounds are formed, which collectively result in the typical bread aroma. The formation of flavor compounds occurs to a greater extent in the crust region than in the crumb. The higher temperature of the surface compared with the crumb allows nonenzymatic browning reactions to occur in the crust region, whereas the aroma of the crumb arises from the fermentation [65]. These flavor compounds are volatile, and with time they are lost. Even during the cooling phase, considerable bread aroma is lost. The proportions of the various flavor compounds change. With time the stale flavor emerges. Some of the compounds have evaporated whereas others may have complexed with starch [72], as indicated by the fact that some of the bread aroma is regained by reheating.

2. Staling of the Crumb

Whereas the staling of the crust is related to a change in water content, staling of the crumb occurs without any change in water content. Nevertheless it causes a change in texture and a loss of aroma. Some of the changes that occur during staling can be reversed if the bread is reheated to temperatures of 50–70°C.

The use of a sensory panel is perhaps the most sensitive method for detecting the changes connected with staling [62, 66, 73, 74]. However, there are also several instrumental methods that have been used to follow the staling of bread. Among these methods are firmness measurements, X-ray diffraction measurements, thermal methods, and the determination of enzymatic susceptibility and the extractability of starch from bread [65].

a. Firmness

In crumb firmness measurements, a slice of bread is compressed and the force necessary to achieve the preselected compression is recorded. The measurements can also be performed in reverse, with the elasticity modulus or Young's modulus being calculated [75]. Rheological results obtained with different types of instruments might not be directly comparable. Deformation under constant load seems to give results that are well correlated to subjective estimates of staleness [76]. These measurements show that the crumb firmness increases over time [45, 73–75, 77–79]; an example is given in Fig. 13. It is evident that the changes are most pronounced during the first day, after which they occur at a much slower rate [64]. The results obtained depend on several factors; for example, the loaf volume affects these measurements—the lower the specific loaf volume, the higher the modulus, or the firmness, of the loaf will be [80]. It is therefore difficult to compare different types of bread by using this kind of measurement. The bread that is softest according to firmness measurements is not necessarily the freshest. The firmness value depends on the temperature, and at –20°C firming is arrested [76]. The maximum firming rate is at temperatures around –2°C, and the bread will be softer for a longer period of time if it is kept at a higher temperature [76, 80]. If stale bread is heated to temperatures of 60–100°C it will be refreshed and almost as soft as it was when it was fresh [76, 81]. However, the refreshed bread will become firm at a much faster rate. A bread that was stored for 5 days, reheated to 80°C, and then stored for 2 more days showed the same firmness value as a corresponding bread stored for 7 days [81].

b. Soluble Starch

If bread crumb is extracted in distilled water, a certain amount of water-solubles is extracted [82]. The amount extracted depends on the protein content of the flour, the storage temperature, and the age of the crumb. For one type of bread a decrease from 3.34% water-soluble starch in fresh bread to 1.22% at day 5 was measured [82]. The composition of the soluble starch also changes with time and

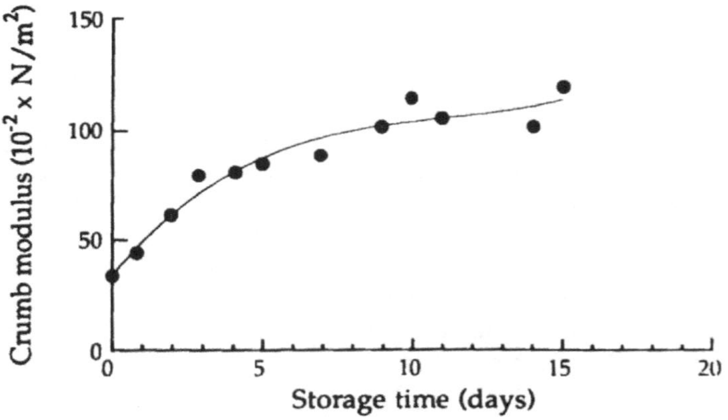

Fig. 13. Changes in the firmness of wheat bread stored at 21°C. (Adapted from [45].)

temperature. The proportion of amylose in the soluble starch decreased from 15.6% in the fresh bread to 8.2% at day 5, and consequently the proportion of amylopectin in the soluble starch increased [82]. A more detailed analysis showed that the amylose content in the soluble starch decreased from 23.9% after 0.1 h to 16.7% after 2 h at room temperature [82]. Further storage caused a decrease to around 12% after 5 h, but thereafter the proportion of amylose did not decrease with longer storage time. In another study the amount of extractable carbohydrates decreased from 0.47 g on day 2 to 0.41 g on day 4 [83]. Amylose was not affected after the first day of storage. The results were interpreted as being due to protein–starch interactions.

c. X-Ray Measurements

The X-ray scattering pattern of a fresh bread indicates an amorphous structure (except for the V-pattern discussed below), but with time an X-ray diffraction pattern of the B type (see Section IV of Chapter 2) emerges [84]. Similar results are obtained for starch pastes, indicating the involvement of crystallization of starch in the aging process [66, 84, 85]. The gelatinized starch in the bread also gives rise to an X-ray diffraction pattern of the V type [66] due to the amylose–lipid complex. The pattern does not change during the storage of the bread. The addition of emulsifiers will certainly affect the crystallization, and this will be discussed later in this chapter. The change in the relative crystallinity of starch during storage of bread has been estimated to be 0.32 at day 0, 0.51 at day 1, and 0.62 at day 7 [85]. When bread is reheated, the B-type X-ray diffraction pattern is lost, and the amorphous pattern (together with the pattern of V-amylose) corresponding to the fresh bread is obtained [66].

d. Thermal Analysis

Thermal analysis with DTA or DSC on fresh bread results only in the thermal transition of the amylose–lipid complex [45]. The gelatinization of starch that occurs during baking is, of course, irreversible. If the bread is stored prior to thermal analysis, however, a staling endotherm is recorded [45, 86–90]. This endotherm is due to the melting of recrystallized amylopectin, as illustrated in Fig. 14 [89]. The size of the endotherm increases with time, and the enthalpy calculated from the endotherm (ΔH_c) thus corresponds to the melting of recrystallized amylopectin. One advantage with the DSC technique is that the influence of loaf volume is avoided. When bread crumb is reheated, DSC analysis shows that the reheating has decreased the endotherm [81]. For a bread stored 5 days and then reheated to 80°C and stored for 2 more days, the ΔH_c value corresponds to the value obtained for a corresponding bread stored for only 2 days [81].

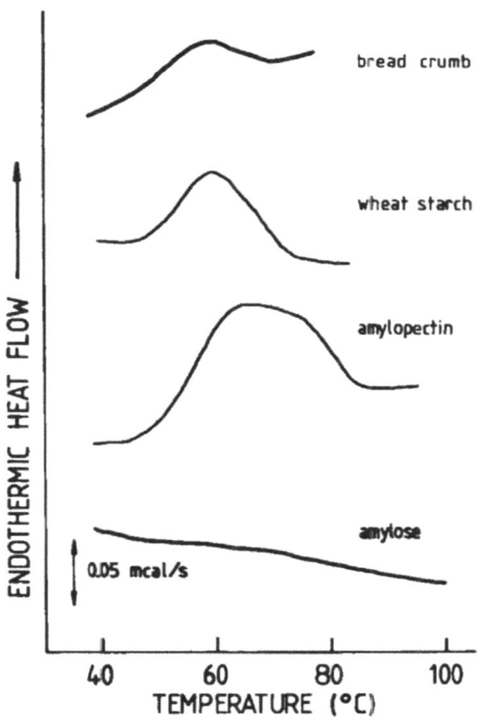

Fig. 14. The melting endotherm of crystallized amylopectin compared with those of bread crumb, starch gel, and amylose. (Reprinted with permission from [89].)

B. Causes of Staling

There are two different mechanisms that are commonly referred to as the causes of staling. These are the *retrogradation (recrystallization) of starch* and the *redistribution of water* between starch and gluten. Of these two mechanisms, the retrogradation is the process that is most easily followed by experiments, and we thus know a lot more about retrogradation than about water transport. A change in gluten gel firmness with time has also been observed [75], thus offering a further cause for the staling of bread. Interactions between the components could also influence the staling.

1. Retrogradation of Starch

The same methods have been used to follow the staling of bread and the retrogradation of starch. Often it is assumed that the staling of bread is a direct reflection of the retrogradation of starch. In this subsection the same type of results that were reported previously for bread crumbs are described for starch gels. The retrogradation of starch was also described in Chapter 2, and here we compare the results obtained for starch gels with the corresponding results for bread crumb.

a. X-Ray Diffraction Pattern

The A-pattern of the native wheat starch is lost during gelatinization, and instead the V-pattern is recorded [66]. If the starch gel is stored, the B-pattern emerges, and the intensity of this pattern increases with time [66, 91, 92]. The B-pattern also develops with time in bread crumb [66], supporting the hypothesis that starch retrogradation is involved in the staling of bread.

b. Thermal Analysis and Bread Firmness Measurements

When bread crumbs is heated in the calorimeter, an endotherm is obtained at temperatures around 60°C (Fig. 14). This endotherm increases in size with time in a way corresponding to the staling of bread. A constant value is obtained after about 7 days at 20°C for a starch gel of 50% (w/w) concentration [93]. When a wheat starch gel is reheated in the calorimeter pan, a melting endotherm is obtained in about the same temperature range as for the bread crumb [89].

A wheat starch gel increases in gel strength (expressed as Young's modulus) during storage, and if the gel is reheated the initial value of the modulus is obtained [75]. The increase in Young's modulus is highest during the first 24 h; thereafter it does not change very much for a starch gel at low concentration. When the starch concentration is 40–50%, the modulus continues to increase for more than 7 days at 20°C [93]. It may, of course, be asked whether the firmness measurements directly reflect the crystallization measured by DSC or X-ray diffraction or if other changes are involved.

Both compressibility measurements and DSC measurements have been analyzed according to the Avrami equation [45, 86, 94]

$$\Phi = (A_L - A_t)/(A_L - A_0) = e^{-kt^n} \qquad (1)$$

where Φ is the fraction of the total change in the measured property (e.g., elastic modulus measured in compressibility measurements or ΔH_c measured by DSC) still to occur, k is a rate constant, and n is the Avrami exponent. A_L, A_t, and A_0 are the experimental values at infinite time (A_L), time t (A_t), and time zero (A_0), respectively. The Avrami exponent, n, should ideally be an integer; it denotes the type of nucleation and crystal growth [86]. In analysis of the staling of bread and starch gels, $n = 1$ is often obtained, denoting rodlike growth from instantaneous nuclei. However, values deviating considerably from $n = 1$ have frequently also been measured [45, 87, 92, 95, 96]. When DSC data (ΔH_c) for four different starch gels were fitted to the Avrami equation, a better fit was obtained with $n = 0.62$ than with $n = 1.0$ [92]. The rate constants calculated from DSC measurements and compressibility measurements are quite similar [45]. However, there are so many uncertainties in using the Avrami model (e.g., determination of limiting values and the meaning of n in a heterogeneous system such as bread) that this model is now used mostly only for comparing data and not for interpretations [92, 95].

A similarity in the rate constants for firmness measurements and DSC data is seen at low and ambient storage temperatures [82, 86, 97]. At higher temperatures, however, the staling rate was found to be higher for bread than for a starch gel [82]. It was also found that the firmness values level off after a shorter period of time in the case of a starch gel than for bread [75]. The difference is attributed to the aging effects of gluten.

c. Amylose and Amylopectin

The different roles of amylose and amylopectin in the retrogradation of starch were discussed in Chapter 2, Section IV. A bread made with waxy barley starch is initially softer than a similar bread made with wheat starch or barley starch [81]. After 3–5 days of storage, all three will show similar firmness values. This result indicates that amylose contributes to the formation of the crumb structure but does not contribute to long-term staling.

d. Temperature

The storage temperature of bread or of starch gels influences crystallization and firmness. The type of X-ray diffraction pattern is the same regardless of storage temperature [98]. The development of the staling endotherm and the transition temperature of this endotherm depend on storage temperature. The endotherm develops much more quickly for starch gels stored at temperatures of −1°C to 10°C than at temperatures above or below this range [87, 89, 98]. Similar results are obtained for cooked wheat grains [97] and for bread [98]. The maximum of the melting endotherm (T_c) increases with increasing storage temperature. For bread stored 3 days at 4°C, T_c was 50°C; for bread stored at 25°C it was 54.5°C; and for bread stored at 40°C it was 64°C [98]. The change in T_c indicates that more perfect crystals are formed at the higher storage temperatures, as a narrower and better defined staling endotherm is obtained after storage at higher temperatures [98].

e. Water Content

The crystallization of starch as measured by the X-ray diffraction technique or DSC depends on the water content of the starch gel [96, 99–101]. The optimum water content for retrogradation is close to the water content in the bread, that is, 40–50% (see Fig. 15). The water content of importance to retrogradation is the water content during retrogradation, not the water content during gelatinization [101].

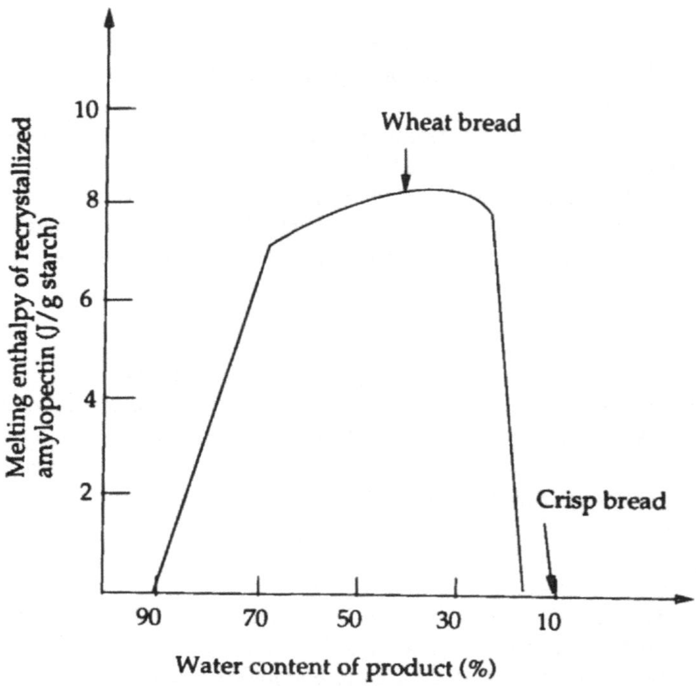

Fig. 15. The influence of water content on the retrogradation of amylopectin. The water contents of two types of bread are indicated. (The enthalpy curve is adapted from [100].)

Firmness measurements do not show such a trend; the more concentrated gel shows the higher elasticity modulus [79]. Dynamic rheological measurements show that the storage modulus G' increases with increasing starch concentration, at least at concentrations of 5–40% [102]. Moreover, an equilibrium value is reached after a shorter storage period at the lower concentrations.

Because the development of crystallinity measured by DSC and the development of firmness show different dependencies on the water content comparison of staling of different breads might give results in which the bread with the highest firmness value has the lowest amount of recrystallized starch measured by DSC [79].

2. Changes in Gluten

The protein content of bread or starch-based gels is of importance for the changes during storage. A higher protein content often results in a lower staling rate [94]. There are several reasons for this. In firmness measurements a higher protein content can result in a higher loaf volume, and thus a softer crumb [80]. The presence of different amounts of protein can also "isolate" the starch granules to varying degrees, and in this way the retrogradation of starch is reduced. The protein quality seems to be unimportant in staling [94].

The melting endotherm of recrystallized amylopectin is affected by the addition of gluten [96]. This could be due to "isolation" effects as well as to influences on the water distribution in the starch gel.

The gluten gel itself also undergoes changes during storage. If a gluten gel is stored it will increase in firmness with time; G' increases whereas G'' (the loss modulus) decreases [75, 103]. However, there is one important difference between the firming of a starch gel or bread and the firming of a gluten gel. If bread crumb or a starch gel is reheated to temperatures of 50–70°C the firmness of the gel decreases, and its rheological properties become similar to those of the fresh gel. If a gluten gel is reheated in the same way, no such change in firmness occurs [75]. The aging rate of the gluten gel increases with temperature, an effect that is especially evident for short storage periods and at temperatures above 10°C [82, 103]. The modulus of a heated gluten gel is strongly related to the water content; at 10% water it was reported to be 3×10^3 MPa; at 30% water, 10 MPa; and at 60% water, 0.03 MPa [22]. It is suggested that the gluten gel passes through a glass transition when the water content is changed from about 20% to about 40%.

3. Redistribution of Water

It has been known for a long time that the changes that occur during the staling of bread are not dependent on the loss of water [65]. The staling can, however, be increased if water losses also occur during storage [62]. Increased baking absorption can increase the softness and decrease the firming rate.

Although no loss of water occurs during staling, water may be redistributed during storage. The redistribution of water between crumb and crust has already been discussed. A redistribution of water between components (gluten and starch) has often been discussed, with some studies reporting a transport from starch to gluten or from gluten to starch and others reporting no transport of water at all [104].

In some measurements, greater water-holding capacity has been observed in stale bread crumb and retrograded starch [105]. This would, of course, indicate water transport from gluten. A water transport from gluten to starch during baking is suggested; this transport would continue during storage [76]. The effect of gluten on the gelatinization temperature of starch has been interpreted as due to water transport from the gluten to the starch [29]. In other investigations a decrease in the water-holding capacity of starch was observed, whereas the water-holding capacity of gluten was unchanged [106].

During storage of a starch gel in close contact with a gluten gel, the water content of the starch gel increases and the water content of the gluten gel decreases [107]. It was noted that an increase in water absorption above the optimum, determined by the farinograph, would not result in a higher water content of the gluten [76]. The optimum water absorption thus corresponds to the saturation of gluten, that is, to its maximum swelling. The total amount of water that takes part in this transport is small. It was estimated that only 3% of the total water is redistributed [107]. In these experiments it was further noted that an antifirming agent works by decreasing the rate of water redistribution. In another investigation it was concluded that emulsifiers do not influence the staling by their effect on the availability of water [101].

C. How to Avoid or Reduce Staling

Earlier, several ways to decrease retrogradation were described. If a relationship exists between starch retrogradation and staling, the same methods should also be effective in decreasing staling. In the previous sections in this chapter it has become evident that the keeping qualities of bread can be improved by choosing the correct storage temperature. Room temperature or freezing temperature is preferred over the temperature in the refrigerator. Added ingredients can influence the keeping qualities, as well as the native flour components. Some of the components that are thought to affect staling are listed in Table 3. In this section the influences of enzymes, emulsifiers, and fats are discussed.

1. Enzymes

The addition of α-amylases to the dough decreases the staling rate of bread [62, 65]. Bacterial α-amylase affects the firmness more than fungal α-amylase or malted wheat flour. The use of enzymes to retard staling is somewhat risky; if the

Table 3 Some Components That Are Thought to Affect Staling of Bread

Component	Effect	Reference
Amylopectin	Increase	81
	Decrease	63, 108
Amylose	Decrease	108
	Increase	92, 110
Monoacyl lipids	Decrease	45, 109
Triglycerides	Decrease	63, 112, 113
Sugar	Decrease	63
Amylases	Decrease	90, 111, 118
Protein	Decrease	63, 94
Pentosans	Decrease	114
	No effect	115
Salt, 0.4–1.0%	Increase	116
Salt, >2%	Decrease	116
Water	Decrease	79, 112
	Increase	Fig. 15.

dosage is too high the bread crumb becomes sticky because of the formation of dextrins, and unacceptable [117]. As much as 20% of the initial enzyme activity can still remain after baking [62]. The influence of the added enzymes is related to their heat stability, which increases in the order fungal amylases < cereal (malt) amylases < bacterial amylases [62]. Bacterial amylases are also more effective when the degree of dextrinization is the same [118].

The presence of extra enzyme activity has also been observed as a decrease in the staling endotherm measured by DSC [90]. When bread was baked from sound flour and from malted flour, the staling endotherm developed much more slowly in the bread made with malted flour. The mechanism of the antistaling function of α-amylases is thought to be that starch chains that connect different crystalline domains are broken [66]. The crystalline network will then be of a smaller dimension.

The X-ray diffraction pattern increases in intensity after supplementation with bacterial α-amylase [85]. α-Amylase-supplemented bread was not as firm as the control bread. This was especially true when bacterial α-amylase was used. The relative crystallinity, on the other hand, was greater, and this was especially true in the bread supplemented with bacterial α-amylase. The crystallization of starch thus, did not follow the change in firmness. An interesting observation was that in the bread supplemented with bacterial α-amylase, the A-pattern was obtained during staling, whereas in the other supplemented breads the B-pattern was also obtained.

It was concluded that the different unit cells could influence the water distribution in the bread differently and in this way account for the discrepancies between X-ray diffraction data and firmness results [85]. The unit cell of the A-pattern contains eight water molecules, whereas the unit cell of the B-pattern holds 36 water molecules. If only the B-crystallites are formed, more water is transformed from gluten to starch than if a mixture of A- and B-crystallites is formed, and even more if only the A-crystallites are formed. The increased water transport from gluten to starch can then result in a firmer bread crumb even though the relative crystallinity is lower.

2. Fats and Shortenings

Fats and shortenings decrease the development of firmness during staling of bread [79, 81, 101]. An addition of 3% fat or shortening has an effect, but if the level is increased to 6 or 12% no further softening effect is obtained [79]. However, the addition of 0.5% monoglycerides decreases the staling further. It might then be concluded that fats and polar lipids affect the staling by different mechanisms.

It has been observed that bread baked from delipidized flour shows a slower increase in firmness than bread made with native flour [79]. The addition of shortening to the delipidized flour has no effect.

The staling endotherm, measured by DSC, is decreased by shortenings [101]. However, the results are not unambiguous; another shortening had no effect at all. The retrogradation of waxy maize starch is not affected by the addition of 1.4 g soybean oil/100 g starch [119]. Variations in the results might be due to different free fatty acid content (a contaminant in oils that can complex amylose/amylopectin).

3. Emulsifiers

When certain lipids (emulsifiers, surfactants) are added to the dough, bread staling decreases. Among the substances that are effective in this respect are monoglycerides, sodium stearoyl-2-lactylate (SSL), and diacetyl tartaric acid ester of monoglycerides (DATEM) [45, 109]. Not all emulsifiers show the same efficiency; for example, monoglycerides are much more effective than DATEM [109]. For the effect of an added lipid on staling, the physical state of the lipid is of utmost importance. It has been shown that in the case of saturated monoglycerides the α-crystalline gel phase is much more effective than, for example, a β-crystalline hydrate [120]. If monoglycerides are added in such a way as to avoid the influence of the different phase behavior (i.e., if they are added as sodium cholate micelles), there are still differences in effects depending on chain length and degree of unsaturation. Monopalmitate was found to have the greatest improving effect on loaf volume and also on the softness of the crumb [121].

If the native wheat flour lipids are extracted and then added back, they increase the staling at a low level of addition (0.4%), but decrease the staling when added at a higher level (1.6%) [79]. The added emulsifiers affect the firmness values as well as the ΔH_c of the crumb measured by DSC [109]. In fact, there is an excellent correlation between these values; correlation coefficients around 0.998 are reported [109]. The effects of added emulsifiers on bread firmness have sometimes been explained by an initial softer crumb due to the additive, but otherwise the staling rate is the same as for the control bread. In some studies, softer breads were obtained when monoglycerides were added [121], but in other studies this was not the case [45]. It could be argued that if amylose is important for the bread crumb structure, then the addition of emulsifiers (which are believed to complex with amylose) could be expected to result in an initially softer crumb [121] in the same way as baking with waxy barley does [81].

Avrami analysis of firmness measurements on bread with and without monoglycerides shows that both A_L and the rate constant decrease in the presence of monoglycerides [45]. The decrease in A_L could be due to a larger loaf volume, but not the decrease in k. Investigations with the X-ray diffraction technique show that emulsifiers influence the starch crystallization directly, and the results obtained in firmness measurements do not simply reflect changes in bread volume and/or initial crumb softness. In the fresh bread, as already discussed, only the V-pattern is present. With time, the B-pattern emerges [66]. However, if emulsifiers are added, the V-pattern increases in intensity but the B-pattern shows a lower intensity than the control.

When emulsifiers are added to a starch gel or to bread, the DSC endotherm due to the retrogradation of amylopectin is decreased [45, 96, 109]. The Avrami analysis shows that A_L and n are independent of the presence of emulsifiers, whereas k decreases [45]. The final extent of crystallization is thus not changed by the addition of a monoglyceride, but it takes a longer time to reach this final value. If the bread is toasted or heated to a temperature above the melting of recrystallized amylopectin, both firmness and the ΔH value decrease [109].

The enthalpy of the endotherm due to the transition of the amylose–lipid complex increases in proportion to the amount of lipid added [96, 109]. The size of this endotherm depends on the type and amount of lipid added, but it does not change during the storage of bread [109]. The added lipid evidently forms the amylose–lipid complex, but it is difficult to rationalize how this complex inhibits the crystallization of amylopectin. There is no simple relation between the intensity of the V-pattern and that of the B-pattern [66]. Neither has a simple relation been found between the amylose-complexing ability of a particular emulsifier and its antistaling effect [66, 122].

As discussed in Chapter 3, added lipids can also form a complex with amylopectin. If amylose and amylopectin are both present, a complex will first form between amylose and the added lipid [110]. When the amylose is saturated with lipids, then a complex forms with amylopectin. Similar conclusions have been drawn from experiments involving bread with added monoglycerides [109]. At low monoglyceride levels, the endotherm due to the transition of the amylose–lipid complex increases in proportion to the amount of monoglyceride, but after about 1% monoglyceride the enthalpy of the amylose–lipid complex is very much the same. The influence of monoglyceride on the retrogradation of amylopectin is quite low until the enthalpy of the complex endotherm has leveled off; then ΔH_c drops (Fig. 16).

The fact that amylopectin definitely also complexes lipids, which will hinder crystallization, and the situation in the bread crumb illustrated in Fig. 9 (with starch forming the continuous phase) provide strong evidence for concluding that amylopectin is the main factor in the staling of bread. The continuous medium in the bread is starch (Fig. 9), and any textural change in the core of the pore system must have strong overall effects.

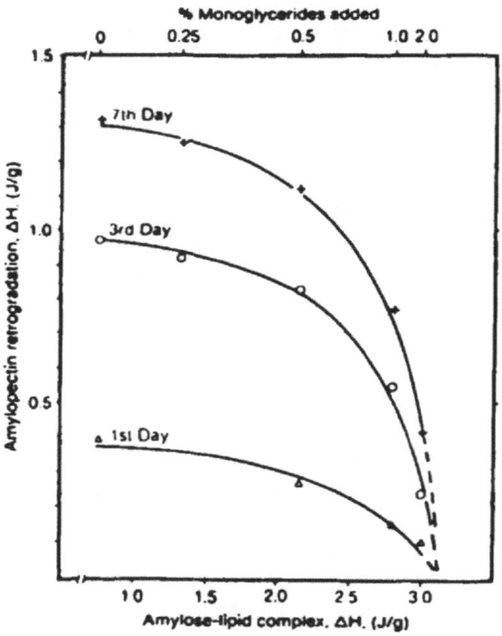

Fig. 16. The relationship between retrogradation of amylopectin and the amount of amylose–lipid complex formed. Breads with different amounts of added monoglyceride were stored at 22°C. (Reprinted with permission from [109].)

REFERENCES

1. Hoseney, R. C, Dreese, P. C., Doescher, L. C., and Faubion, J. M. (1987). Thermal properties of gluten, in *Proceedings of the 3rd International Workshop on Gluten Proteins*, R. Lasztity and F. Bekés, Eds., World Scientific, Singapore, p. 518.
2. Junge, R. C., and Hoseney, R. C. (1981). A mechanism by which shortening and certain surfactants improve loaf volume in bread, *Cereal Chem.*, 58: 408.
3. Rask, C. (1988). The heat transfer in a convection oven—influence on some product characteristics, in *Cereal Science and Technology in Sweden*, N.-G. Asp, Ed., Lund University, Lund, p. 148.
4. Hallström, B. (1988). Heat and mass transfer in baking ovens, in *Cereal Science and Technology in Sweden*, N.-G. Asp, Ed., Lund University, Lund, p. 137.
5. de Vries, U., Sluimer, P., and Bloksma, A. H. (1988). A quantitative model for heat transport in dough and crumb during baking, in *Cereal Science and Technology in Sweden*, N.-G. Asp, Ed., Lund University, Lund, p. 174.
6. Czuchajowska, Z., Pomeranz, Y., and Jeffers, H. C. (1989). Water activity and moisture content of dough and bread, *Cereal Chem.*, 66: 128.
7. Larsen, N. G., and Greenwood, D. R. (1991). Water addition and the physical properties of mechanical dough development doughs and breads, *J. Cereal Sci.*, 13: 195.
8. Bloksma, A. H. (1990). Rheology of the breadmaking process, *Cereal Foods World*, 35: 228.
9. Moore, W. R., and Hoseney, R. C. (1985). The leavening of bread dough, *Cereal Foods World*, 30: 791.
10. van Dam, H. W. (1986). The biotechnology of baker's yeast: old or new business? in *Chemistry and Physics of Baking*, J. M. V. Blanshard, P. J. Frazier, and T. Galliard, Eds., The Royal Society of Chemistry, London, p. 117.
11. Kulp, K. (1988). Bread industry and processes, in *Wheat: Chemistry and Technology*, Y. Pomeranz, Ed., American Association of Cereal Chemists, St. Paul, Minn. p. 371.
12. Kruger, J. E., and Reed, G. (1988). Enzymes and color, in *Wheat: Chemistry and Technology*, Y. Pomeranz, Ed., AACC, St. Paul, Minn., p. 441.
13. Fox, P. F., and Mulvihill, D. M. (1986). Enzymes in wheat, flour, and bread, in *Advances in Cereal Science and Technology*, Y. Pomeranz, Ed., AACC, St. Paul, Minn., p. 107.
14. Audidier, Y. (1968). Effects of thermal kinetics and weight loss kinetics on biochemical reactions in dough, *Baker's Dig.*, 42(5): 36.

15. Privalov, P. L., and Khechinashvili, N. N. (1974). A thermodynamic approach to the problem of stabilization of globular protein structure: a calorimetric study, *J. Mol. Biol.*, 86: 665.
16. Eliasson, A.-C., and Hegg, P.-O. (1981). Thermal stability of gluten, *Cereal Chem.*, 57: 436.
17. Hoseney, R. C., Zeleznak, K., and Lai, C. S. (1986). Wheat gluten: a glassy polymer, *Cereal Chem.*, 63: 285.
18. Tatham, A. S., and Shewry, P. R. (1985). The conformation of wheat gluten proteins. The secondary structures and thermal stabilities of α-, β-, γ-gliadins, *J. Cereal Sci.*, 3: 103.
19. Tatham, A. S., Field, J. M., Smith, S. J., and Shewry, P. R. (1987). The conformation of wheat gluten proteins. II. Aggregated gliadins and low molecular weight subunits of glutenin, *J. Cereal Sci.*, 5: 203.
20. Ablett, S., Barnes, D. J., Davies, A. P., Ingman, S. J., and Patient, D. W. (1988). ^{13}C and pulse nuclear magnetic resonance spectroscopy of wheat proteins, *J. Cereal Sci.*, 7: 11.
21. Schofield, J. D., Bottomley, R. C., Timms, M. F., and Booth, M. R. (1983). The effect of heat on wheat gluten and the involvement of sulphydryl–disulphide interchange reactions, *J. Cereal Sci.*, 1: 241.
22. Attenburrow, G., Barnes, D. J., Davies, A. P., and Ingman, S. J. (1990). Rheological properties of wheat gluten, *J. Cereal Sci.*, 12: 1.
23. Bale, R., and Muller, H. G. (1970). Application of the statistical theory of rubber elasticity to the effect of heat on wheat gluten, *J. Food Technol.*, 5: 295.
24. Jeanjean, M. F., Damidaux, R., and Feillet, P. (1980). Effect of heat treatment on protein solubility and viscoelastic properties of wheat gluten, *Cereal Chem.*, 57: 325.
25. Eliasson, A.-C., and Silverio, J. (1990). Interfacial behaviour of gluten proteins after heat treatment, in *Gluten Proteins 1990*, W. Bushuk and R. Tkachuk, Eds., AACC, St. Paul, Minn., p. 11.
26. MacRitchie, F. (1987). Evaluation of contributions from wheat protein fractions to dough mixing and breadmaking, *J. Cereal Sci.*, 6: 259.
27. Lineback, D. R., and Wongsrikasem, E. (1980). Gelatinization of starch in baked products, *J. Food Sci.*, 45: 71.
28. Greenwood, C. T. (1976). Starch, in *Advances in Cereal Science and Technology*, Y. Pomeranz, Ed., AACC, St. Paul, Minn. p. 119.
29. Eliasson, A.-C. (1983). Differential scanning calorimetry studies on wheat starch–gluten mixtures. I. Effect of gluten on the gelatinization of wheat starch, *J. Cereal Sci.*, 1: 199.
30. Eliasson, A.-C. (1989). Some physico-chemical properties of wheat starch, in *Wheat End-Use Properties: Wheat and Flour Characterization for Specific End-Use*, H. Salovaara, Ed., Helsinki University, p. 355.

31. Eliasson, A.-C. (1985). Starch gelatinization in the presence of emulsifiers. A morphological study of wheat starch, *Starch/Stärke*, 37: 411.
32. Willhoft, E. M. A. (1971). Bread staling. I. Experimental study, *J. Sci. Food Agric.*, 22: 176.
33. Dreese, P. C., Faubion, J. M., and Hoseney, R. C. (1988). Dynamic rheological properties of flour, gluten, and gluten-starch doughs. I. Temperature-dependent changes during heating, *Cereal Chem.*, 65: 348.
34. Eliasson, A.-C. (1986). Viscoelastic behavior during the gelatinization of starch. II. Effects of emulsifiers, *J. Text. Stud.*, 17: 357.
35. Sandstedt, R. M. (1961). The function of starch in the baking of bread, *Baker's Dig.*, 35(3): 36.
36. Hoseney, R. C., Finney, K. F., Pomeranz, Y., and Shogren, M. D. (1971). Functional (breadmaking) and biochemical properties of wheat flour components. VIII. Starch, *Cereal Chem.*, 48: 191.
37. Williams, M. R., and Bowler, P. (1982). Starch gelatinization: a morphological study of triticeae and other starches, *Starch/Stärke*, 34: 211.
38. Bowler, P., Williams, M. R., and Angold, R. E. (1980). A hypothesis for the morphological changes which occur on heating lenticular wheat starch in water, *Starch/Stärke*, 32: 186.
39. Varriano-Marston, E., Ke, V., Huang, G., and Ponte, J., Jr. (1980). Comparison of methods to determine starch gelatinization in bakery food, *Cereal Chem.*, 57: 242.
40. Varriano-Marston, E., Zeleznak, K., and Nowotna, A. (1985). Structural characteristics of gelatinized starch, *Starch/Stärke*, 37: 326.
41. Dennett, K., and Sterling, C. (1979). Role of starch in bread formation, *Starch/Stärke*, 31: 209.
42. Seguchi, M., and Yamada, Y. (1988). Hydrophobic character of heat-treated wheat starch, *Cereal Chem.*, 65: 375.
43. Eliasson, A.-C., and Tjerneld, E. (1990). Adsorption of wheat proteins on wheat starch granules, *Cereal Chem.*, 67: 366.
44. Eliasson, A.-C. (1986). On the effects of surface active agents on the gelatinization of starch—a calorimetric investigation, *Carbohydr. Polym.*, 6: 463.
45. Russell, P. L. (1983). A kinetic study of bread staling by differential scanning calorimetry and compressibility measurements. The effect of added monoglycerides, *J. Cereal Sci.*, 1: 297.
46. DeStefanis, V. A., Ponte, J. G., Chung, F. H., and Ruzza, N. A. (1977). Binding of crumb softeners and dough strengtheners during breadmaking, *Cereal Chem.*, 54: 13.
47. Soulaka, A. B., and Morrison, W. R. (1985). The bread baking quality of six wheat starches differing in composition and physical properties, *J. Sci. Food Agric.*, 36: 719.

48. Moore, W. R., and Hoseney, R. C. (1986). The effects of flour lipids on the expansion rate and volume of bread baked in a resistance oven, *Cereal Chem.*, 63: 172.
49. Bloksma, A. H. (1990). Dough structure, dough rheology, and baking quality, *Cereal Foods World*, 35: 237.
50. Junge, R. C., Hoseney, R. C., and Varriano-Marston, E. (1981). Effect of surfactants on air incorporation in dough and the crumb grain of bread, *Cereal Chem.*, 58: 338.
51. MacRitchie, F. (1983). Role of lipids in baking, in *Lipids in Cereal Technology*, P. J. Barnes, Ed., Academic, London, p. 165.
52. Marion, D, Le Roux, C., Akora, S., Tellier, C., and Gallant, D. (1987). Lipid–protein interactions in wheat gluten: a phosphorous nuclear magnetic resonance spectroscopy and freeze-fracture electron microscopy study, *J. Cereal Sci.*, 5: 101.
53. Westerlund, E., Theander, O., and Åman, P. (1989). Effects of baking on protein and aqueous ethanol-extractable carbohydrate in white bread fractions, *J. Cereal Sci.*, 10: 139.
54. Westerlund, E., Theander, O., Andersson, R., and Åman, P. (1989). Effects of baking on polysaccharides in white bread fractions, *J. Cereal Sci.*, 10: 149.
55. Andersson, S., Hyde, S. T., Larsson, K., and Lidin, S. (1988). Minimal surfaces and structures: from inorganic and metal crystals to cell membranes and biopolymers, *Chem. Rev.*, 88: 221.
56. Kulp, K. (1991). Breads and yeast-leavened bakery foods, in *Handbook of Cereal Science and Technology*, K. J. Lorenz and K. Kulp, Eds., Marcel Dekker, New York, p. 639.
57. Siljeström, M., and Asp, N.-G. (1985). Resistant starch formation during baking—effect of baking time and temperature and variations in the recipe, *Z. Lebensm. Unters. Forsch.*, 181: 4.
58. Berry, C. S. (1986). Resistant starch: formation and measurement of starch that survives exhaustive digestion with amylolytic enzymes during the determination of dietary fibre, *J. Cereal Chem.*, 4: 301.
59. Gaillard, T. (1989). Baking industry response to consumer demand. A. General aspects and B. Wholemeal flour quality, in *Cereal Science and Technology in Sweden* N.-G. Asp, Ed., Lund University, Lund, Sweden. p. 55.
60. He, H., and Hoseney, R. C. (1991). Gas retention of different cereal flours, *Cereal Chem.*, 68: 334.
61. Eliasson, A.-C., Tjerneld, E., and Larsson, K. (1985). Development of a gluten-free bread (in Swedish), *Livsmedelsteknik*, 27: 150.
62. Herz, K. O. (1965). Staling of bread—a review, *Food Technol.*, 19: 90.

63. Maga, J. A. (1975). Bread staling, *CRC Crit. Rev. Food Technol.*, 29: 443.
64. Knightly, W. H. (1977). The staling of bread, *Baker's Dig.*, 51(5): 52.
65. Kulp, K., and Ponte, Jr., J. G. (1981). Staling of white pan bread: fundamental causes, *CRC Crit. Rev. Food Sci. Nutr.*, 15: 1.
66. Zobel, H. F. (1973). A review of bread staling, *Baker's Dig.*, 47: 52.
67. Czuchajowska, Z., and Pomeranz, Y. (1989). Differential scanning calorimetry, water activity, and moisture contents in crumb center and near-crust zones of bread during storage, *Cereal Chem.*, 66: 305.
68. Sperling, L. H. (1986). *Introduction to Physical Polymer Science*, Wiley, New York.
69. Levine, H., and Slade, L. (1990). Influences of the glassy and rubbery states on the thermal, mechanical, and structural properties of doughs and baked products in *Dough Rheology and Baked Product Texture*, H. Faridi and J. M. Faubion, Eds. Van Nostrand Reinhold, New York, p. 157.
70. Zeleznak, K. J., and Hoseney, R. C. (1987). The glass transition in starch, *Cereal Chem.*, 64: 121.
71. Slade, L., and Levine, H. (1988). Non-equilibrium behavior of small carbohydrate–water systems, *Pure Appl. Chem.*, 60: 1841.
72. Schoch, T. J. (1965). Starch in bakery products, *Baker's Dig.*, 39: 48.
73. Maleki, M., and Seibel, W. (1972). Über das Altbackenwerden von Brot. 1. Mitteilung: Beziehungen zwischen Sensorik, Penetrometer und Panimeter beim Messen des Altbackenwerdens von Weizenbrot, *Getreide, Mehl Brot*, 26: 58.
74. Stöllman, U., and Lundgren, B. (1987). Texture changes in white bread: effects of processing and storage, *Cereal Chem.*, 64: 230.
75. Cluskey, J. E., Taylor, N. W., and Senti, F. R. (1959). Relation of the rigidity of flour, starch and gluten gels to bread staling, *Cereal Chem.*, 36: 236.
76. Willhoft, E. M. A. (1973). Recent developments on the bread staling problem, *Bakers' Dig.*, 47(6): 14.
77. Knorr, D., Klaushofer, H., and Kleemair, F. (1976). Texturmessungen an Weissbrot. 1. Die lagerungszeitabhängige Veränderung der Festigkeit, Weichheit und Elastizität von Weissbrotkrume, *Lebensm. Wiss. Technol.*, 9: 59.
78. Baker, A. E., Walker, C. E., and Kemp, K. (1988). An optimum compression depth for measuring bread crumb firmness, *Cereal Chem.*, 65: 302.
79. Rogers, D. E., Zeleznak, K. J., Lai, C. S., and Hoseney, R. C. (1988).

Effect of native lipids, shortening, and bread moisture on bread firming, *Cereal Chem.*, 65: 398.

80. Axford, D. W. E., Colwell, K. H., Cornford, S. J., and Elton, G. A. H. (1968). Effect of loaf specific volume on the rate and extent of staling in bread, *J. Sci. Food Agric.*, 19: 95.
81. Ghiasi, K., Hoseney, R. C., Zeleznak, K., and Rogers, D. E. (1984). Effect of waxy barley starch and reheating on firmness of bread crumb, *Cereal Chem.*, 61: 281.
82. Kim, S. K., and D'Appolonia, B. L. (1977). Bread staling studies. II. Effect of protein content and storage temperature on the role of starch, *Cereal Chem.*, 54: 216.
83. Lin, W., and Lineback, D. R. (1990). Changes in carbohydrate fractions in enzyme-supplemented bread and the potential relationship to staling. *Starch/Stärke*, 42: 385.
84. Katz, J. R. (1934). X-ray investigation of gelatinization and retrogradation of starch in its importance for bread research, *Bakers' Weekly*, 81: 34.
85. Dragsdorf, R. D., and Varriano-Marston, E. (1980). Bread staling: X-ray diffraction studies on bread supplemented with alfa-amylses from different sources, *Cereal Chem.*, 57: 310.
86. McIver, R. G., Axford, D. W. E., Colwell, K. H., and Elton, G. A. H. (1968). Kinetic study of the retrogradation of gelatinised starch, *J. Sci. Food Agric.*, 19: 560.
87. Colwell, K. H., Axford, D. W. E., Chamberlain, N., and Elton, G. A. H. (1969). Effect of storage temperature on the ageing of concentrated wheat starch gels, *J. Sci. Food Agric.*, 20: 550.
88. Fearn, T., and Russell, P. (1982). A kinetic study of bread staling by differential scanning calorimetry. The effect of loaf specific volume, *J. Sci. Food Agric.*, 33: 537.
89. Eliasson, A.-C. (1985). Retrogradation of starch as measured by differential scanning calorimetry, in *New Approaches to Research on Cereal Carbohydrates*, R. D. Hill and L. Munck, Eds., Elsevier, Amsterdam, p. 93.
90. Siljeström, M., Björck, I., Eliasson, A.-C., Lönner, C., Nyman, M., and Asp, N.-G. (1988). Effects on polysaccharides during baking and storage of bread—in vitro and in vivo studies, *Cereal Chem.*, 65: 1.
91. Ring, S. G. (1985). Observations on the crystallization of amylopectin from aqueous solution, *Int. J. Biol. Macromol.*, 7: 253.
92. Russell, P. L. (1987). The ageing of gels from starches of different amylose/amylopectin content studied by differential scanning calorimetry, *J. Cereal Sci.*, 6: 147.
93. Roulet, P., Raemy, A., and Wuersch, P. (1987). Retrogradation kinetics

of concentrated gelatinized wheat starch in gels or powders; a comparative study by rigidity modulus, differential scanning calorimetry (DSC) and X-ray diffraction, *Food Hydrocolloids*, 1: 575.
94. Kim, S. K., and D'Appolonia, B. L. (1977). Bread staling studies. I. Effect of protein content on staling rate and bread crumb pasting properties, *Cereal Chem.*, 54: 207.
95. Wong, R. B. K., and Lelievre, J. (1982). Effects of storage on dynamic rheological properties of wheat starch pastes, *Starch/Stärke*, 34: 231.
96. Eliasson, A.-C. (1983). Differential scanning calorimetry studies on wheat starch–gluten mixtures. II. Effect of gluten and sodium stearoyl lactylate on starch crystallization during ageing of wheat starch gels, *J. Cereal Sci.*, 1: 207.
97. Jankowski, T., and Rha, C. K. (1986). Retrogradation of starch in cooked wheat, *Starch/Stärke*, 38: 6.
98. Zeleznak, K. J., and Hoseney, R. C. (1987). Characterization of starch from bread aged at different temperatures, *Starch/Stärke*, 39: 231.
99. Hellman, N. N., Fairchild, B., and Senti, F. R. (1954). The bread staling problem. Molecular organization of starch upon aging of concentrated starch gels at various moisture levels. *Cereal Chem.*, 31: 495.
100. Longton, J., and LeGrys, G. A. (1981). Differential scanning calorimetry studies on the crystallization of ageing wheat starch gels, *Starch/Stärke*, 33: 410.
101. Zeleznak, K. J., and Hoseney, R. C. (1986). The role of water in the retrogradation of wheat starch gels and bread crumb, *Cereal Chem.*, 63: 407.
102. Biliaderis, C. G., and Zawistowski, J. (1990). Viscoelastic behavior of aging starch gels: effects of concentration, temperature, and starch hydrolysates on network properties, *Cereal Chem.*, 67: 240.
103. Mita, T. (1990). Effect of aging on the rheological properties of gluten gel, *Agric. Biol. Chem.*, 54: 927.
104. Kim, S. K., and D'Appolonia, B. L. (1977). The role of wheat flour constituents in bread staling, *Baker's Dig.*, 51(1): 38.
105. Bachrach, H. L., and Briggs, D. R. (1947). Studies on bread staling. II. Water relationships during staling of bread crumb and the retrogradation of starch, *Cereal Chem.*, 24: 492.
106. Wynne-Jones, S., and Blanshard, J. M. V. (1986). Hydration studies of wheat starch, amylopectin, amylose gels and bread by proton magnetic resonance, *Carbohydr. Polym.*, 6: 289.
107. Breaden, P. W., and Willhoft, E. M. A. (1971). Bread staling. III. Measurement of the re-distribution of moisture in bread by gravimetry, *J. Sci. Food Agric.*, 22: 647.
108. Slade, L., and Levine, H. (1987). Recent advances in starch retrograda-

tion, in *Industrial Polysaccharides: The Impact of Biotechnology and Advanced Methodologies*, S. S. Stivala, V. Crescenzi, and I. C. M. Dea, Eds., Gordon and Breach, New York, p. 387.

109. Krog, N., Olesen, S. K., Toernaes, H., and Joensson, T. (1989). Retrogradation of the starch fraction in wheat bread, *Cereal Foods World*, 34: 281.

110. Gudmundsson, M., and Eliasson, A.-C. (1990). Retrogradation of amylopectin and the effects of amylose and added surfactants/emulsifiers, *Carbohydr. Polym.*, 13: 295.

111. Maleki, M., Schulz, A., and Brümmer, J.-M. (1972). Über das Altbackenwerden von Brot. II. Frischhalteeffekt von Bakterien-, Pilz- und Getreide-Alpha-Amylasen, *Getreide, Mehl Brot*, 26: 221.

112. Maleki, M., and W., S. (1972). Über das Altbacken von Brot. 3. Der Einfluss von Fett, Emulgatoren und Schüttwassermenge, *Getreide, Mehl Brot*, 26: 258.

113. Cornford, S. J., Axford, D. W. E., and Elton, G. A. H. (1964). The elastic modulus of bread crumb in linear compression in relation to staling, *Cereal Chem.*, 41: 216.

114. Kim, S. K., and D'Appolonia, B. L. (1977). Bread staling studies. III. Effect of pentosans on dough, bread, and bread staling rate, *Cereal Chem.*, 54: 225.

115. Gudmundsson, M., Eliasson, A.-C., Bengtsson, S., and Åman, P. (1991). The effects of water soluble arabinoxylan on gelatinization and retrogradation of starch, *Starch/Stärke*, 43: 5.

116. Russell, P. L., and Oliver, G. (1989). The effect of pH and NaCl content on starch gel ageing. A study by differential scanning calorimetry and rheology, *J. Cereal Sci.*, 10: 123.

117. Tipples, K. H. (1969). The relation of starch damage to the baking performance of flour, *Baker's Dig.*, 43(6): 28.

118. Martin, M. L., and Hoseney, R. C. (1991). A mechanism of bread firming. II. Role of starch hydrolyzing enzymes, *Cereal Chem.*, 68: 503.

119. Eliasson, A.-C., and Ljunger, G. (1988). Interactions between amylopectin and lipid additives during retrogradation in a model system. *J. Sci. Food Agric.*, 44: 353.

120. Krog, N., and Nybo-Jensen, B. (1970). Interaction of monoglycerides in different physical states with amylose and their anti-firming effects in bread, *J. Food Technol.*, 5: 77.

121. Riisom, T., Krog, N., and Eriksen, J. (1984). Amylose complexing capacities of *cis*- and *trans*-unsaturated monoglycerides in relation to their functionality in bread, *J. Cereal Sci.*, 2: 105.

122. Osman, E. M., Leith, S. J., and Fles, M. (1961). Complexes of amylose with surfactants, *Cereal Chem.*, 38: 449.

Index